Tribological Aspects of Additive Manufacturing

Tribological Aspects of Additive Manufacturing provides a technical discussion on the roles of the 3D printing process in processing polymeric-, metallic-, and ceramics-based additive manufactured products in order to improve the tribological properties. It explores design flexibility, waste minimization, and cost reduction.

Emphasizing the various types of additive manufacturing technologies, this book demonstrates how these can effectively influence the tribological properties of additively manufactured components. It examines 3D printing process parameters, carbon fiber reinforcement, natural fiber reinforcement, and surface structure on tribological properties of 3D-printed parts. This book also covers wear and friction resistance of additively manufactured parts prepared with natural fiber and carbon fiber.

This book will be a useful reference for undergraduate and graduate students and academic researchers in the fields of materials science, tribology, additive manufacturing, maintenance engineering, and 3D printing.

Emerging Materials and Technologies

Series Editor: Boris I. Kharissov

The *Emerging Materials and Technologies* series is devoted to highlighting publications centered on emerging advanced materials and novel technologies. Attention is paid to those newly discovered or applied materials with potential to solve pressing societal problems and improve quality of life, corresponding to environmental protection, medicine, communications, energy, transportation, advanced manufacturing, and related areas.

The series takes into account that, under present strong demands for energy, material, and cost savings, as well as heavy contamination problems and worldwide pandemic conditions, the area of emerging materials and related scalable technologies is a highly interdisciplinary field, with the need for researchers, professionals, and academics across the spectrum of engineering and technological disciplines. The main objective of this book series is to attract more attention to these materials and technologies and invite conversation among the international R&D community.

Impedance Spectroscopy and its Application in Biological Detection
Edited by Geeta Bhatt, Manoj Bhatt, and Shantanu Bhattacharya

Nanofillers for Sustainable Applications
Edited by N.M Nurazzi, E. Bayraktar, M.N.F. Norrrahim, H.A. Aisyah, N. Abdullah, and M.R.M. Asyraf

Chemistry of Dehydrogenation Reactions and its Applications
Edited by Syed Shahabuddin, Rama Gaur, and Nandini Mukherjee

Biosorbents
Diversity, Bioprocessing, and Applications
Edited by Pramod Kumar Mahish, Dakeshwar Kumar Verma, and Shailesh Kumar Jadhav

Principles and Applications of Nanotherapeutics
Imalka Munaweera and Piumika Yapa

Energy Materials
A Circular Economy Approach
Edited by Surinder Singh, Suresh Sundaramuthy, Alex Ibhadon, Faisal Khan, Sushil Kansal, and S.K. Mehta

Tribological Aspects of Additive Manufacturing
Edited by Rashi Tyagi, Ranvijay Kumar, and Nishant Ranjan

For more information about this series, please visit: www.routledge.com/Emerging-Materials-and-Technologies/book-series/CRCEMT

Tribological Aspects of Additive Manufacturing

Edited by
Rashi Tyagi, Ranvijay Kumar, and Nishant Ranjan

CRC Press
Taylor & Francis Group
Boca Raton London New York

CRC Press is an imprint of the
Taylor & Francis Group, an **informa** business

Designed cover image: Shutterstock

First edition published 2024
by CRC Press
2385 NW Executive Center Drive, Suite 320, Boca Raton FL 33431

and by CRC Press
4 Park Square, Milton Park, Abingdon, Oxon, OX14 4RN

CRC Press is an imprint of Taylor & Francis Group, LLC

ISBN: 978-1-032-50975-4 (hbk)
ISBN: 978-1-032-50977-8 (pbk)
ISBN: 978-1-003-40052-3 (ebk)

DOI: 10.1201/9781003400523

Typeset in Times
by codeMantra

Contents

                and Friction Resistance 76

                Rajnish P. Modanwal, Dan Sathiaraj, Pradeep K. Singh,
                Rashi Tyagi, and Ashwath Pazhani

                6.1   Introduction 76
                6.2   Natural Fibers 79
                6.3   Tribology 82
                       6.3.1   Pin on Drum 82
                       6.3.2   Pin on Disk 83
                       6.3.3   Block on Ring 83
                       6.3.4   Block on Disk 84
                       6.3.5   Linear Tribo Machine 84
                       6.3.6   Dry Sand Rubber Wheel 85
                6.4   Description of AM 3DP Technique 85
                       6.4.1   Fused Filament Fabrication 86
                       6.4.2   Direct Write 87
                       6.4.3   Stereolithography 87
                       6.4.4   Selective Laser Sintering 88
                       6.4.5   Binder Jetting 89
                6.5   Wear Performance of 3D AM Composites 89
                       6.5.1   Biogenic Carbon/PLA Composite 89
                       6.5.2   Flex Yarn/PLA Composite 90
                       6.5.3   Grewia/Nettle/Sisal/PLA Composite 91
                       6.5.4   Corn Cob/PLA Composite 91
                       6.5.5   Date Particle/PLA Composite 91
                6.6   Conclusion 93
                Acknowledgments 94
                References 94

Chapter 7   Study on the Effect of Carbon-Fiber-Reinforced
                Composites on Tribological Properties 97

                Shalini Mohanty, Adrian Murphy, and Rashi Tyagi

                7.1   Introduction 97
                7.2   Tribological Analysis of Carbon-Fiber-Reinforced
                       Composites 99
                7.3   Case Studies 101
                7.4   Applications and Future in 3D-Printed Carbon Fiber
                       Composites 103
                7.5   Conclusions 104
                Acknowledgments 105
                References 105

Preface

The book entitled *Tribological Aspects of Additive Manufacturing* aims to deliver various additive manufacturing processes to study the tribological properties (wear, friction, and lubrication) of additively manufactured parts, suitable for demanding industrial and biomedical applications. This book provides a comprehensive overview of the tribological aspects of AM. It covers the fundamentals of tribology, the tribological behavior of AM-produced materials, and the design and manufacturing of AM components for tribological applications. This book is intended for a wide audience, including engineers, scientists, and students who are interested in learning more about the tribological aspects of AM. The purpose of this book is to emphasize the role of various types of additive manufacturing technologies that can effectively influence the tribological properties of additively manufactured components. This book provides a collection of roles for 3D printing process parameters, carbon fiber reinforcement, natural fiber reinforcement, and surface structure on tribological properties of 3D-printed parts. The literature, methodology, experimental results, and theoretical aspects of tribology in 3D printing are discussed in the chapters of this book. This book is designed to cover the significant contribution of the research fraternity, across the world, for different classes of 3D printing process roles in tribology field and end applications. This book is designed as a key source of information on the tribological aspects of 3D-printed parts.

Dr. Rashi Tyagi
Chandigarh University, India

Dr. Ranvijay Kumar
Chandigarh University, India

Dr. Nishant Ranjan
Chandigarh University, India

Editors' Brief Bios

Dr. Rashi Tyagi is currently working as an assistant professor in University Centre for Research and Development at Chandigarh University. Dr. Tyagi has won CII MILCA AWARD in the field of tribology of electrical discharge coating in 2022. She has completed her PhD in mechanical engineering from the Indian Institute of Technology (Indian school of mines), Dhanbad, India. Her PhD work was focused on surface modification by an electrical discharge process for solid lubrication and enhanced tribological performance. She has completed her MTech from IIT(ISM), Dhanbad, India. She has done her MTech project on the fiber laser cladding of TiN+SS 304 powder to enhance tribological performance. She has published over 21 SCI articles in peer-reviewed international journals, conference proceedings, and book chapters as a first author in the field of 3D printing and tribology of coatings. She has published two articles in *Tribology International*. She is also a reviewer in several SCI and Scopus indexed journals. She is currently working on tribology of fiber reinforce composite prepared by fused filament fabrication, fused deposition modeling, thermoplastic polymers, and natural and synthetic composites.

Dr. Ranvijay Kumar is an assistant professor in University Centre for Research and Development, Chandigarh University. He has received a PhD in mechanical engineering from Punjabi University, Patiala. Additive manufacturing, shape memory polymers, smart materials, friction-based welding techniques, advance materials processing, polymer matrix composite preparations, reinforced polymer composites for 3D printing, plastic solid waste management, thermosetting recycling, and destructive testing of materials are the skills of Dr. Kumar. He has won prestigious CII MILCA award 2020. He has co-authored more than 55 research papers in science citation indexed journals and 38 book chapters and has presented 20 research papers in various national/international level conferences. He has contributed extensively in Additive Manufacturing literature with publications appearing in the *Journal of Manufacturing Processes, Composite Part: B, Rapid Prototyping Journal*, the *Journal of Thermoplastic Composite Materials, Measurement, Proceedings of the Institution of Mechanical Engineers, Part C (iMeche Part C), Proceedings of the Institution of Mechanical Engineers, Part H: Journal of Engineering in Medicine*, the *Journal of Thermoplastic Composite Materials, Materials Research Express, Proceedings of the National Academy of Sciences, India Section A: Physical Sciences*, the *Journal of Central South University*, the *Journal of the Brazilian Society of Mechanical Sciences and Engineering*,

Composite Structures, CIRP Journal of Manufacturing Science and Technology, etc.
He is the editor of book *Additive Manufacturing for Plastic Recycling: Efforts in
Boosting A Circular Economy* publishing in CRC Press (Taylor and Francis).

Dr. Nishant Ranjan is working as Assistant Professor
at University Centre for Research and Development of
Chandigarh University. Fused deposition modeling; extrusion;
thermoplastic polymers; composition of thermoplastic polymers; natural and synthetic biopolymers; scaffolds printing;
3D printing technology; thermal, mechanical, morphological,
and chemical properties of thermoplastic polymers; biocompatible and biodegradable fillers; and reinforcement of materials are the main focused area of Dr. Nishant Ranjan. He has
co-authored more than 20 research papers in science citation
indexed journal, 1 book, and 27 book chapters and has presented 14 research papers
in various international/national level conferences. He has been reviewing research
articles of various peer-reviewed SCI and Scopus indexed journals.

List of Contributors

Mukul Anand
Department of Mechanical Engineering
Indian Institute of Technology (Indian
    School of Mines, Dhanbad)
Dhanbad, India

Harish Bishwakarma
Department of Mechanical Engineering
Indian Institute of Technology (Indian
    School of Mines, Dhanbad)
Dhanbad, India

Harpreet Kaur Channi
Department of Electrical Engineering
Chandigarh University
Mohali, India

Jasgurpeet Singh Chohan
Department of Mechanical Engineering
University Centre for Research and
    Development
Chandigarh University
Mohali, India

Alok Kumar Das
Department of Mechanical Engineering
Indian Institute of Technology (Indian
    School of Mines, Dhanbad)
Dhanbad, India

Amit Rai Dixit
Department of Mechanical Engineering
Indian Institute of Technology (Indian
    School of Mines, Dhanbad)
Dhanbad, India

Suryank Dwivedi
Department of Mechanical Engineering
Indian Institute of Technology (Indian
    School of Mines, Dhanbad)
Dhanbad, India

Sehra Farooq
Department of Mechanical Engineering
Chandigarh University
Mohali, India

Alireza Hajialimohammadi
Faculty of Mechanical Engineering
Semnan University
Semnan, Iran

Nitesh Kumar
Department of Mechanical Engineering
Indian Institute of Technology (Indian
    School of Mines, Dhanbad)
Dhanbad, India

Ranvijay Kumar
Department of Mechanical Engineering
University Centre for Research and
    Development
Chandigarh University
Mohali, India

Raushan Kumar
Department of Mechanical Engineering
Indian Institute of Technology (Indian
    School of Mines, Dhanbad)
Dhanbad, India

Vinay Kumar
Department of Mechanical Engineering
University Centre of Research and
    Development
Chandigarh University
Mohali, India

Vishal Kumar
Indian Institute of Technology
    (Indian School of Mines, Dhanbad)
Dhanbad, India

Rajnish P. Modanwal
Advanced Forming Lab
Department of Mechanical Engineering
Indian Institute of Technology, Indore
Indore, India

Annada Prasad Moharana
Indian Institute of Technology (Indian
    School of Mines, Dhanbad)
Dhanbad, India

Shalini Mohanty
School of Mechanical and Aerospace
    Engineering
Queen's University
Belfast, United Kingdom

Adrian Murphy
School of Mechanical and Aerospace
    Engineering
Queen's University
Belfast, United Kingdom

Gaurav Parmar
Department of Mechanical Engineering
Indian Institute of Technology (Indian
    School of Mines, Dhanbad)
Dhanbad, India

Ashwath Pazhani
School of Mechanical, Aerospace and
    Automotive Engineering
College of Engineering, Environment
    and Science
Coventry University
Coventry, United Kingdom

Ratnesh Raj
Indian Institute of Technology (Indian
    School of Mines, Dhanbad)
Dhanbad, India

Nishant Ranjan
Department of Mechanical Engineering
University Centre for Research and
    Development
Chandigarh University
Mohali, India

Dan Sathiaraj
Advanced Forming Lab
Department of Mechanical Engineering
Indian Institute of Technology Indore
Indore, India

Pratik Kumar Shaw
Department of Mechanical Engineering
Indian Institute of Technology (Indian
    School of Mines, Dhanbad)
Dhanbad, India

Ankan Shrivastava
Department of Mechanical Engineering
University Centre for Research and
    Development
Chandigarh University
Mohali, India

Pradeep K. Singh
Department of Mechanical Engineering
Sant Longowal Institute of Engineering
    and Technology
Longowal, India

Rupinder Singh
Department of Mechanical Engineering
Chandigarh University
Mohali, India

Vishal Thakur
Department of Mechanical Engineering
Chandigarh University
Mohali, India

Rashi Tyagi
Department of Mechanical Engineering
University Centre for Research and
    Development
Chandigarh University
Mohali, India

Kumar Ujjwal
Department of Mechanical Engineering
Indian Institute of Technology (Indian
    School of Mines, Dhanbad)
Dhanbad, India

1 Tribological Study of 3D-Printed Thermoplastic Polymers

Vishal Thakur, Rupinder Singh, and Ranvijay Kumar
Chandigarh University

1.1 INTRODUCTION

Additive manufacturing (AM) is a technique that has the capability to manufacture 3D-printed parts with customization in mechanical properties, but there are some factors that affect the tribological properties of the manufactured parts. Tribology is a study related to friction, wear, and lubrication properties at the interacting surfaces. Adsorbed, self-assembled, or functionally grafted molecular structures that develop near or on the surface as the outcome of physical and chemical processes are referred to as interfacial molecular films (Jabbarzadeh, 2018). Numerous tribological applications including extremely precise ball bearings, shock absorption components, and bushings in diesel fuel injection pumps can be made with polymers and polymeric composites (Roy and Mukhopadhyay, 2021). Acrylonitrile butadiene styrene (ABS) materials were used for the fabrication of the 3D-printed parts by using the fused deposition modelling (FDM) process. The process parameters used for the fabrication of parts are layer height, orientation, infill density, raster angle, width of the raster, and bed temperature, and these process parameters influenced the tribological properties of the ABS-printed parts or structures. The addition of graphite, carbon fiber, and calcium carbonate is also beneficial for the enhancement of the tribological properties of ABS (Kaur et al., 2022). 3D-printed parts were manufactured by FDM process with composites of polyamide (PA6) blended with 10, 20, and 30 wt.% of titanium dioxides (TiO_2). The pin-on-desk tribometer was used for the investigation by varying loads of 5, 10, 15, and 20 N at sliding speeds of 1.256 m/s for 5- and 10-minute run times. The wear rate was observed minimum for PA6-30wt.% TiO_2 (Soundararajan et al., 2019). In order to conduct tribological testing against a water-sensitive skin model, two low-friction 3D printing materials such as thermoplastic polyurethane (TPU) and polyamide (TPA) were used. A gelatine-based model constructed with cotton and crosslinked with glutaraldehyde was used to create the skin model. TPU/TPA was tested tribologically against a skin model in both dry and wet circumstances. In comparison to the dry condition, the wet condition showed a greater coefficient of friction (COF). To lower the greater friction, heat pressing was used to create TPA/TPU-sodium polyacrylate composites, which considerably reduced the COF of TPU and TPA in wet circumstances by around 40% and 75%, respectively (Kasar et al., 2022). The polycarbonate (PC)-ABS composite structure was fabricated by using FDM with varying process parameters consisting of layer thickness, air gap, raster angle, build orientation, road width, and the number of contours. The previous studies have reported the impact of various FDM fabrication parameters on the tribological behaviour and wear behaviour of manufactured prototypes was investigated. The layer thickness of 0.127 mm, air gap of 0.00275 mm, raster angle of 81°, build orientation of 9°, road width of 0.4693 mm, and five contours were the optimum input parameters that yielded the lowest wear rate (Mohamed et al., 2017). Due to self-lubrication, corrosion resistance, and vibration-dampening capacity properties, the polymeric materials have employed for tribological applications (Briscoe and Sinha, 2008). PLA, high-temperature PLA (HT-PLA), and Polyethylene Terephthalate Glycol (PETG) were used for the manufacturing of test specimens for the investigation of tribological properties. HT-PLA observed low COF and wear, which confirms its potential to be used for tribological applications as compared to PLA and PETG thermoplastic materials (Hanon et al., 2020b). Carbon nanotubes (CNTs) were used to strengthen polyphenylene sulfide (PPS), and the FDM method was used to create composite structures. The mechanical properties of printed parts are enhanced by CNT, which also plays a significant role in load transfer. Tribological performance was greatly modified by CNT reinforcement (Pan et al., 2021). ABS material was used for the fabrication of structures to explore the friction and wear behaviour in context to the effect of process parameters such as load, speed, and orientation. Friction is more influenced by sliding speed than by orientation under a constant load. Wear patterns, such as abrasion, fatigue, corrugation, erosion, and cavitation, often harm contact surfaces. Due to changes in the spaces between the solids in contact, abrasive wear frequently causes irreversible alterations to the body's shapes. The wear rate typically oscillates between a minimum value and a value where it stabilizes and follows a fixed linear rate (Gurrala and Regalla, 2014). The changes in the mechanical and tribological properties are dependent on the process parameters and materials being used for 3D printing (Noraniet al., 2021). Although the 3D-printed surface had a lower surface pressure than the turned surface, the adhesion wear mechanism on the turned surface initially altered to smoother and stronger than that on the manufactured area. The previous study has revealed the comparison between the turned part and the 3D-printed part on the tribological viewpoint (Hanon et al., 2019). The 3D printing materials used for manufacturing of structures/components for tribological applications such as bearings, brakes, rolling or sliding components, gear drives and other innovative applications (Friedrich, 2018; Gbadeyan et al., 2021; Hanon et al., 2020a, Slapnik et al., 2020).

1.2 BACKGROUND

The database, which is available on the www.webofknowledge.com website, has been used for checking the background of the study related to 3D printing and tribological properties. There are 175 results observed on the keywords "3D printing" and "tribological properties" in the last 9 years (2015–2023). The research work related to the tribological properties of 3D printing is increasing significantly. In 2020, the maximum number of research papers were reported related to the tribological properties of 3D-printed structures as shown in Figure 1.1.

For the purpose of analyzing the gaps in the previously published literature, the VosViewer software program has been utilized. The 175 research papers contained a total of 4,467 terms; by maintaining the first five occurrences of each term, 104 were filtered out, and the most pertinent terms were chosen for the relationship diagram (Figure 1.2).

Figure 1.3 has been drawn with the help of Figure 1.2 to reflect the technological gap between the reported terms. The partially hidden term in Figure 1.3 shows the studies related to tribological properties which have been less reported. Figure 1.3 shows that most of the studies reported tribological properties of the materials such as wear characteristics and friction resistance. So, there is a scope to extend the study of tribological properties related to 3D-printed structures or the materials used in 3D printing of the composite structures, which is beneficial to enhance the application fields of the thermoplastic materials. Table 1.1 shows the relevance score of selected terms for the development of a relationship diagram.

1.3 TRIBOLOGICAL PROPERTIES OF 3D-PRINTED POLYMERS

The properties of the material used in 3D printing that are associated with friction, wear, and lubrication are referred to as its tribological properties. These properties are crucial when 3D-printed parts or components are put through motion or sliding contact.

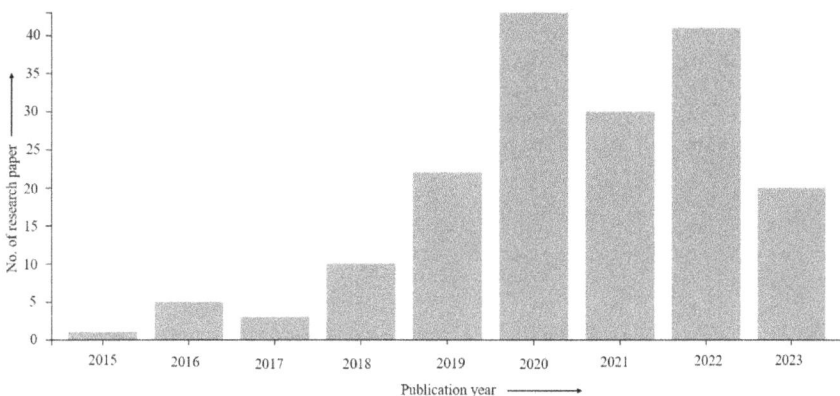

FIGURE 1.1 Number of research papers reported for tribological properties of 3D printing materials.

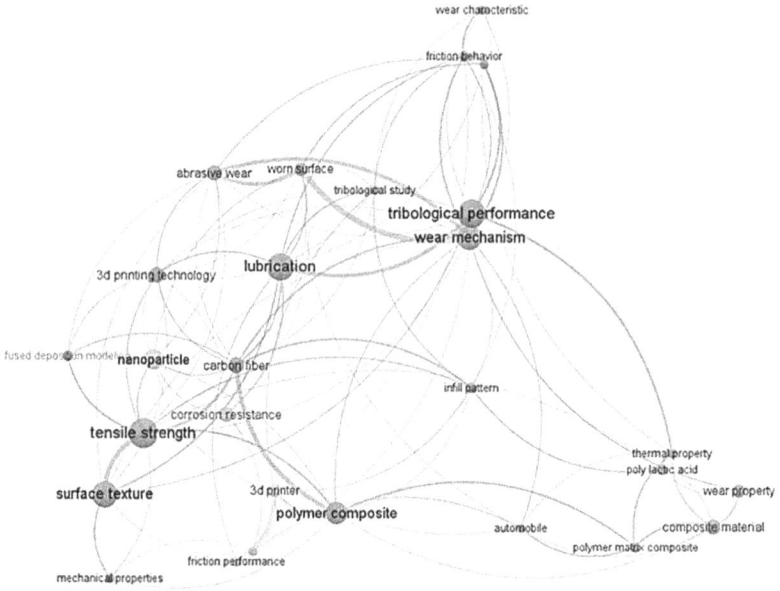

FIGURE 1.2   Relationship diagram of terms from reported literature of tribological properties related to 3D printing of materials.

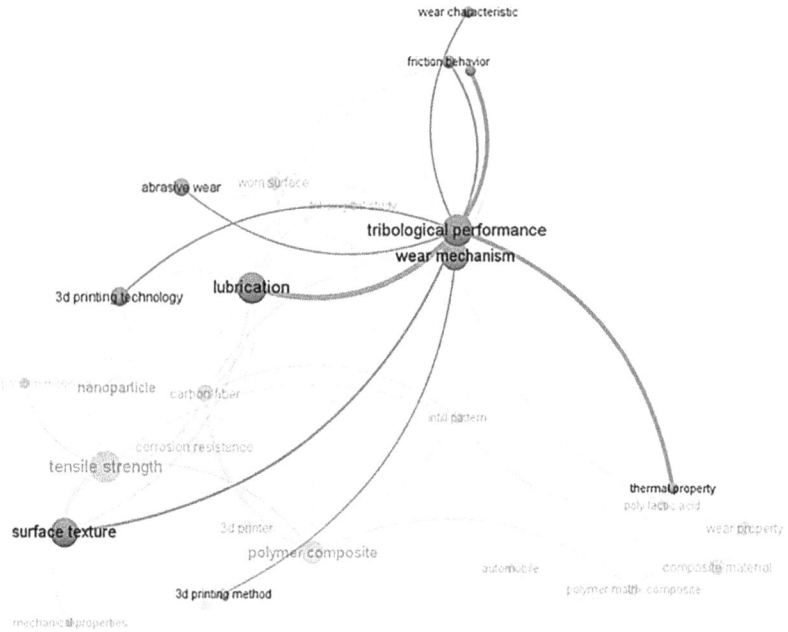

FIGURE 1.3   Gaps in the reported previous literature.

TABLE 1.1

Relevance Score of Selected Terms for the Development of Relationship Diagram

Id	Term	Occurrences	Relevance Score
1	3d printer	8	1.49
2	3d Printing method	5	1.33
3	3d Printing technology	11	0.27
4	Abrasive wear	10	0.52
5	Automobile	5	1.05
6	Carbon fiber	11	0.49
7	Composite material	10	2.09
8	Corrosion resistance	11	0.74
9	Disc tribometer	5	0.64
10	Friction behaviour	6	0.77
11	Friction performance	6	2.33
12	Fused deposition modelling	6	0.96
13	Infill pattern	7	0.57
14	Lubrication	23	0.60
15	Mechanical properties	5	1.13
16	Nanoparticle	15	0.76
17	Poly lactic acid	6	1.39
18	Polymer composite	17	0.71
19	Polymer matrix composite	6	1.83
20	Surface texture	21	0.79
21	Tensile strength	26	0.75
22	Thermal property	5	1.48
23	Tribological performance	24	0.58
24	Tribological study	5	0.53
25	Wear characteristic	5	1.26
26	Wear mechanism	20	0.43
27	Wear property	8	1.81
28	Worn surface	9	0.54

The composition of the material, COF, wear resistance, surface roughness, lubrication, and environmental conditions are the main factors to take into account. Different 3D printing materials have different tribological properties, and elements like surface roughness and lubrication contribute to minimizing wear and friction. The tribological behaviour of the materials can also be impacted by environmental factors. The tribological properties and prospective applications of various thermoplastics materials is shown in Table 1.2.

TABLE 1.2

Tribological Properties and Prospective Applications of Various Thermoplastic Materials

Sr. No.	Polymer	Reinforced material	Findings	Prospective applications	Ref.
1.	ABS and PLA	–	Effect of process parameter on wear	Windshields, door hinges, and hoses	Roy and Mukhopadhyay (2021)
2.	Polyester ether ketone (PEEK)	–	PEEK is a good polymer for tribological applications	Medical devices	Singh et al. (2019)
3.	Polypropylene (PP)	CNT	Wear resistance increases	Packaging, laboratory equipment, and automobile parts	Gandhi et al. (2013)
4.	PP	Carbonized bone	Wear rate decreased	Automobile parts	Asuke et al. (2014)
5.	Thermoplastic polyurethane (TPU)	Titanium micro powder	Wear rate increases with the increment of titanium wt.%, Shore hardness increases	Drive belts, automotive instrument panels, and medical devices	Banoriya et al. (2020)
6.	PC	ABS	Wear resistance improved by reduction in layer thickness and orientation	Automotive parts, industrial equipment, and electronic housings	Mohamed et al. (2018)
7.	ABS	Calcium carbonate (CaCO$_3$)	Optimum specific wear rate has been observed	Electronic components, packaging	Sudeepan et al. (2014)
8.	ABS	Nano-zirconia and PTFE	Tribological properties have improved	Engine components, surgical components	Amrishraj and Senthilvelan (2018)

1.4 LUBRICATION TECHNIQUES TO REDUCE WEAR RATE AND FRICTION BEHAVIOUR

1.4.1 SOLID LUBRICANTS

The wear and friction caused by thermoplastics during friction phenomena can be significantly reduce by using solid lubricants. Graphite, molybdenum disulfide (MoS$_2$), and polytetrafluoroethylene (PTFE) are common examples of solid lubricants. By providing a lubricated coating on the surface exposed to wear, such additives can reduce the wear and physical contact between connecting components (Aderikha et al., 2014).

1.4.2 Liquid Lubricants

Liquid lubrications applied to thermoplastic surfaces exposed to friction can reduce friction as well as wear. Oils, greases, and silicone-based lubricant are some common examples of the liquid lubricants. The liquid lubricants act as the thin coating between the solid surfaces to minimize direct contact (Zhang et al., 2020). The use of liquid lubricants is helpful in reducing the wear rate caused due to friction and heat.

1.4.3 Self-lubricating Thermoplastic Materials

Some of the thermoplastics materials are made with the purpose of having built-in self-lubricating properties. The self-lubricating thermoplastics uses fillers or additives materials that release lubricants to create a lubricated layer between the surfaces when friction occurs. The example involves 3D printed and oil impregnated porous polyimide for self lubricating property (Yang et al., 2021).

1.4.4 Use of Composite Materials

The wear resistance properties of thermoplastic materials can be tuned by adding fibers or fillers/additives that resist abrasion during the friction. The mechanical properties of the material are improved by using additives like glass fibers and carbon fibers, which decrease wear rate and increase the durability of the materials as well as products (Jesthi et al., 2018).

1.5 SUSTAINABILITY ASPECTS RELATED TO TRIBOLOGICAL PROPERTIES OF 3D-PRINTED POLYMERS

The sustainability of the 3D printed parts (especially thermoplastics) in tribological applications are dependent on some of the factors including circular economy/recyclability, mechanical properties, energy consumption and economy of the processes (Friedrich, 2018; Singh et al., 2019; Gbadeyan et al., 2021). The wear rate and frictional force depended upon the materials selected in such applications. Using the materials which exhibit low COF can have good wear resistance properties. Norani et al. (2021) have reported that parameters of 3D printing process are significantly affect the wear properties. The tribological properties can be improved of 3D-printed parts to function more effectively in tribological applications, especially for less energy consumption. Reducing the wasteful energy consumption in tribological applications can promote the sustainability of 3D printed parts in tribological applications. The selection of right materials in AM with good tribological properties can help in improving the efficiency of manufactured components and decreasing the negative environmental effects. For example, selecting materials that apply in dry environment or with less lubrication might reduce the requirement of lubricants and the chances of waste generation. The materials selected for the 3D printing should have recyclable properties and suitable for the tribological applications. Specifically, the good wear resistance properties of 3D printed parts reduce the materials losses. In such applications, the 3D printing processes are helpful in promoting the sustainability by optimizing the use of raw materials and waste generation. In recent times, the use of biodegradable and bio-compatible materials (e.g. PEEK polymer) have increased by 3D printing processes due to their excellent wear properties in biomedical applications (Singh et al., 2019). Using such biodegradable and biocompatible materials, the sustainability of the processes and products may be ascertained. It is possible to improve sustainability by decreasing the consumption of energy, increasing the life of 3D printed parts, increasing material effectiveness, introducing the recycling, and selecting the sustainable materials. The 3D printing process is capable of manufacturing such sustainable components/parts in tribological applications (Singh et al., 2019; Kaur et al., 2022).

1.6   CONCLUSION

This chapter has reviewed the tribological aspects of the 3D printing process for polymers, their applications and related technical details. The 3D printing processes play key role in the various fields such as the automotive industry, aerospace industry, and nowadays medical fields. The applications of the 3D printing processes have advanced greatly as a result of research into tribological properties. It may be expected that the effectiveness, economy, and sustainability of the 3D printed parts in tribological applications will continue to be increased. The recent developments of 3D printing for tribological applications have balanced need for energy efficiency, decreased maintenance costs, and enhanced system reliability by minimizing friction and wear rate. Specifically, with the development of energy-effective systems (such as 3D printing processes), and the use of sustainable lubricants, the applications domain of the tribology may be extended in the fruitful way.

ACKNOWLEDGEMENTS

The authors are thankful to University Centre for Research and Development, Chandigarh University for technical support.

REFERENCES

Aderikha VN, Krasnov AP, Shapovalov VA, Golub AS. (2014) Peculiarities of tribological behaviour of low-filled composites based on polytetrafluoroethylene (PTFE) and molybdenum disulfide. *Wear*. 320:135–42. https://doi.org/10.1016/j.wear.2014.09.004.

Amrishraj D, Senthilvelan T. (2018) Dry sliding wear behaviour of ABS composites reinforced with nano Zirconia and PTFE. *Materials Today: Proceedings*. 5(2):7068–77. https://doi.org/10.1016/j.matpr.2017.11.371.

Asuke F, Abdulwahab M, Aigbodion VS, Fayomi OS, Aponbiede O. (2014) Effect of load on the wear behaviour of polypropylene/carbonized bone ash particulate composite. *Egyptian Journal of Basic and Applied Sciences*. 1(1):67–70. https://doi.org/10.1016/j.ejbas.2014.02.002.

Banoriya D, Purohit R, Dwivedi RK. (2020) Wear performance of titanium reinforced biocompatible TPU. *Advances in Materials and Processing Technologies*. 6(2):284–91. https://doi.org/10.1080/2374068X.2020.1731232.

Briscoe BJ, Sinha SK. (2008) Tribological applications of polymers and their composites: Past, present and future prospects. *Tribology and Interface Engineering Series*. 55:1–14. https://doi.org/10.1016/S1572-3364(08)55001-4.

Friedrich K. (2018) Polymer composites for tribological applications. *Advanced Industrial and Engineering Polymer Research*. 1(1):3–9. https://doi.org/10.1016/j.aiepr.2018.05.001.

Gandhi RA, Palanikumar K, Ragunath BK, Davim JP. (2013) Role of carbon nanotubes (CNTs) in improving wear properties of polypropylene (PP) in dry sliding condition. *Materials & Design*. 48:52–7. https://doi.org/10.1016/j.matdes.2012.08.081.

Gbadeyan OJ, Mohan TP, Kanny K. (2021) Tribological properties of 3D printed polymer composites-based friction materials. *Tribology of Polymer and Polymer Composites for Industry*. 4:161–191. https://doi.org/10.1007/978-981-16-3903-6_9.

Gurrala PK, Regalla SP. (2014) Friction and wear behaviour of abs polymer parts made by fused deposition modeling (FDM). *Technology Letters*. 1:13–7.

Hanon MM, Alshammas Y, Zsidai L. (2020a) Effect of print orientation and bronze existence on tribological and mechanical properties of 3D-printed bronze/PLA composite. *The International Journal of Advanced Manufacturing Technology*. 108:553–70.

Hanon SM, Kovács M, Zsidai L. (2019) Tribological behaviour comparison of ABS polymer manufactured using turning and 3D printing. *International Journal of Engineering and Management Sciences*. 4(1):46–57. https://doi.org/10.21791/IJEMS.2019.1.7.

Hanon MM, Marczis R, Zsidai L. (2020b) Impact of 3D-printing structure on the tribological properties of polymers. *Industrial Lubrication and Tribology*. https://doi.org/10.1108/ILT-05-2019-0189.

Jabbarzadeh A. (2018) "Tribological Properties of Interfacial Molecular Films." In *Encyclopedia of Interfacial Chemistry: Surface Science and Electrochemistry*, edited by Klaus Wandelt, 864–874. https://doi.org/10.1016/B978-0-12-409547-2.13128-2.

Jesthi DK, Mandal P, Rout AK, & Nayak RK. (2018). Enhancement of mechanical and specific wear properties of glass/carbon fiber reinforced polymer hybrid composite. *Procedia Manufacturing*, 20:536–541. https://doi.org/10.1016/j.promfg.2018.02.080

Kasar AK, Chan A, Shamanaev V, Menezes PL. (2022) Tribological interactions of 3D printed polyurethane and polyamide with water-responsive skin model. *Friction*. 10:159–66. https://doi.org/10.1007/s40544-020-0472-2.

Kaur G, Singari RM, Kumar H. (2022) A review of fused filament fabrication (FFF): Process parameters and their impact on the tribological behaviour of polymers (ABS). *Materials Today: Proceedings*. 51:854–60. https://doi.org/10.1016/j.matpr.2021.06.274.

Mohamed OA, Masood SH, Bhowmik JL. (2018) Analysis of wear behaviour of additively manufactured PC-ABS parts. *Materials Letters*. 230:261–5. https://doi.org/10.1016/j.matlet.2018.07.139.

Mohamed OA, Masood SH, Bhowmik JL, Somers AE. (2017) Investigation on the tribological behaviour and wear mechanism of parts processed by fused deposition additive manufacturing process. *Journal of Manufacturing Processes*. 29:149–59. https://doi.org/10.1016/j.jmapro.2017.07.019.

Norani MN, Abdullah MI, Abdollah MF, Amiruddin H, Ramli FR, Tamaldin N, Tunggal D, dan Pembuatan FT, Malaysia UT. (2021) Mechanical and tribological properties of FFF 3D-printed polymers: A brief review. *Jurnal Tribologi*. 29:11–30.

Pan S, Shen H, Zhang L. (2021) Effect of carbon nanotube on thermal, tribological and mechanical properties of 3D printing polyphenylene sulfide. *Additive Manufacturing*. 47:102247. https://doi.org/10.1016/j.addma.2021.102247.

Roy R, Mukhopadhyay A. (2021) Tribological studies of 3D printed ABS and PLA plastic parts. *Materials Today: Proceedings*. 41:856–62. https://doi.org/10.1016/j.matpr.2020.09.235.

Singh S, Prakash C, Ramakrishna S. (2019) 3D printing of polyether-ether-ketone for bio-medical applications. *European Polymer Journal*. 114:234–48. https://doi.org/10.1016/j.eurpolymj.2019.02.035.

Slapnik J, Stiller T, Wilhelm T, Hausberger A. (2020) Influence of solid lubricants on the tribological performance of photocurable resins for vat photopolymerization. *Lubricants*. 8(12):104. https://doi.org/10.3390/lubricants8120104.

Soundararajan R, Jayasuriya N, Vishnu RG, Prassad BG, Pradeep C. (2019) Appraisal of mechanical and tribological properties on PA6-TiO2 composites through fused deposition modelling. *Materials Today: Proceedings*. 18:2394–402. https://doi.org/10.1016/j.matpr.2019.07.084.

Sudeepan J, Kumar K, Barman TK, Sahoo P. (2014) Study of tribological behaviour of ABS/CaCO3 composite using grey relational analysis. *Procedia Materials Science*. 6:682–91. https://doi.org/10.1016/j.mspro.2014.07.084.

Yang C, Jiang P, Qin H, Wang X, Wang Q. (2021) 3D printing of porous polyimide for high-performance oil impregnated self-lubricating. *Tribology International*. 160:107009. https://doi.org/10.1016/j.triboint.2021.107009.

Zhang Y, Dong Z, Li C, Du H, Fang NX, Wu L, Song Y. (2020) Continuous 3D printing from one single droplet. *Nature Communications*. 11(1):4685. https://doi.org/10.1038/s41467-020-18518-1.

2 Investigation on Tribology of Additively Manufactured Metal Part

Kumar Ujjwal and Raushan Kumar
Indian Institute of Technology (ISM Dhanbad)

Rashi Tyagi
Chandigarh University

Alok Kumar Das
Indian Institute of Technology (ISM Dhanbad)

2.1 INTRODUCTION

The term additive manufacturing (AM), often used synonymously with 3D printing, refers to the process where the material is selectively deposited in a layer-by-layer fashion to form a component from the computer-aided design (CAD) model data (Ranjan et al., 2023a; Tyagi et al., 2023). The term prototype refers to the 1:1 scale working model of the component or product in the development phase. So, in RP, the prototype of the component is made rapidly to have a physical feel of the component and to test and modify the product's design (Ranjan et al., 2023b; Tyagi & Tripathi, 2023). Whereas the term 3D printing refers to the process of forming the component using 3D printers from the CAD data. In contrast to conventional manufacturing processes, which follow the top-down approach, where the sequence of operations is laid out to manufacture the part, AM utilizes the bottom-up approach, joining the material in a layered manner to form the final component. As per ASTM standards, the AM has been classified into seven categories (ASTM E384, 2002), viz. DED, PBF, material extrusion, binder jetting (BJ), material jetting, sheet lamination, and vat photopolymerization, as depicted in Figure 2.1.

2.2 DIFFERENT AM PROCESSES

2.2.1 DIRECTED ENERGY DEPOSITION (DED)

Directed energy deposition (DED) utilizes energy in the form of a laser, electron beam, or electric arc to melt and deposit the material selectively with the help of CNC (computer numerical control) or robot, and the feedstock material can be powder or wire, as depicted in Figure 2.2. It is suitable for printing large metallic components

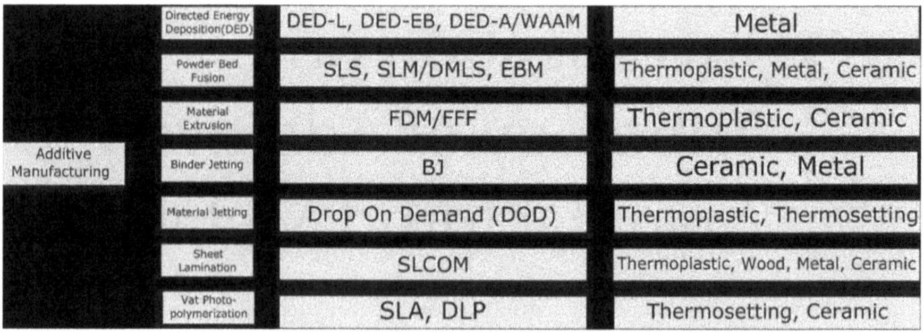

Additive Manufacturing	Directed Energy Deposition(DED)	DED-L, DED-EB, DED-A/WAAM	Metal
	Powder Bed Fusion	SLS, SLM/DMLS, EBM	Thermoplastic, Metal, Ceramic
	Material Extrusion	FDM/FFF	Thermoplastic, Ceramic
	Binder Jetting	BJ	Ceramic, Metal
	Material Jetting	Drop On Demand (DOD)	Thermoplastic, Thermosetting
	Sheet Lamination	SLCOM	Thermoplastic, Wood, Metal, Ceramic
	Vat Photo-polymerization	SLA, DLP	Thermosetting, Ceramic

FIGURE 2.1   AM classification.

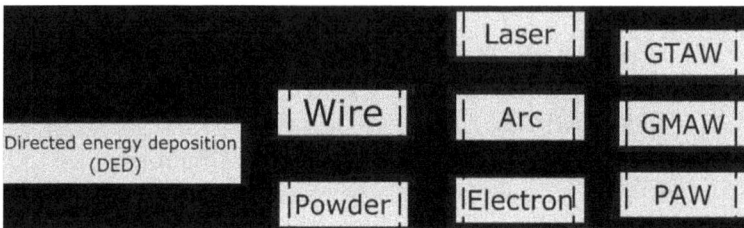

Directed energy deposition (DED)	Wire	Laser	GTAW
		Arc	GMAW
	Powder	Electron	PAW

FIGURE 2.2   DED classification.

with medium to low level of complexity in their features. Generally, in laser-based DED or DED-L, if the feedstock is powder, it is fed co-axially through the nozzle. In the case of wire as a feedstock, the wire is fed with the help of an automatic wire feeder, as shown in Figure 2.3. Similarly, in the case of an electron beam DED process, the laser source is replaced by the e-beam. In addition, if the heat source is in the form of an electric arc and the fed material is a wire, it is referred to as arc-based DED and also known as wire arc additive manufacturing (WAAM).

2.2.2 POWDER BED FUSION (PBF)

Powder bed fusion (PBF) was the first among all AM techniques to be commercialized (Gibson et al., 2015). Figure 2.4 presents the schematics of a typical PBF setup. It consists of a laser as an energy source, and a thin layer of metallic powder is laid onto the substrate with the help of a roller. The laser is scanned over this bed of powder to fuse the powder together. Later, the stage is lowered, a new layer of powder is laid onto the stage, and again laser is scanned. This process continues until the complete part is printed. It is suitable for printing thermoplastics, metals, and ceramics.

2.2.3 MATERIAL EXTRUSION

This process is often used for the printing of thermoplastics. Fused deposition modelling (FDM) is one of the most popular material extrusion AM techniques. Figure 2.5 shows the basic material extrusion setup. The material in the form of filament is fed

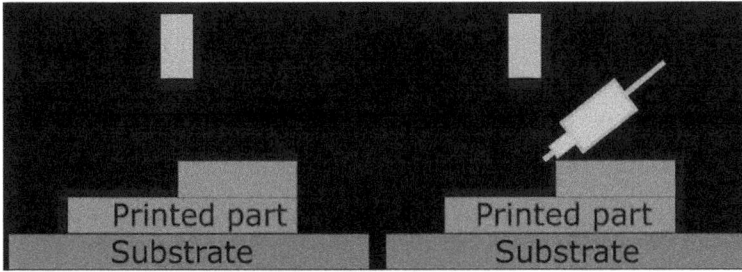

FIGURE 2.3 Laser-based DED.

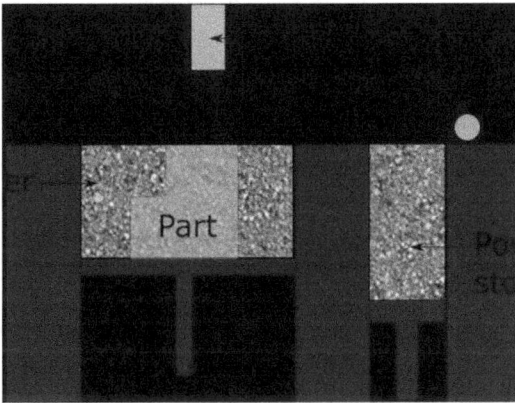

FIGURE 2.4 PBF schematics.

FIGURE 2.5 Material extrusion setup.

into the liquefier chamber with the help of a pinch roller. The heating element heats up the filament and liquefies it (Stevens & Covas, 2012). This molten material is forced through the nozzle and allowed to deposit in a selective manner.

2.2.4 BINDER JETTING

Binder jetting technology is suitable for printing sand moulds, metals, and ceramics. The idea for the working principle was inspired by the inkjet printers. Figure 2.6 demonstrates the working of the binder jetting AM setup. The setup is similar to PBF, but here instead of a laser (energy source), the binders are selectively sprayed over the material layer, which is in the form of powder. Following the binder spray, the heating element evaporates the excess binder, and the process goes on until the required part is built (Sachs et al., 1990). Further, the component is extracted from the powder bed and sintered to ensure the strength of the component.

2.2.5 MATERIAL JETTING

Material jetting, also referred to as drop on demand (DOD), is suitable for printing of waxy polymers, ceramics, and metals. The idea was inspired by the available ink-jet printer technology, where the printable ink is sprayed onto the surface and dried. Figure 2.7 presents the schematics of the material jetting AM setup. The photopolymer material is heated by the heating coil and selectively sprayed, followed by ultraviolet (UV) curing of the polymer. Then the build plate is lowered, and again the next layer is deposited; this process continues until the final part is built (Gibson et al., 2015). Later in the post-processing, the support structure is removed from the part and cleaned.

2.2.6 SHEET LAMINATION

Sheet lamination or laminated object manufacturing (LOM) uses thin sheets of material to make 3D parts. Figure 2.8 shows the typical LOM setup in which the material to be printed is fed in the form of sheets (Wimpenny et al., 2003). The part

FIGURE 2.6   Binder jetting setup.

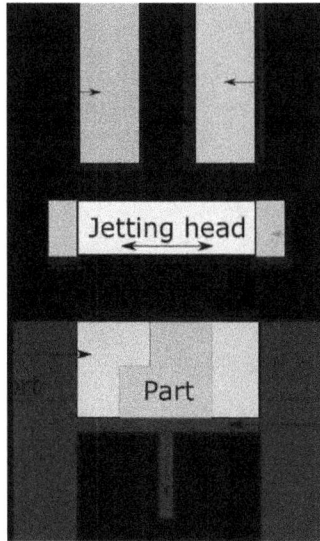

FIGURE 2.7   Material jetting setup.

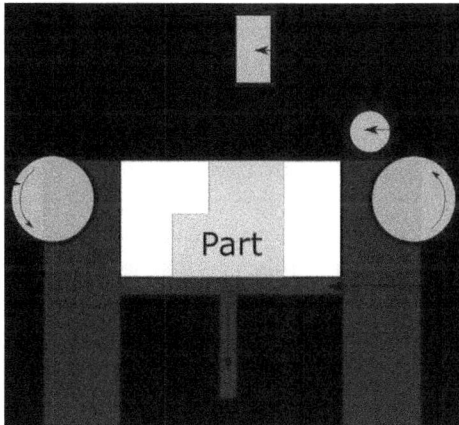

FIGURE 2.8   Laminated object manufacturing setup.

to be printed is sliced into thin cross-sections; each cross-section is cut with the help of a laser, and the excess material is cross-hatched. The layers are joined together with the help of adhesive.

2.2.7   VAT PHOTOPOLYMERIZATION

Photopolymers are polymers that turn into solids when exposed to light due to the chemical reaction known as photopolymerization. Vat photopolymerization AM takes advantage of this phenomenon to print polymer prototypes. Figure 2.9 shows

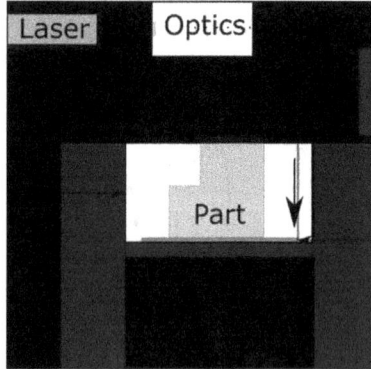

FIGURE 2.9   Vat photopolymerization setup.

the typical vat photopolymerization setup. The vat consists of the photopolymer, and the laser or UV light scans over the platform dipped in the vat. Subsequently, the vat is lowered based on the cross-section thickness of the component, and these steps are repeated until the final component is built.

2.3   BASIC TRIBOLOGY

Tribology is the branch of science that deals with the relative motion between two rubbing surfaces. The design, usage, and maintenance of tribological systems are the focus of tribology in engineering. Any system where two or more surfaces are moving relative to one another is referred to as a tribological system. Examples include bearings, gears, locks or seals, and brakes. The important pillars of engineering tribology are surface-dominant properties such as friction, wear, and lubrication. Engineering tribology aims to prolong the lifespan of tribological systems while reducing friction and wear. Several strategies can be used to do this, such as selecting the appropriate materials for the surfaces coming into contact, designing the system to reduce contact pressure, using lubrication, applying surface coatings, keeping an eye out for deterioration, and monitoring the system.

Even while engineering tribology is a convoluted and challenging field, it is also significantly rewarding. Engineers may develop and operate improved, reliable, and durable tribological systems by having a solid understanding of their underlying concepts. Tribology can also contribute to improved reliability by lowering the risk of wear-related failure in machinery and devices. Numerous engineering fields depend heavily on the topic of engineering tribology. It is necessary for the design and functioning of machinery and gadgets with moving parts. Tribology is also used extensively in the medical, aerospace, and auto industries.

2.3.1   SURFACES AND CONTACTS

Engineering tribology is a complex and difficult field, but it is also a very significant one. It studies surfaces and interactions. Engineers can design materials and components with low friction and wear by understanding the principles of surface and contact tribology, which can result in longer component life and better performance. In engineering tribology, surfaces and contacts represent important considerations. Where a substance interacts with its surroundings is on its surface, and friction and wear happen when two surfaces come into contact. Surface roughness has an impact on how much wear and friction it experiences, as well as how effectively it adheres to other materials. The choice of method for measuring surface roughness depends on the application and the required accuracy. For example, if the surface roughness is critical for the performance of a component, then a more accurate method, such as a stylus instrument or AFM, may be required. However, if the surface roughness is not critical, then a less accurate method, such as a profilometer, may be sufficient.

There are three main categories of methods for measuring surface roughness:

1. Direct measurement methods use a physical probe to contact the surface and measure the height of the peaks and valleys.
2. Comparison-based techniques compare the surface to a known standard. This can be done visually or with instruments such as a Wallace surface dynamometer.
3. Non-contact methods do not touch the surface at all. These methods use optical, acoustic, or electron beam techniques.

Surface contacts are the points at which two surfaces come into contact with each other. In tribology, the study of surfaces in relative motion, the type of surface contact plays a critical role in determining the friction and wear characteristics of the system. The area of contact, the normal load, and the coefficient of friction are used to characterize the surface contact.

There are two main types of surface contacts:

- **Asperity Contact:** This is the most common type of surface contact. It occurs when the peaks of the two surface's asperities come into contact with each other.
- **Adhesion Contact:** This kind of surface contact happens when adhesive forces are used to bind the surfaces together. Adhesion contact is typically found in systems where the surfaces are made of similar materials or where the surfaces are contaminated with a third substance. Adhesion contact can lead to high friction and wear, and it can also be difficult to break apart.

The type of surface contact has a significant impact on the friction and wear characteristics of a tribological system. The surface roughness is the most important factor, as it determines the size of the asperities and the true area of contact. The materials of the surfaces also play a role, as some materials are more prone to adhesive contact than others. Asperity contact typically leads to lower friction and wear than adhesion contacts. However, adhesion contact can be more difficult to break apart, which can lead to problems in some applications.

2.3.2 FRICTION

The force of resistance to motion created by solid surfaces, fluid layers, and moving material components is known as friction (Hanaor et al., 2016). The friction force needed to start sliding is known as the static friction force, and the friction force needed to maintain sliding is known as the kinetic friction force. For the same combination of material and other parameters, the latter (μ_k) usually has a lower value than the former (μ_s).

The importance of friction may be classified into four categories:

  i. The inclusion of a sufficient amount of friction for activities like walking, grasping objects, and operating machinery;
 ii. the reduction of friction in situations requiring the use of machinery;
iii. the maintenance of friction within specific limits for activities like metal rolling industries, and precision equipment; and
 iv. the provision of essential requirements for friction-caused oscillation for operations like stringed musical instruments.

2.3.2.1 The Coefficient of Friction

The resistance of the two surfaces to sliding past one another is measured by the coefficient of friction (COF), a dimensionless quantity. Consider a body on a horizontal plane subjected to a normal load W, and progressively, a force F is applied parallel to the plane in order to set it into motion and increase its speed from 0 to s. A friction force F in the plane, acting in opposition to the sliding motion, will be produced as a result of this displacement. We can then compute the COF, which is denoted by Equation (2.1), using W and F as experimental inputs:

$$F = \mu W \tag{2.1}$$

where F represents force of friction and W represents the normal load. The proportionality constant μ, denoted by the symbol, is also referred to as the friction coefficient. The COF is written as μ_s, μ_k, or μ_r, depending on whether the body is in a static, sliding motion, or rolling motion.

2.3.2.2 Measurement

In order to measure the tangential force restraining motion, two specimens must be placed together, moved relative to one another, and subjected to a defined Normal load.

  i. **Inclined Plane Rig**

   A specimen is set down on a flat surface as shown in Figure 2.10 whose angle of inclination with respect to the horizontal increases gradually till the specimen on the surface begins to slide. If the inclination at this moment is α, then $\mu_s = \tan \alpha$. It is clear that this method cannot determine friction in a continuous sliding action.

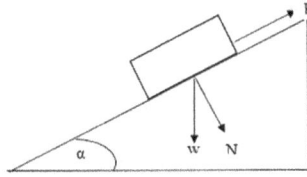

FIGURE 2.10   Measuring friction by an inclined.

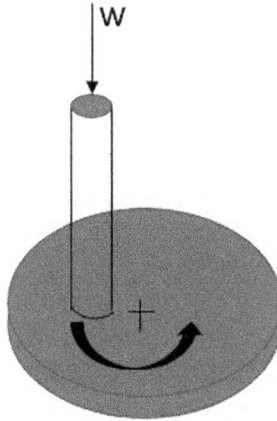

FIGURE 2.11   Pin-on-disc device.

ii. **Pin-on Disc Rig**

The pin-on-disc setup rig-based (Figure 2.11), which is used in continuous sliding cases, uses a pin that is held static while the disc is made to revolve at a normal load. With the aid of a calibrated tangential motion of a capacitive or inductive transducer installed on the stationary sample from the centre of the disc, the friction force is recorded; however, in a single-pass setup, the transducer is moved radially during the experiment. Figure 2.12 displays more common configurations, such as pin-on-cylinder, crossed-cylinders, and reciprocating configurations.

2.3.2.3   Friction Theories

a. **Bowden and Tabor's Simple Adhesion Theory**

As was previously mentioned, when two surfaces are loaded, intimate contacts form at the asperity peaks, and the actual area of contact is much less than the nominal area. The asperity tip of a softer material deforms plastically and widens the total contact area due to the strong pressure at the actual points of contact. Two factors contribute to this expansion of the contact area: (a) individual contact sites and (b) the beginning of new contacts.

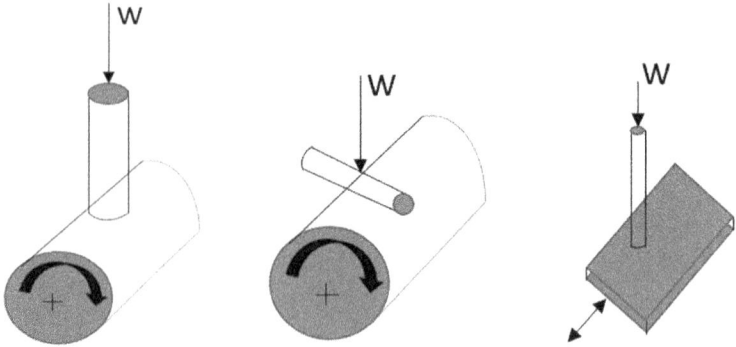

FIGURE 2.12 Friction measuring devices: (a) pin-on-cylinder, (b) crossed cylinders, and (c) reciprocating rig.

When the total actual surface area of contact is enough to support the load elastically till the process is repeated. The normal load W for elastic-plastic material can then be represented as

$$W = A \cdot P_o$$

Hence, A is the real area of contact and P_o is the yield pressure of softer material, which is very close to the hardness, H.

$$W = A \cdot H$$

Asperity joints are cold welded and strong adhesive bonds are created as the plastic deformation is higher. If the friction force from adhesion is F_{ad}, which is the force necessary to induce the shear failure of the Junctions, then

$$F_{ad} = A \cdot \Pi$$

where Π is the shear strength of the softer material. Thus, the coefficient of friction due to adhesion is

$$\mu_{ad} = F_{ad}/W = \frac{\Pi}{H}$$

b. **Deformation Theory: Ploughing**

When the adhesion is little, ploughing takes over as the dominant source of friction. If the surface during sliding contact is harder compared with the another and when shear strength is surpassed, then the asperities of the harder surface may plough through the softer surface and generate grooves. Ploughing can also happen when wear particles are impacted or entrapped. When two rough surfaces meet, mechanical interlocking may occur, which leads to one of the surfaces being ploughed during sliding. In addition to raising friction, ploughing also produces wear particles, which raise subsequent friction and wear.

Four such model asperities are considered to calculate the ploughing component: (a) conical (b) spherical, (c) upright cylindrical, and (d) transverse cylindrical. During sliding only, the front surface of the asperity comes into contact with the softer body. Therefore, the normal load W is supported by the vertical projection of the asperity contact area (A_V) and the friction force (F) is supported by the horizontal projection of the asperity contact area (A_H). Considering the yield pressure of the solid as P_O and the isotropic yielding

$$W = P_O \cdot A_V$$

$$F = P_O \cdot A_H$$

Thus, the coefficient of friction due to ploughing is given as follows:

$$\mu_P = \frac{F}{W} = A_V / A_H$$

Therefore, by choosing materials with greater, less, or the same hardness as well as by removing wear particles from the contact friction component, wear particles can be minimized.

2.3.3 WEAR

Wear is the removal or loss of material moving relative to two solid surfaces (sliding, rolling, or impacting). It is a system reaction that is strongly influenced by the operating conditions. Surface damage due to material displacement with no net change in volume or weight is also called wear. Desirable cases of wear include machining, polishing, shearing, and writing with a pencil, whereas undesirable cases include almost all machine applications such as bearings, gears, cams, and seals.

Friction and wear are often correlated, but they are not the same thing. Friction is the force that opposes the relative motion of two surfaces in contact, while wear is the loss of material from a surface due to friction. High friction can sometimes lead to increased wear rates, but this is not always the case. Friction is caused by the microscopic interlocking of asperities (roughness) on the two surfaces, as well as by adhesive forces between the surfaces. Type of loading (impact, static, or dynamic), type of motion (slide, rolling), temperature, and lubrication all have a significant effect on wear rate. So, the study of wear is important for designing and optimizing machine parts.

Wear can be caused by a number of mechanisms, including abrasion, adhesion, fatigue, and corrosion. Abrasion is the most common type of wear, and it occurs when hard particles or grit are dragged across a surface, causing the surface to be scratched or pitted. Adhesion is the process by which two surfaces stick together, and it can lead to wear when the two surfaces are pulled apart. Fatigue wear occurs when a surface is repeatedly stressed, causing tiny cracks to form that eventually break apart.

2.3.3.1   Wear Types and Mechanisms

Relative motion, the nature of the mechanism, and whether it has an impact on the base layer or a layer that is capable of self-regeneration are used to classify different types of wear (Varenberg, 2013). Synergistic interactions between wear mechanisms as well as sub-mechanisms frequently occur, leading to a higher rate of wear than the total number of individual wear mechanisms (Williams, 2005).

a. **Adhesive Wear**

Adhesive wear is the unintended transfer of material compounds and wear debris from one surface to another while the surfaces are in frictional contact. Adhesive wear occurs when the atomic forces between the materials in contacting surfaces under relative pressure outweigh the inherent material characteristics of each surface.

There are two types of this wear:

1. Relative motion, "direct contact," and plastic deformation, which result in wear debris as well as material transfer from one surface to another, are the causes of adhesive wear.
2. Cohesive adhesive forces, with or without actual material transfer, hold two surfaces together even though they are separated by a measurable distance.

    Typically, material transfer is facilitated when two bodies slide over or are pressed into one another, which causes adhesive wear. Strong adhesive forces between atoms (Rabinowicz, 1995) and energy buildup in the plastic zone among asperities while in motion both contribute to the severity of how oxide fragments are pulled off and added to the other surface, which is affected by the asperities or surface roughness present on each surface.

    However, naturally occurring contaminants, lubricants, and oxidation films typically prevent adhesion (Stachowiak & Batchelor, 2005). Adhesive wear can cause the surface to become rougher and to develop protrusions (also known as lumps) above the surface.

b. **Abrasive Wear**

Abrasive wear occurs when a hard-rough surface slides across a softer surface (Rabinowicz, 1995). 2-body and 3-body abrasive wear are the types of abrasive wear. Hard particles such as Grits take material from the opposing surface, causing 2-body wear. The typical comparison is during a cutting or ploughing process involving the removal or shifting of material. When particles are unrestrained and allowed to freely roll and glide along a surface, 3-body wear happens. Abrasive wear is commonly classified according to the type of contact and the contact environment (Coeffi, 2011). The type of wear—open or closed—is determined by the contact environment. The Taber Abrasion Test can be used to measure abrasive wear as a loss of mass according to ASTM D 4060 or ISO 9352.

The three most frequently recognized mechanisms of abrasive wear are ploughing, cutting, and fragmentation. Ploughing happens when material is moved to the side, away from the wear particles, creating grooves without actually removing any material. When abrasive particles are later introduced, the ridges and grooves created by the displaced material can be removed. Cutting happens when there is little to no material displaced to the sides of the grooves and material is separated from the surface in the form of the main debris or microchips. This mechanism has a lot in common with traditional machining. When material is cut away from a surface, fragmentation happens because the wear material is locally fractured by the indenting abrasive. Then, as these cracks spread locally and freely around the wear groove, more material is removed by spalling (Coeffi, 2011).

c. **Surface Fatigue Wear**

Surface fatigue is a type of general fatigue of material in which the surface of a material wears out due to cyclic loading. When the wear particles are separated from the surface by cyclic microcrack formation, fatigue wear is created. There are two types of microcracks: superficial cracks and subsurface cracks.

d. **Fretting Wear**

Fretting wear is caused by repeated cycle rubbing of two surfaces. Fretting over a period of time removes material from one or both of the surfaces in contact. It typically happens in bearings. When the airborne oxidation of the small particles is eliminated by wear, this issue arises. Wear increases as the tougher particles progressively roughen the metal surfaces, because oxides tend to be harder than the base metal.

e. **Erosive Wear**

Erosive wear is a sliding motion that happens very quickly over a short time. Erosive wear occurs due to solid or liquid particles impacting an object's surface (Davis, 2001). The impacting particles repeatedly deform due to cutting actions on the surface, eroding material from it over time (Padhy & Saini, 2008).

Along with the characteristics of the surface being eroded, the primary factors that affect are the material characteristics of the particles, such as their shape, hardness, impact velocity, and impingement angle. One of the most crucial elements is the impingement angle, which is widely acknowledged in literature (Sinmazçelik & Taşkiran, 2007). For non-ductile materials, the maximum wear rate is observed when the impingement angle is normal to the surface, whereas the wear rate for ductile materials is found when the angle is around 30° (Sinmazçelik & Taşkiran, 2007).

f. **Corrosion and Oxidation Wear**

Corrosion wear occurs when a surface is exposed to a corrosive environment, such as acids or chemicals. When the corroding media and the worn material interact chemically, it causes corrosion and oxidation wear. Tribocorrosion is another name for wear which arises due to the combined effect of tribological stresses and corrosion.

2.4 INFLUENCE OF DIFFERENT FACTORS ON TRIBOLOGICAL PROPERTIES OF ADDITIVELY MANUFACTURED (AMED) MATERIALS

2.4.1 THE INFLUENCE OF SURFACE FINISH

Wear may be facilitated by the additively manufactured (AMed) parts due to low surface finish. In general, wear is worse as surface roughness increases (Özel et al., 2011). The metallic AMed parts have a roughness of microns range (Lekkala et al., 2011). Although the printing conditions can be carefully adjusted to reduce this roughness, it is still too high for parts in contact (Chae et al., 2006). In order to obtain optimum lubrication, this high roughness not only increases the rate of wear during dry sliding (Edkinsa et al., 2014; Le Coz et al., 2020; Özel et al., 2011), but also needs lubricating oil with higher viscosity during lubrication (Sharman et al., 2015). Initial wear is linearly related to surface roughness in the case of steel (Bonaiti et al., 2017).

Some surface finishing processes can be used to enhance the AMed part's surface finish. Surface roughness can be significantly decreased by using mechanical polishing techniques like shot peening and barrel finishing (Le Coz et al., 2017). The metallic pin used in the ultrasonic burnishing process is made of robust materials. This pin is repeatedly forced against the printed metal surfaces by ultrasonic vibration (Bruschi et al., 2016; Le Coz et al., 2017). For AMed AlSi10Mg alloy, roughness Ra gets reduced from 18 to 3.5 μm. The material can only be polished on the outside using this approach. If the AMed item is made of a non-magnetic material, it can be polished using a magneto-rheological fluid that is propelled by magnetism. This can penetrate inside surfaces and reduce surface roughness to a level comparable to the ultrasonic approach (Bruschi et al., 2016). The tribological and wear behaviour of a part can also be improved by printing patterns on the surface (Hojati et al., 2020).

2.4.2 THE INFLUENCE OF MICROSTRUCTURE

Microstructure has an important role in factors that influence the wear behaviour of an AMed part. Due to its resistance to oxidation, stainless steel can experience severe adhesive wear through metal transfer and adhesion (Biswal et al., 2019; Wu et al., 2017). Additionally, the homogeneous austenite structure promotes the development of adhesive wear (Wu et al., 2017). Compared with wrought or cast components of AMed stainless steel, it has fine as well as inhomogeneous microstructure (Christopherson, 2015; Tutunea-Fatan et al., 2011; Varghese & Mujumdar, 2021) (Figure 2.13). Because of this microstructure, AMed parts are twice as hard as wrought steel (Varghese & Mujumdar, 2021), which reduces both abrasive and adhesive wear. AMed SS alloys are more resistant to wear when its hardness is comparable to that of wrought stainless steel (Fakhri et al., 2013). Despite being negative for corrosion resistance, chromium carbide segregation significantly increases wear resistance. In comparison to the cast alloy, the AMed aluminium alloy AlMgSc exhibits superior wear resistance (Gorsse et al., 2017). In this instance, the cellular structure reduces wear.

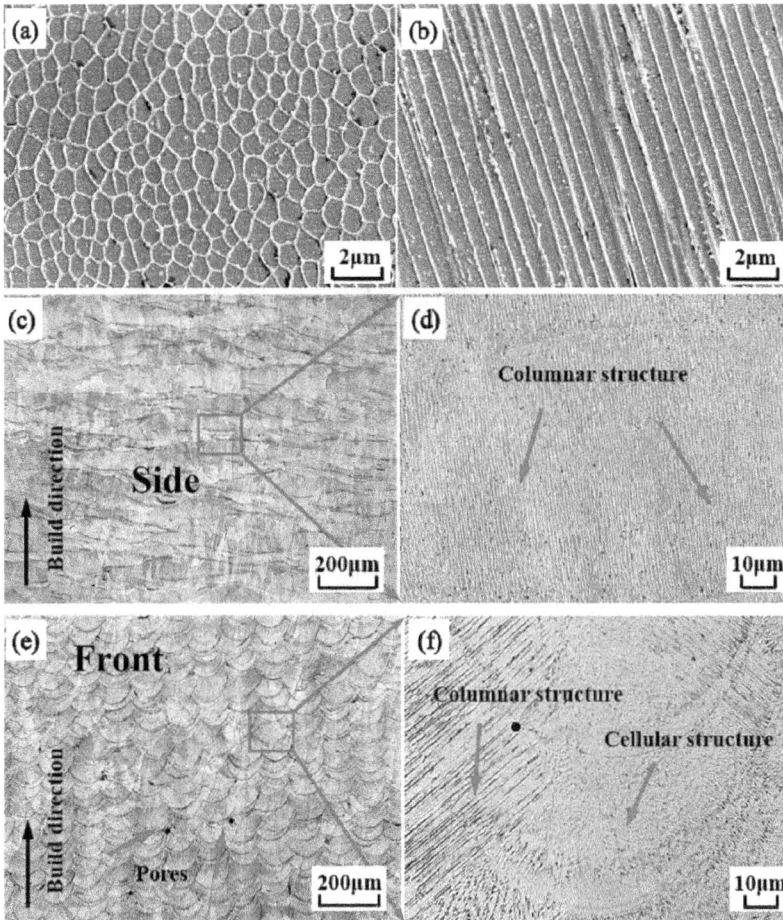

FIGURE 2.13   The microstructure SLM 316L stainless steel (Varghese & Mujumdar, 2021).

The AMed material has anisotropic wear performance due to its unique micro-structure (Baufeld et al., 2010; Vrancken et al., 2012). When the direction of sliding coincides with the printing direction (cellular structure's direction), either in the case of aluminium alloys or stainless steel, the rate of wear rises. Materials are more susceptible to adhesive wear in that direction because they also have the maximum friction. The effect is demonstrated in Figure 2.14.

Unfavourable microstructures in the AMed metal component can make it more prone to wear. Comparison with printed parts with optimal printing parameters, the wear rate of pores generated with unfavourable printing conditions for Inconel 718 can be doubled (Mok et al., 2008). The AMed TiC/Inconel 718 experiences increased wear due to the low-density porous structure as well (Mok et al., 2008). The hardness and wear resistance of stainless steel is reduced due to the larger grains created by the arc additive manufacturing process (Brandl et al., 2010). Repetitive melting can cause intergranular segregation of chromium carbide ($Cr_{23}C_6$) in stainless steel or

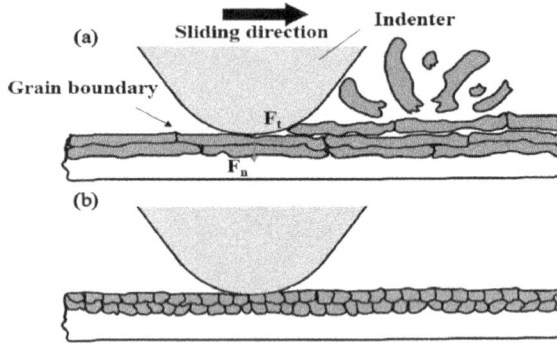

FIGURE 2.14 Schematic of wear resistance mechanism: (a) sliding and (b) perpendicular direction (Vrancken et al., 2012).

Co-Cr alloys. The hardness of materials is reduced by these separated carbide particles, and third body wear may occur (M'Saoubi et al., 2008).

The anti-wear performance of AMed items can be enhanced by heat treatment after printing. The Co alloy's metal matrix can have its carbide removed using the solutioning method (M'Saoubi et al., 2008; Shunmugavel et al., 2017). For printed Co-Cr-Mo alloy, this procedure can reduce wear by one-third. By producing a hard coating, the laser-treated Co-Cr-Mo alloy also has better wear resistance (Chen et al., 2017). By adding some alloying elements, the microstructure can also be improved.

2.5 TRIBOLOGICAL BEHAVIOUR OF ADDITIVELY MANUFACTURED TITANIUM ALLOYS

This section discusses the properties, wear behaviour, and effective ways to enhance the wear resistance of Ti alloys that is specific to the AM process. Titanium alloys are a group of materials that combine the element titanium with one or more other elements, often including aluminium, vanadium, nickel, or other metals. Titanium alloys are versatile materials prized for their exceptional properties, including strength, lightweight nature, corrosion resistance, biocompatibility, and high-temperature performance. Also, the titanium alloys are very formable and are simple to weld. High cost, difficult machinability, and limited availability are the challenges associated with titanium alloys. Titanium alloys have extensive applications in various industries, including aerospace, automotive, medical devices, and sports equipment, due to their unique characteristics.

2.5.1 CHARACTERISTICS OF TITANIUM ALLOYS

Strength and Lightweight: Titanium alloys possess a remarkable strength-to-weight ratio, making them a great choice for uses where strength and lightweight are important. This attribute is particularly crucial in aerospace engineering, where the need for strong yet lightweight materials is paramount.

Corrosion Resistance: Titanium alloys have an inherent resistance to corrosion, even in aggressive environments such as seawater or acidic conditions. This property makes them suitable for applications in marine equipment, chemical processing, and medical implants, where corrosion can compromise performance and safety.

Biocompatibility: The biocompatibility of titanium alloys makes them suitable for medical applications. They are often used in surgical implants, dental prosthetics, and medical instruments due to their ability to integrate well with living tissues and bones without causing adverse reactions.

High-Temperature Performance: Certain titanium alloys can keep their mechanical properties unchanged even at elevated temperatures, making them suitable for applications in jet engines, gas turbines, and other high-temperature environments.

Low Thermal Expansion: Titanium alloys exhibit a low coefficient of thermal expansion, which makes them valuable in applications where dimensional stability across a broad temperature range is essential.

2.5.2 CATEGORIES AND GRADES

- Four main categories of Ti alloys:

    α Alloys: This type of Ti alloy exclusively contains α stabilizers (such as aluminium or oxygen) and/or neutral alloying elements (like tin). These cannot be heat treated. Examples of α alloys are Ti-8Al-1Mo-1V and Ti-5Al-2Sn-.

    Near-α Alloys: There is a trace quantity of ductile β-phase in Near-α alloys. Near-α alloys contain 1–2% β-phase stabilizers, such as Mo, Si, or V, in addition to α-phase stabilizers. Ti-6Al-2Sn-4Zr-2Mo, Ti-5Al-5Sn-2Zr-2Mo, IMI 685, and Ti 1100 are few examples.

    α and β Alloys: These Ti alloys are heat treatable, metastable, and typically contain a combination of both α and β stabilizers. Ti-6Al-4V, Ti-6Al-4V-ELI, Ti-6Al-6V-2Sn, Ti-6Al-7Nb, and Ti62A are few examples.

    β and Near-β Alloys: These alloys may be solution treated and aged to increase strength, are metastable, and include enough β stabilizers (such as Mo, Si, and V) to preserve the β phase when quenched. β alloys show the BCC allotropic form of Ti which contains one or more non-titanium elements such as Cu, Fe, Cr, Ni, Co, Mo, V, Nb, Ta, Zr, and Mn in variable proportions. Several examples include Ti-10V-2Fe-3Al, Ti-29Nb-13Ta, and Ti-4.6Zr; Beta C, Ti-15-3, Ti-8Mo-8V-2Fe-3Al, and Ti-13V-11Cr-3Al.

- **Ti Grades:**

    There are 38 grades of titanium alloys available for application purposes and also mentioned in the ASTM International standard. Grades 5, 23, 24, 25, 29, 35, or 36 are annealed or aged; Grades 9, 18, 23, 28, or 29 are transformed-β conditions; Grades 9, 18, 28, or 38 are cold-worked and annealed or stress-relieved; and Grades 19, 20, or 21 are solution-treated and/or aged.

    Note: According to a user association's assessment of more than 5,200 test reports for commercial grades 2, 7, 16, and 26, more than 99% of them met the minimal UTS of 58 ksi. Also, the H grades were introduced in response to their request having the exception of the higher guaranteed minimum UTS. The H grades (like 7H, 26H, 2H, and 16H) are typically used in pressure vessels.

2.5.3 FACTORS INFLUENCING THE ADDITIVE MANUFACTURING OF TI ALLOYS

Five of the most commonly utilized MAM techniques, including SLM, EBM, LAD, WAAM, and CSAM, were found among the various AM procedures used to create Ti components. Depending on the size of their print chambers, the SLM and EBM procedures may create small components with dimensions between a few millimetres and a few centimetres, while the LAD, WAAM, and CSAM techniques can create large-scale components with dimensions between a few centimetres and a few metres.

Due to their design freedom, the PBF-based AM techniques SLM and EBM are most commonly used to fabricate objects made of Ti alloys. These AM technologies are suitable for products with small dimensions and modest production volumes, but they are constrained by the diameters of their machine chambers. Alternative AM techniques including LAD, WAAM, and CSAM are being used more frequently to create Ti components with greater dimensions and rapid production rates. However, these procedures result in products that need additional post-process machining.

The harsh cutting conditions directly cause a higher tool wear rate while machining, which further compromises the workpieces' surface integrity and results in greater roughness of the surface and plastically deformed subsurface layers that have elevated hardness (Bordin et al., 2015). AMed Ti component's machinability is greatly impacted by the porosities and anisotropy of the material characteristics (Bordin et al., 2015). Moreover, post-processing or heat treatment after AM of materials also causes variations in microstructural properties, further dynamically affecting the machining or wear response.

2.5.4 WEAR ANALYSIS OF TI ALLOYS

The community of AM is very interested in titanium alloys and titanium-based composite materials. It is a highly attractive option for applications including aviation, engines, and medical applications because of its strong corrosion resistance, biocompatibility (Al-Rubaie et al., 2020), low density, and outstanding mechanical behaviour in a higher-temperature range. However, popular engineering Ti alloys have poor wear resistance (Bruschi et al., 2017). When using titanium alloys in tribological contacts, the use of anodization and coating is recommended (Bruschi et al., 2017). The anti-wear properties of Ti alloys like Ti-6A-4V are only marginally improved by the SLM technique (Kaymakci et al., 2012; Niu et al., 2013). This may be due to the SLM process's improved hardness being compensated by the high roughness. The changes in wear as a result of different AMed Ti alloys combined with a range of fillers are shown in Table 2.1 which were taken from different literature.

TABLE 2.1
Wear Resistance Change and Compositing of Ti, Al, and Steel Alloys (Renner et al., 2021: 189)

AM Process	Alloy	Matrix	Reinforcement	Wear-Rate (WR) Reduction (%)
LENS	Ti-alloy	Ti-6Al-4V	Si-hydroxyapatite	70.8
Laser cladding	Ti-alloy	Ti-CN	CeO_2 NP	~14
E-beam melting	Ti-alloy	Ti-6Al-4V	$TiC + Ti3SiC2 + Ti5Si3$	~29.4
SLM	Ti-alloy	Ti-6Al-4V	Boron, TiO_2	~40
SLM	Al-alloy	AlSi10Mg	SiC	63.4
SLM	Al-alloy	Al2024	Al_2O_3	38
SLM	Al-alloy	Al4SiC	SiC	66
SLM	Al-alloy	AlSi10Mg	ALN	60
SLM	Steel	316L SS	TiC	10
SLM	Steel	P20	Al_2O_3	33
LMD	Steel	316L	Cr_3C_2	73

2.6 TRIBOLOGICAL BEHAVIOUR OF ADDITIVELY MANUFACTURED ALUMINIUM ALLOYS

Aluminium alloys are a diverse group of materials formed by combining aluminium with various other elements, resulting in improved mechanical properties, corrosion resistance, and other desirable characteristics. These alloys are widely used in industries ranging from aerospace and automotive to construction and electronics. For example, 6000 series alloys are often used in structural applications due to their combination of strength and formability, while 5000 series alloys have good corrosion resistance and are used in marine and automotive applications. Depending on the specific alloying elements and their proportions, aluminium alloys exhibit a wide range of properties suited for different applications. Their versatility, strength, lightweight nature, corrosion resistance, and electrical conductivity have made them indispensable in modern industries, contributing to innovations in transportation, construction, electronics, and beyond.

2.6.1 CATEGORIES

Each category of aluminium alloy has its own unique properties and advantages, making aluminium alloys versatile and adaptable to a wide range of applications across industries such as aerospace, automotive, construction, and electronics. Aluminium alloys are also categorized based on a four-digit numbering system, where the first digit represents the major alloying element or group and the subsequent digits provide more specific information about the alloy composition. The following are different categories of aluminium alloys:

a. **Wrought Alloy:**

1xxx Series: These alloys are nearly pure aluminium and have high electrical conductivity and good corrosion resistance. They are often used for electrical conductors and components where corrosion resistance is essential. For example, 1420.

2xxx Series: The 2xxx series is a copper alloy that is able to precipitation hardened to have steel-like strengths. They were formerly known as duralumin and were the most popular aerospace alloys, but because they were prone to stress corrosion cracking, the 7xxx series is progressively taking their place in modern designs. For example, 2008, 2036, 2090, 2124, 2324.

3xxx Series: These alloys have manganese as their main alloying element, which imparts moderate strength and excellent formability. They are commonly used for applications requiring good formability and corrosion resistance, such as cookware and architectural components. For example, 3003, 3004.

4xxx Series: The 4xxx series has Si as an alloying element and variations of Al-Si alloys intended for casting (and therefore not included in the 4xxx series) are also known as silumin. For example, 4047.

5xxx Series: Alloys in this category contain magnesium as the primary alloying element. They exhibit good weldability, moderate strength, and excellent corrosion resistance. They find use in structural components, marine applications, and automotive parts. For example, 5052, 5059, 5083, 5086.

6xxx Series: These alloys have Mg and Si as their main alloying elements. They offer good formability, weldability, and moderate strength. They are often used for extrusions, architectural components, and automotive parts. For example, 6013, 6061, 6063, 6113, 6951.

7xxx Series: Alloys in this category are primarily strengthened by zinc and can also contain magnesium and copper. They have higher strength and are used in aerospace applications and structural components requiring high strength-to-weight ratios. For example, 7010, 7050,7150, 7475.

8xxx Series: These alloys are not commonly used for structural applications but rather for specialized purposes. They may contain lithium for enhanced properties in certain aerospace applications. For example, 8009.

b. **Cast Alloy:**

For cast alloy, designation is generally based on AA (Aluminium association) system in which the second two digits indicate the minimum percentage of aluminium. For example, 150.x indicates a minimum of 99.50% aluminium. The digit following the decimal point indicates casting or ingot, which has a value of 0 or 1, respectively. The following are the primary alloying components of the AA system:

1xx.x series: These alloys are having minimum 99% aluminium.

2xx.x Series: These alloys are primarily aluminium-copper alloys and offer excellent casting characteristics and good corrosion resistance. They are used for various cast parts.

3xx.x Series: Cast alloys in this category are aluminium-silicon alloys, with added Cu and/or Mg commonly used in automotive applications due to their good castability and wear resistance.

Other Series: 4xx.x, 5xx.x, 7xx.x, and 8xx.x alloys having silicon, magnesium, zinc, and tin as alloying elements. These cast alloys are modified versions of their wrought counterparts and are used for specialized applications where the cast form provides specific advantages.

c. **Based on Heat Treatment**:

Heat-Treatable Alloy: Alloys in this category can be heat-treated to enhance their mechanical properties. Examples include the 2xxx, 6xxx, and 7xxx series, which are commonly used for aerospace and structural applications.

Non-heat Treatable Alloy: These alloys are not subjected to heat treatment for strengthening. Examples include the 1xxx, 3xxx, and 5xxx series, which are often used for applications where formability and corrosion resistance are key criteria.

2.6.2 WEAR ANALYSIS OF ALUMINIUM ALLOYS

Despite having some advantageous characteristics, such as low weight and a high strength-to-weight ratio, alloys of aluminium have low wear resistance. However, in the 1970s, the use of Al alloys was started to produce Metal Matrix Composite (MMC) in order to enhance its tribological behaviour (Gurusamy & Rao, 2017; Kiran Sagar et al., 2019; Melkote et al., 2015). The potential application of Al MMC in sectors like automotive parts is quite promising.

The tribological characteristics of the Al MMC material produced by AM have improved. With this enhancement and the increased hardness brought on by the AM method, aluminium alloys can finally be employed in moving parts and bearings. Table 2.1 displays the literature results of SLM-fabricated alloys of aluminium combined with various fillers and the related wear-rate decrease.

2.7 TRIBOLOGICAL BEHAVIOUR OF ADDITIVELY MANUFACTURED STAINLESS STEEL

Stainless steel is renowned for its exceptional resistance to corrosion, durability, and versatility. It has wide applications due to its unique combination of properties. Stainless steel is primarily composed of iron, with the addition of Cr and often other alloying elements, such as Ni, Mo, or Ti, to enhance its properties. Stainless steel's diverse properties and applications have led to its use in various industries, including construction, architecture, automotive, aerospace, medical, food processing, and more.

2.7.1 Properties and Characteristics

2.7.1.1 Properties

Conductivity: Similar to steel, stainless steels have a comparatively poor electrical conductivity that are lower than those of copper. Particularly, the dense protective oxide layer of stainless steel causes non-electrical contact resistance (ECR), which restricts its usability in applications as electrical connectors. However, stainless steel connectors are used in circumstances when corrosion resistance is necessary but ECR is a lesser design criterion such as in hot and oxidizing conditions.

Melting Point: The melting point of SS is represented as a range of temperatures rather than as a single value like most alloys. Depending on the unique consistency of the temperature, temperatures range from 1,400°C to 1,530°C (2,550°F–2,790°F; 1,670–1,800 K; 3,010°R–3,250°R).

Hardness: Stainless steel is a very robust metal with an outstanding level of hardness due to Cr and Ni. On the metal's surface, chromium creates an oxide layer that shields it from corrosion and wear, while nickel increases the material hardness by improving its ductility and strength.

Magnetism: While martensitic, duplex, and ferritic stainless steels are magnetic, austenitic stainless steel is normally non-magnetic. Ferritic steel's magnetism is due to the bcc crystal structure, in which Fe atoms are arranged in cubes (with one at each of the corners and another iron atom in the centre). The magnetic characteristics of ferritic steel are due to this core iron atom, and due to this, the steel can only absorb about 0.025% of carbon.

Wear: Galling, also known as cold welding, is a type of adverse adhesive wear that can happen when two metal surfaces are in close association or in relative motion to one another and are being put under heavy strain. Fasteners made of austenitic stainless steel are especially sensitive to thread galling, but other alloys that produce a protective oxide film on their own are also problematic.

Density: The density of stainless steel can be somewhere between 7,500 and 8,000 kg/m^3 depending on the alloy.

2.7.1.2 Characteristics

Corrosion Resistance: One of the most significant features of stainless steel is its resistance to corrosion and rust. The Cr content creates a layer of passive protection on the surface known as a chromium oxide layer. This layer prevents oxygen and moisture from coming into direct contact with the underlying metal, thereby inhibiting the development of corrosion.

Strength and Durability: Stainless steel is a strong and durable material that is capable of withstanding high mechanical stresses and impacts. Different grades of stainless steel offer varying levels of strength and are used for various applications, from structural components to precision instruments.

Heat Resistance: Many stainless steel grades retain their strength and corrosion resistance at higher temperatures when used for applications involving heat exposure. This property is particularly advantageous in industries such as manufacturing, chemical processing, and automotive.

Hygiene and Cleanliness: Stainless steel is non-porous and easy to clean, making it a preferred choice in environments that require high levels of hygiene, such as hospitals, kitchens, and pharmaceutical facilities.

Aesthetic Appeal: Stainless steel's shiny and reflective surface, along with its ability to resist discolouration and staining, makes it popular in architectural and decorative applications. It's commonly used in building facades, interior design, and modern appliances.

Formability and Fabrication: Stainless steel can be easily formed, welded, and fabricated into various shapes and sizes, making it adaptable to a wide range of manufacturing processes.

Biocompatibility: Some stainless steel grades are biocompatible, meaning they are well-tolerated by the human body and can be used in medical implants and devices.

Recyclability: Stainless steel is highly recyclable, making it an environmentally friendly choice. It can be recycled and reused without losing its inherent properties.

Cost-Effectiveness: While some specialized stainless steel grades can be expensive, there are many general-purpose grades that offer an excellent balance of properties and affordability.

2.7.2 TYPES AND GRADES

There are around 150 recognized grades of stainless steel, with 15 being the most popular. There are other grading systems in use, such as the US SAE steel grades. The ASTM created the UNS, or Unified Numbering System, for Metals and Alloys, in 1970. EN 10088 has been adopted by the Europeans. The choice of a specific stainless steel grade depends on the intended application, the required properties, and the environmental conditions in which it will be used.

Stainless steel grades are classified into five main families based on their microstructure, which in turn influences their mechanical properties and corrosion resistance:

  i. **Austenitic:** The largest family of stainless steels, comprising around two-thirds of all stainless steel production, is austenitic stainless steel which has FCC crystal. Subgroups of austenitic stainless steel, 200 series and 300 series (most popular). The 200 series of Cr, Mn, and Ni alloys make the best utilization of nitrogen and manganese while using the least amount of nickel possible. They have a yield strength that is roughly 50% more than 300-series SS sheets because of the nitrogen addition. The austenitic microstructure of the 300-series chromium-nickel alloys is almost entirely produced by nickel alloying; sometimes less Ni is needed due to the addition of nitrogen during extreme alloying. For example, Type 201, 202, 304, 316.
 ii. **Ferritic:** Ferritic SS contains between 10.5% and 27% Cr, with little or no Ni, and has a bcc crystal structure known as the ferrite microstructure. Due to the Cr addition, this microstructure is present at all temperatures, making

heat treatment unable to harden them. In comparison to austenitic stainless steels, they can't be as much strengthened through cold work. They attract each other. Niobium (Nb), titanium (Ti), and zirconium (Zr) additions to type 430 enable effective welding. They are less costly than austenitic steels and are used in numerous goods because of the almost complete lack of nickel in them.

iii. **Martensitic:** Martensitic SS are employed as creep-resistant steels, tool steels, and engineering steels because of their wide range of characteristics and bcc crystal structure. Due to their low Cr concentration, they are magnetic and have a lower level of corrosion resistance than ferritic and austenitic stainless steels. They have four categories which include Fe-Cr-C grades, Fe-Cr-Ni-C grades, precipitation hardening grades, and Creep-resisting grades.

iv. **Duplex:** The optimal ratio for austenite and ferrite in duplex stainless steels is 50:50, while commercial alloys may have ratios as low as 40:60. They differ from austenitic SS in having higher Cr (19–32%) and Mo (up to 5%) levels and lower Ni contents. The yield strength (YS) of duplex SS is nearly double that of austenitic SS. Compared to austenitic SS grades 304 and 316, its mixed microstructure offers better resistance to chloride stress corrosion cracking. Lean duplex, standard duplex, and super duplex are the three subgroups of duplex grades that are typically separated depending on how well they resist corrosion.

v. **Precipitation Hardening:** Despite the fact that it can be precipitation-hardened to even greater strengths compared to martensitic grades, their corrosion resistance is equal comparable to that of austenitic kinds. There are three types of precipitation-hardening stainless steels:
- Martensitic 17-4 PH (AISI 630 EN 1.4542): It contains about 17% Cr, 4% Ni, 4% Cu, and 0.3% Nb.
- Semi-austenitic 17-7 PH (AISI 631 EN 1.4568): It contains about 17% Cr, 7.2% Ni, and 1.2% Al.
- Austenitic A286 (ASTM 660 EN 1.4980): It contains about Cr 15%, Ni 25%, Ti 2.1%, Mo 1.2%, V 1.3%, and B 0.005%.

2.7.3 Wear Analysis of AMed Steel Alloys

Most significant technical materials that are commonly manufactured by the AM method is steel alloys, particularly stainless steel. Steel alloys have a substantially higher hardness than their forge or cast counterparts, unlike titanium alloys. If advantageous AM parameters, such as energy density and scanning speed, are employed, the hardness increases and can reduce the wear rate (see Table 2.1).

Despite having a strong corrosion resistance, austenitic stainless steel produced additively has a low wear resistance. Due to its low hardness and uniform microstructure, it is more vulnerable to adhesive wear than to abrasive wear (Wu et al., 2017). It also works well to increase its wear resistance by incorporating composite materials. SiC-reinforced SS316L MMCs achieve a high strength up to 1.3 GPa and a lower WR of 0.77×10^{-5} mm^3/Nm (Zou et al., 2021).

2.8 CONCLUSIONS

AM has reported the fast growth by making it possible to fabricate complex shapes with better design flexibility. The present study presents a summarized review of investigation into the tribological characteristics of metal components fabricated via AM techniques. These studies reported the mechanism of various available AM processes for fabricating metal parts and the effects of major AM parameters on the tribological properties of additively manufactured metal parts. By studying the results, it can be concluded that a range of common metals and alloys are used in AM, and tribology behaviour was evaluated under different operating conditions and loads, which led to the excellent improvement in wear and friction properties. This technique gets attention in various industrial applications so that it can be said that the study of tribological behaviour of additively manufactured metal parts is important to ensure their reliable performance in real-world applications.

ACKNOWLEDGEMENTS

The authors are thankful to the Department of Mechanical Engineering, Indian Institute of Technology (ISM) Dhanbad and University Centre for Research and Development, Chandigarh University for the technical support.

REFERENCES

Al-Rubaie, K. S., Melotti, S., Rabelo, A., Paiva, J. M., Elbestawi, M. A., & Veldhuis, S. C. (2020). Machinability of SLM-produced Ti6Al4V titanium alloy parts. *Journal of Manufacturing Processes*, *57*, 768–786. https://doi.org/10.1016/J.JMAPRO.2020.07.035.

ASTM E384. (2002). Standard test method for microindentation hardness of materials ASTM E384. *ASTM Standards*, *14*, 1–24.

Baufeld, B., Van Der Biest, O., & Dillien, S. (2010). Texture and crystal orientation in Ti-6Al-4V builds fabricated by shaped metal deposition. *Metallurgical and Materials Transactions A: Physical Metallurgy and Materials Science*, *41*(8), 1917–1927. https://doi.org/10.1007/S11661-010-0255-X/FIGURES/12.

Biswal, R., Zhang, X., Syed, A. K., Awd, M., Ding, J., Walther, F., & Williams, S. (2019). Criticality of porosity defects on the fatigue performance of wire + arc additive manufactured titanium alloy. *International Journal of Fatigue*, *122*, 208–217. https://doi.org/10.1016/J.IJFATIGUE.2019.01.017.

Bonaiti, G., Parenti, P., Annoni, M., & Kapoor, S. (2017). Micro-milling Machinability of DED Additive titanium Ti-6Al-4V. *Procedia Manufacturing*, *10*, 497–509. https://doi.org/10.1016/J.PROMFG.2017.07.104.

Bordin, A., Bruschi, S., Ghiotti, A., & Bariani, P. F. (2015). Analysis of tool wear in cryogenic machining of additive manufactured Ti6Al4V alloy. *Wear*, *328–329*, 89–99. https://doi.org/10.1016/J.WEAR.2015.01.030.

Brandl, E., Baufeld, B., Leyens, C., & Gault, R. (2010). Additive manufactured Ti-6Al-4V using welding wire: Comparison of laser and arc beam deposition and evaluation with respect to aerospace material specifications. *Physics Procedia*, *5*(PART 2), 595–606. https://doi.org/10.1016/J.PHPRO.2010.08.087.

Bruschi, S., Bertolini, R., & Ghiotti, A. (2017). Coupling machining and heat treatment to enhance the wear behaviour of an additive manufactured Ti6Al4V titanium alloy. *Tribology International*, *116*, 58–68. https://doi.org/10.1016/J.TRIBOINT.2017.07.004.

Bruschi, S., Tristo, G., Rysava, Z., Bariani, P. F., Umbrello, D., & De Chiffre, L. (2016). Environmentally clean micromilling of electron beam melted Ti6Al4V. *Journal of Cleaner Production*, *133*, 932–941. https://doi.org/10.1016/J.JCLEPRO.2016.06.035.

Chae, J., Park, S. S., & Freiheit, T. (2006). Investigation of micro-cutting operations. *International Journal of Machine Tools and Manufacture*, *46*(3–4), 313–332. https://doi.org/10.1016/J.IJMACHTOOLS.2005.05.015

Chen, Q., Guillemot, G., Gandin, C. A., & Bellet, M. (2017). Three-dimensional finite element thermomechanical modeling of additive manufacturing by selective laser melting for ceramic materials. *Additive Manufacturing*, *16*, 124–137. https://doi.org/10.1016/J.ADDMA.2017.02.005.

Christopherson, D. (2015). Machinability of powder metallurgy steels. *Powder Metallurgy*, *7*, 384–394. https://doi.org/10.31399/ASM.HB.V07.A0006103

Coeffi, R. F. (2011). Standard test method for linearly reciprocating ball-on-flat sliding wear 1. *Lubrication*, *05*(Reapproved 2010), 1–10. https://doi.org/10.1520/G0133-05R10.2.

Davis, J. R. (2001). Weld-overlay coatings. *Surface Engineering for Corrosion and Wear Resistance*. https://www.asminternational.org/surface-engineering-for-corrosion-and-wear-resistance/results/-/journal_content/56/06835G/PUBLICATION/.

Edkinsa, K. D., Van Rensburga, N. J., & Laubscher, R. F. (2014). Evaluating the subsurface microstructure of machined Ti-6Al-4V. *Procedia CIRP*, *13*, 270–275. https://doi.org/10.1016/J.PROCIR.2014.04.046.

Fakhri, M. A., Bordatchev, E. V., & Tutunea-Fatan, O. R. (2013). Framework for evaluation of the relative contribution of the process on porosity-cutting force dependence in micromilling of titanium foams. *Proceedings of the Institution of Mechanical Engineers, Part B: Journal of Engineering Manufacture*, *227*(11), 1635–1650. https://doi.org/10.1177/0954405413491243/ASSET/IMAGES/LARGE/10.1177_0954405413491243-FIG18.JPEG.

Gibson, I., Rosen, D., & Stucker, B. (2015). *Additive manufacturing technologies: 3D printing, rapid prototyping, and direct digital manufacturing*, pp. 1–498. Springer. https://doi.org/10.1007/978-1-4939-2113-3/COVER.

Gorsse, S., Hutchinson, C., Gouné, M., & Banerjee, R. (2017). Additive manufacturing of metals: A brief review of the characteristic microstructures and properties of steels, Ti-6Al-4V and high-entropy alloys. *Science and Technology of Advanced Materials*, *18*(1), 584–610. https://doi.org/10.1080/14686996.2017.1361305.

Gurusamy, M. M., & Rao, B. C. (2017). On the performance of modified Zerilli-Armstrong constitutive model in simulating the metal-cutting process. *Journal of Manufacturing Processes*, *28*, 253–265. https://doi.org/10.1016/J.JMAPRO.2017.06.011.

Hanaor, D. A., Gan, Y., & Einav, I. (2016). Static friction at fractal interfaces. *Tribology International*, *93*, 229–238.

Hojati, F., Daneshi, A., Soltani, B., Azarhoushang, B., & Biermann, D. (2020). Study on machinability of additively manufactured and conventional titanium alloys in micro-milling process. *Precision Engineering*, *62*, 1–9. https://doi.org/10.1016/J.PRECISIONENG.2019.11.002.

Kaymakci, M., Kilic, Z. M., & Altintas, Y. (2012). Unified cutting force model for turning, boring, drilling and milling operations. *International Journal of Machine Tools and Manufacture*, *54–55*, 34–45. https://doi.org/10.1016/J.IJMACHTOOLS.2011.12.008.

Kiran Sagar, C., Kumar, T., Priyadarshini, A., & Kumar Gupta, A. (2019). Prediction and optimization of machining forces using oxley's predictive theory and RSM approach during machining of WHAs. *Defence Technology*, *15*(6), 923–935. https://doi.org/10.1016/J.DT.2019.07.004.

Le Coz, G., Fischer, M., Piquard, R., D'Acunto, A., Laheurte, P., & Dudzinski, D. (2017). Micro cutting of Ti-6Al-4V parts produced by SLM process. *Procedia CIRP*, *58*, 228–232. https://doi.org/10.1016/J.PROCIR.2017.03.326.

Le Coz, G., Piquard, R., D'Acunto, A., Bouscaud, D., Fischer, M., & Laheurte, P. (2020). Precision turning analysis of Ti-6Al-4V skin produced by selective laser melting using a design of experiment approach. *International Journal of Advanced Manufacturing Technology*, *110*(5–6), 1615–1625. https://doi.org/10.1007/S00170-020-05807-8/FIGURES/12.

Lekkala, R., Bajpai, V., Singh, R. K., & Joshi, S. S. (2011). Characterization and modeling of burr formation in micro-end milling. *Precision Engineering*, *35*(4), 625–637. https://doi.org/10.1016/J.PRECISIONENG.2011.04.007.

Melkote, S. N., Liu, R., Fernandez-Zelaia, P., & Marusich, T. (2015). A physically based constitutive model for simulation of segmented chip formation in orthogonal cutting of commercially pure titanium. *CIRP Annals*, *64*(1), 65–68. https://doi.org/10.1016/J.CIRP.2015.04.060.

Mok, S. H., Bi, G., Folkes, J., Pashby, I., & Segal, J. (2008). Deposition of Ti-6Al-4V using a high power diode laser and wire, Part II: Investigation on the mechanical properties. *Surface and Coatings Technology*, *202*(19), 4613–4619. https://doi.org/10.1016/J.SURFCOAT.2008.03.028.

M'Saoubi, R., Outeiro, J. C., Chandrasekaran, H., Dillon, O. W., & Jawahir, I. S. (2008). A review of surface integrity in machining and its impact on functional performance and life of machined products. *International Journal of Sustainable Manufacturing*, *1*(1–2), 203–236. https://doi.org/10.1504/IJSM.2008.019234.

Niu, W., Bermingham, M. J., Baburamani, P. S., Palanisamy, S., Dargusch, M. S., Turk, S., Grigson, B., & Sharp, P. K. (2013). The effect of cutting speed and heat treatment on the fatigue life of Grade 5 and Grade 23 Ti-6Al-4V alloys. *Materials & Design (1980–2015)*, *46*, 640–644. https://doi.org/10.1016/J.MATDES.2012.10.056.

Özel, T., Thepsonthi, T., Ulutan, D., & Kaftanolu, B. (2011). Experiments and finite element simulations on micro-milling of Ti-6Al-4V alloy with uncoated and cBN coated micro-tools. *CIRP Annals*, *60*(1), 85–88. https://doi.org/10.1016/J.CIRP.2011.03.087.

Padhy, M. K., & Saini, R. P. (2008). A review on silt erosion in hydro turbines. *Renewable and Sustainable Energy Reviews*, *12*(7), 1974–1987. https://doi.org/10.1016/J.RSER.2007.01.025.

Rabinowicz, E. (1995). *Friction and Wear of Materials* (2nd edition), p. 336. Wiley. https://www.wiley.com/en-us/Friction+and+Wear+of+Materials%2C+2nd+Edition-p-9780471830849.

Ranjan, N., Tyagi, R., Kumar, R., & Babbar, A. (2023a). 3D printing applications of thermo-responsive functional materials: A review. *Advances in Materials and Processing Technologies*, 1–17. https://www.tandfonline.com/doi/abs/10.1080/2374068X.2023.2205669.

Ranjan, N., Tyagi, R., Kumar, R., & Kumar, V. (2023b). On fabrication of acrylonitrile buta-diene styrene-zirconium oxide composite feedstock for 3D printing-based rapid tooling applications. *Journal of Thermoplastic Composite Materials*, *0*(0), 08927057231186310. https://doi.org/10.1177/089270572311863.

Renner, P., Jha, S., Chen, Y., Raut, A., Mehta, S. G., & Liang, H. (2021). A review on corrosion and wear of additively manufactured alloys. *Journal of Tribology*, *143*(5), 050802. https://doi.org/10.1115/1.4050503/1103641.

Sachs, E., Cima, M., & Cornie, J. (1990). Three dimensional printing: Rapid tooling and prototypes directly from CAD representation. *Journal of Manufacturing Science and Engineering*. https://doi.org/10.15781/T2GT5FZ4F.

Sharman, A. R. C., Hughes, J. I., & Ridgway, K. (2015). The effect of tool nose radius on surface integrity and residual stresses when turning Inconel 718TM. *Journal of Materials Processing Technology*, *216*, 123–132. https://doi.org/10.1016/J.JMATPROTEC.2014.09.002.

Shunmugavel, M., Polishetty, A., Goldberg, M., & Nomani, J. (2017). *Influence of Build Orientation on Machinability of Selective Laser Melted Titanium Alloy-Ti-6Al-4V Machinability Assessment and Surface Integrity Characteristics of Austempered Ductile Iron (ADI) u View project Engineering Education View project.* International Scholarly and Scientific Research & Innovation. https://www.researchgate.net/publication/318352450.

Sinmazçelik, T., & Taşkiran, I. (2007). Erosive wear behaviour of polyphenylenesulphide (PPS) composites. *Materials & Design, 28*(9), 2471–2477. https://doi.org/10.1016/J. MATDES.2006.08.007.

Stachowiak, G. W., & Batchelor, A. W. (2005). *Engineering Tribology,* pp. 1–801. Cambridge University Press. https://doi.org/10.1016/B978-0-7506-7836-0.X5000-7.

Stevens, M., & Covas, J. (2012). *Extruder Principles and Operation.* https://books.google. com/books?hl=en&lr=&id=NM33CAAAQBAJ&oi=fnd&pg=PP8&dq=Stevens+MJ, +Covas+JA+(1995)+Extruder+principles+and+operation,+2nd+edn.+Springer,+Do rdrecht,+494+pp.+ISBN:+0412635909,+9780412635908&ots=GiJ_B8bJEE&sig=K p9K6N6a3HysiHrRbLXhTD1Crrg.

Tutunea-Fatan, O. R., Fakhri, M. A., & Bordatchev, E. V. (2011). Porosity and cutting forces: From macroscale to microscale machining correlations. *Proceedings of the Institution of Mechanical Engineers, Part B: Journal of Engineering Manufacture, 225*(5), 619–630. https://doi.org/10.1177/2041297510394057.

Tyagi, R., Singh. G., Kuma, R., Kumar, V., & Singh, S. (2023). 3D-printed sandwiched acrylonitrile butadiene styrene/carbon fiber composites: Investigating mechanical, morphological, and fractural properties. *Journal of Materials Engineering and Performance,* 1–14. https://link.springer.com/article/10.1007/s11665-023-08292-8.

Tyagi, R., & Tripathi, A. (2023). Coating/cladding based post-processing in additive manufacturing. In: *Handbook of Post-Processing in Additive Manufacturing: Requirements, Theories, and Methods.* Edited by Gurminder Singh, Ranvijay Kumar, Kamalpreet Sandhu, Eujin Pei, & Sunpreet Singh, p. 127. CRC Press.

Varenberg, M. (2013). Towards a unified classification of wear. *Friction, 1*(4), 333–340. https:// doi.org/10.1007/S40544-013-0027-X/METRICS.

Varghese, V., & Mujumdar, S. (2021). Micromilling-induced surface integrity of porous additive manufactured Ti6Al4V alloy. *Procedia Manufacturing, 53,* 387–394. https://doi. org/10.1016/J.PROMFG.2021.06.041.

Vrancken, B., Thijs, L., Kruth, J. P., & Van Humbeeck, J. (2012). Heat treatment of Ti6Al4V produced by selective laser melting: Microstructure and mechanical properties. *Journal of Alloys and Compounds, 541,* 177–185. https://doi.org/10.1016/J. JALLCOM.2012.07.022.

Williams, J. A. (2005). Wear and wear particles-some fundamentals. *Tribology International, 38*(10), 863–870. https://doi.org/10.1016/J.TRIBOINT.2005.03.007.

Wimpenny, D. I., Bryden, B., & Pashby, I. R. (2003). Rapid laminated tooling. *Journal of Materials Processing Technology, 138*(1–3), 214–218. https://doi.org/10.1016/ S0924-0136(03)00074-8.

Wu, X., Li, L., & He, N. (2017). Investigation on the burr formation mechanism in micro cutting. *Precision Engineering, 47,* 191–196. https://doi.org/10.1016/J. PRECISIONENG.2016.08.004.

Zou, Y., Tan, C., Qiu, Z., Ma, W., Kuang, M., & Zeng, D. (2021). Additively manufactured SiC-reinforced stainless steel with excellent strength and wear resistance. *Additive Manufacturing, 41,* 101971. https://doi.org/10.1016/J.ADDMA.2021.101971.

3 Tribological Properties of Polymer-Reinforced Matrix Composite Prepared by Additive Manufacturing

Ankan Shrivastava, Jasgurpeet Singh Chohan,
Ranvijay Kumar, and Vinay Kumar
Chandigarh University

3.1 INTRODUCTION

Manufacturing for the fabrication of several components can be done in a variety of methods. One of the rapidly and continuously developing technologies is 3D printing, which makes it possible to create complicated and flexible structures that would have been challenging for conventional manufacturing processes to construct. 3D printing techniques construct components with the help of CAD models in a layer-upon-layer method (Tambrallimath et al., 2021). One of the particular potential 3D printing techniques is fused deposition modeling (FDM), which has tremendous application potential thanks to its simplicity of use and low cost in producing industrial parts with complex shapes (Zhou et al., 2017). FDM is a filament-based technique in which the heated nozzle is filled with filaments and heated until it melts. Extruding molten material onto the substrate and depositing layers on top of the previous layer, the heated nozzle moves following the design to create the necessary additive components (Li et al., 2018). Polymers have gained popularity as materials in the last ten years for a variety of uses because of their appealing qualities, such as their low weight, production ease of use, and cost-effectiveness. As a result, considerable attempts have been made to implement polymers in many industrial applications, often through adding metal powders, ceramics, etc., or other types of reinforcement into the polymers to improve their tribological qualities (Omrani et al., 2016). Over three decades, the researchers explored and looked at polymer tribology. Tribology is the primary characteristic used to gauge wear and friction in polymer materials and their composites. Typically, the polymer material's friction and wear performance involved sliding across an aggressive material. Solid material surfaces are necessary for the production of the polymer transfer layer (Bahadur and Gong, 1992). Numerous inorganic substances, including metallic powders, and mineral substances have been employed as fillers for polymers over the years. Typically, metal particles are added to polymers as fillers to enhance their electrical, thermal, and tribological properties. When only the filler loading increases above a certain point, these qualities start to improve (Bahadur, 2000). Fillers have a large surface area-to-volume ratio, which allows them to have a considerable impact on the tuning of characteristics even at extremely low filler loadings. Multifunctional fillers increase the potential applications of polymer composites by enabling them to be specifically tailored to a given application, such as tribo-components for extreme operating temperatures, working conditions with unacceptable levels of lubricants, and highly corrosive environments. These polymer composites exhibit intriguing tribological activity, but some of them also possess significant electrical conductivity for application in microsystems or self-healing properties for mechanical components where routine maintenance is challenging (Friedrich, 2018). The wear properties of the plastic-based materials especially composites are dependent upon the composition of the materials. These factors affect the formation of transfer films and the actual contact area, which modifies the coefficient of friction (COF) and wear patterns (Myshkin and Kovalev, 2018). The polymers transfer layer, which frequently forms during polymer-metal or polymer-polymer sliding and ultimately alters the contact surface, is the most crucial component in polymer tribology. Material always transfers from the polymer to the metal during the metal-polymer phase (Chan et al., 2021).

The various investigations that were conducted on polymer composites and how the various fillers affected the tribological properties of the composite material are thoroughly covered in this paper. Further in this study, a novel composite is developed by blending Al powder into the PLA Matrix. The objective of this paper is to examine the tribological properties of PLA-Al-based composite which is manufactured through FDM 3D printing.

3.2 LITERATURE SURVEY ON POLYMER COMPOSITES

The reported research on the tribological behavior of polymer composites is summarized in this section. PTFE, PEEK, and PLA are the polymers that are currently most frequently utilized in tribology research. In high-load bearing systems working in challenging conditions where lubricants have failed, such as in the oil and gas sector and the air-conditioning and refrigeration industries, ABS-based composite materials have also demonstrated significant potential. Table 3.1 shows the tribological characteristics of polymer-based composites and how the reinforcement provided by various fillers influences those characteristics.

According to Table 3.1, the reinforcement of the materials in the polymers leads to decreased wear rate and good tribological properties. This resulted in the composite material having better tribological performance than polymer materials. The proper composition of the filler is highly important for the composite material because if the material is not compatible with each other, different properties can be affected.

TABLE 3.1

Tribological Properties of Polymer Composites

Sr. No.	Polymer	Reinforcement	Findings	Ref.
1.	PTFE	Graphite and bronze fillers	The COF and the graphite filler have been decreased about 27% and 4.2 times, respectively. On the other hand, the bronze filler exhibited a COF increase of up to 5%	Valente et al. (2020)
2.	PLA	Bronze fillers	With a 150 N load, the sample at a 45° angle has the lowest dynamic friction coefficient (0.52). The dynamic friction coefficient for bronze/PLA has a range of 0.52–0.6. Average values for the measured wear depth range from 9.7 to 17.5 m	Hanon et al. (2020)
3.	PLA	Graphite	The wear rate is significantly reduced by 10% of graphite (11 k) and is 65% lower than the wear rate of PLA (32 k)	Przekop et al. (2020)
4.	PLA	MoS_2	The addition of 0.5% MoS_2 results in a significant increase in tribological characteristics of up to 457% with an average wear rate of 180 μm/km	Pawlak et al. (2020)
5.	PLA	Silicon	In comparison to neat PLA, the specific wear and COF of the PLA composite containing 5% silicon particles were 2×10^{-3} mm³/Nm and 0.11, respectively	Vishal et al. (2022)
6.	PLA	Graphene fillers (30 μm)	In comparison to PLA samples, the reinforced PLA samples had lower OF (0.49–0.6). The sample with the lowest value (4.04×10^4 mm³/N·m) consisted of 5% graphene	Al Abir et al. (2023)
7.	PLA	Graphene	The results showed that PLA-graphene had 14% more wear resistance than PLA. PLA-graphene displays a COF in the range of 0.45–0.55, whereas PLA has a greater COF in the range of 0.5–0.6	Bustillos et al. (2018)
8.	ABS	Graphite	The ABS matrix's incorporation of graphite showed weight reductions for the FDM parts in addition to a reduction in the COF. Additionally, the degree of wear and friction loss reduces as the graphite proportion of weight in the ABS matrix increases	Kumar et al. (2022)
9.	PEEK	MWCNT black powder (200–500 μm), GNP (6.99 μm)	The study noticed a decrease in COF of PEEK (0.25) by adding CNT to (0.08) and GNP fillers (0.1) [18]	Arif et al. (2020)

(Continued)

TABLE 3.1 (*Continued*)
Tribological Properties of Polymer Composites

Sr. No.	Polymer	Reinforcement	Findings	Ref.
10.	PLA	GNP (dia, 5–7 nm) MWCNP (dia, 10–30 nm)	The reinforced fillers decreased the COF compared to PLA, which had a COF of (0.15–0.16), (0.11) for GNP/PLA at 12% GNP and (0.10) for MWCNT/PLA at 12% MWCNT	Batakliev (2020)
11.	PLA	Alumina powder (dia, 15–20 nm) Carbon black (dia, 30–60 nm)	The study found that PLA composites had a higher COF of friction. The PLA displayed 0.30 COF, while PLA-Al and PLA/CB displayed 0.39 and 0.31 COF, respectively	Cardoso et al. (2020)
12.	PLA	Copper powder (5–20 μm)	The COF of reinforced copper in PLA matrix has considerable wear characteristics. The highest COF (0.4) was noticed at room temperature sliding conditions and the lowest COF (0.13) was noticed in 70°C	Karabeyoğlu et al. (2021)
13.	Nylon 6	Iron powder (Fe)	The wear rate and COF are decreased when the Fe content increases. The highest wear (0.64 g) was noticed in 50/50 Nylon-Fe content and the lowest wear (0.04) was noticed in 60/40 Nylon-Fe	Garg and Singh (2017)
14.	Nylon 6	Al powder (41–44 μm) Al_2O_3 (125–149 μm)	Composite materials offer greater wear resistance than ABS material because Al and Al_2O_3 are incorporated in the Nylon 6 matrix. The composition PLA 60/Al40/Al_2O_{14} has the lowest wear (100 μm) and $PLA_{60}/Al_{30}/Al_2O_{10}$ has a wear rate of (200 μm) at 10 minutes interval	Boparai et al. (2015)
15.	PLA	Nano silica (20 nm)	The decrement in COF and wear rate was noticed by the addition of nano-silica fillers. The COF and specific wear rate for PLA were 0.254 and 0.00567 mm³/N·m. The result dropped to 0.154 for COF and 0.00189 mm³/N·m for SWR with 6 wt% of nano-silica	Ramachandran and Rajeswari (2022)

3.3 METHODS AND EXPERIMENTATION

3.3.1 MATERIALS

The PLA which was purchased from the Green Dot BioPak in Gujarat, India with 1.24 g/cm³ density has been used as Matrix material with 80 wt%, whereas Al metal powder which was purchased from Shiva Chemicals Ludhiana, India with 20 wt% has been used as reinforced material.

3.3.2 PREPARATION OF COMPOSITE FILAMENT

The 20% Al metal powder and precisely 80% PLA, by weight percentage, combination was chosen as the feedstock filament for the FDM process. Al metal powder was used as filler primarily due to its reasonably robust mechanical properties. A Felfil evo single screw extruder (Make – Italy) was used to create the feedstock filament. The filament with a diameter of 1.750.5 mm was extruded after the extruder's settings of RPM 4 and a temperature of 175°C–180°C were made.

3.3.3 MANUFACTURING OF 3D-PRINTED SAMPLES

The pin-on-disk samples were manufactured with the help of Creality Ender 3 Pro FDM 3D printer which has a nozzle diameter of 0.4 mm. A total of three samples were manufactured: one sample of PLA and two samples of PLA-Al composite. The samples which were manufactured have a diameter of 8 mm and a length of 30 mm. Throughout the entire work, the printing speed of 60 mm/s, print temperature of 60°C, and nozzle temperature of 220°C remain the same. Both samples of PLA-Al composite were constructed using 0.15 mm layer with grid pattern and 60% infill and 0.20 mm layer thickness with grid pattern and 80% infill. The PLA sample with 0.18 mm layer thickness, with linear fill pattern and 40% infill, was fabricated. All the samples are depicted in Figure 3.1. To investigate the tribological characteristics, these samples were developed.

3.3.4 WEAR TESTING

The 3D-printed pin-on-disk samples with 8 mm diameter and 30 mm in length were taken into consideration for wear testing. The wear test was performed at room temperature using a tribometer (model TR-20LE-PHM-400; maximum capacity: 200 N).

FIGURE 3.1 Pin-on-disk samples before wear testing.

Using a track diameter of 80 mm, a track speed of 100 rpm, and an applied load of 20 N for 5 minutes in atmospheric circumstances, the wear tests were carried out. A measurement system was added to the wear test device. On a monitor, an immediate picture of wear (μm), the COF, was shown.

3.4 RESULT AND DISCUSSION

The wear rate and COF of samples of PLA and PLA-Al composites have been examined to determine the impact of the Al powder addition within the PLA matrix on wear characteristics. Figures 3.2–3.4 show that the wear of PLA material is significantly higher than that of composite materials, demonstrating that the recently produced composites are significantly more wear-resistant than PLA. The wear rate and COF of all the samples are mentioned in Table 3.2 in which it can be seen that the PLA has more wear as compared to PLA-Al composite. After the wear test, the PLA demonstrated a wear rate of 1,188 μm and a COF of 1.8. The PLA sample's initial weight was 2.3516 g, but after the wear test, it weighed only 2.2731 g. The weight loss of 0.0785 g means that during the wear test, 0.0785 g of material was removed from the PLA sample after 5 minutes interval which also indicates that the PLA is less durable. Figure 3.2a demonstrates plots of wear rate and Figure 3.2b depicts the COF with time.

Figure 3.3a and b clearly shows that the sample with 0.15 mm layer thickness, 60% infill density, and grid infill pattern has the best wear resistance of all the samples in terms of wear rate and COF. Following the wear test, the sample with a 0.15 mm thickness had a wear rate of 501 μm and a COF of 2.6. And the sample's weight loss during testing was reduced from 2.3874 to 2.3245, meaning that the material loss throughout the wear test was only 0.0629, proving that it has a good capacity for resistance during friction. Mixing Al powder into PLA Matrix improves its abrasion and slide resistance because Al has the ability to avoid seizures. PLA and Al combined in the composite have a synergistic effect that improves wear resistance. Compared

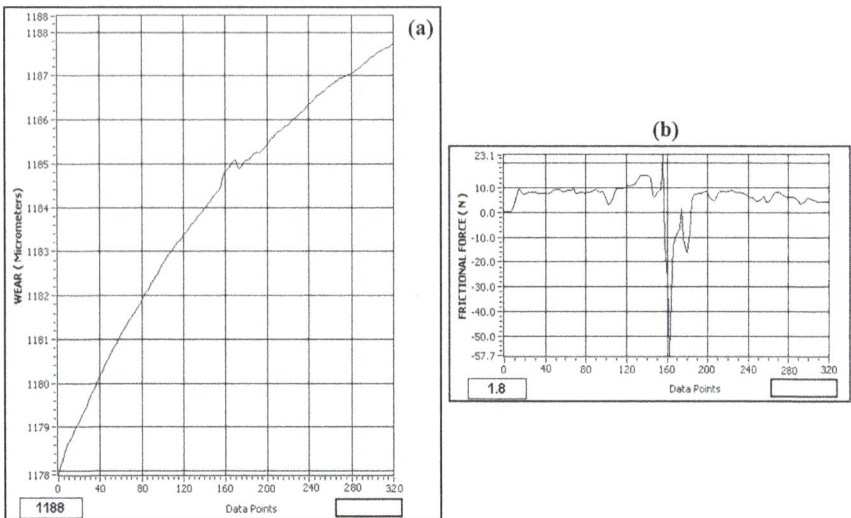

FIGURE 3.2 Wear rate and COF of PLA.

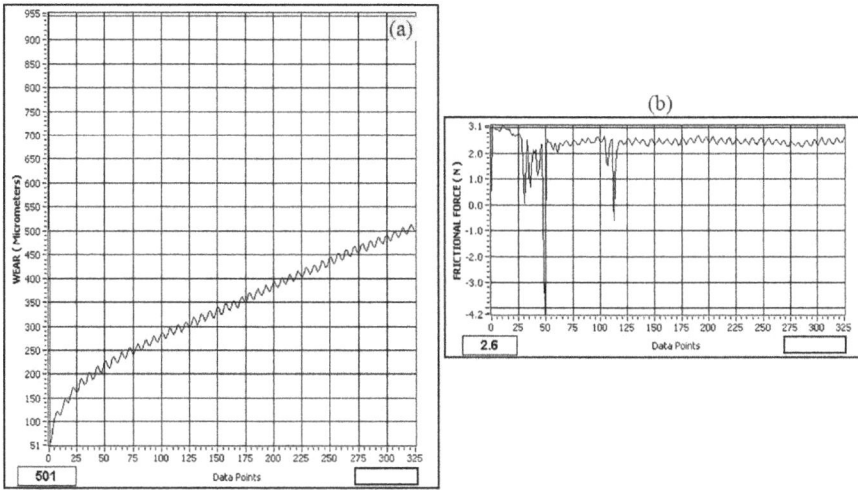

FIGURE 3.3   Wear rate and COF of PLA-Al 0.15 mm layer thickness sample.

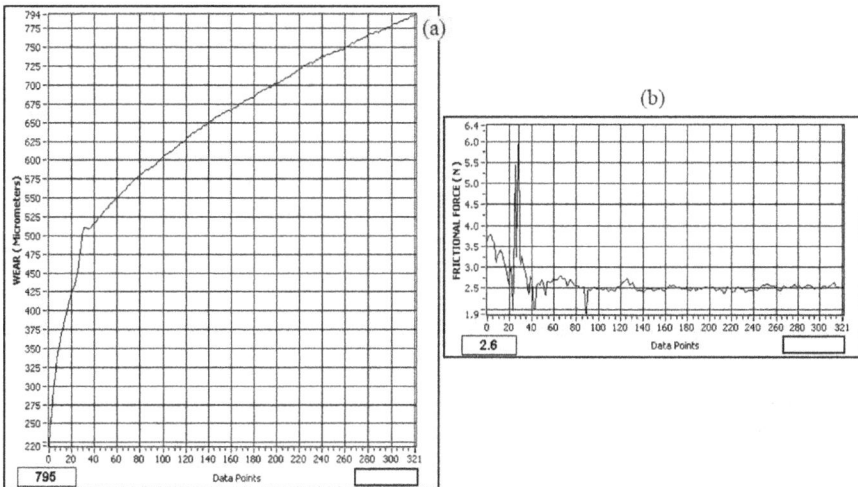

FIGURE 3.4   Wear rate and COF of PLA-Al 0.20 mm layer thickness sample.

TABLE 3.2
Tribological Properties of All Samples

Material	Wear Rate (μm)	COF	Material Loss
PLA	1,188	1.8	0.0785
PLA-Al 0.15 mm layer thickness	501	2.6	0.0629
PLA-Al 0.20 mm layer thickness	795	2.6	0.0279

to PLA, Al is a stronger and more resilient material. In addition to improving the overall composite material and boosting its resistance to wear and abrasion, the use of Al particles improves the load-bearing capacity of the material as compared to PLA. Because of this, composite material exhibits improved wear resistance. On the other hand, the sample with 0.20 mm layer thickness, with zig zag type of patterns and 80% infill density, also demonstrates good wear resistance against the sliding mechanism in which it shows 795 μm wear rate and 2.6 COF as depicted in Figure 3.4. When compared to other samples, the material loss which consisted from 2.3911 to 2.3632, the sample with a 0.20 mm layer thickness experienced the least amount of material loss throughout testing, resulting in a total loss of 0.0279. From the study, it was concluded that the optimized parameters for getting better tribological results were 0.15 mm layer thickness, with grid type of pattern and 60% infill.

3.5 CONCLUSION

This study thoroughly investigated the tribological properties of PLA and composite PLA-Al material samples produced using an FDM 3D printer. The impact of the applied load of 20 N and the runtime of 5 minutes at a speed of 100 rpm was examined using pin-on-disk equipment. The study's conclusions are as follows:

- From the literature, it was noticed that adding fillers into the polymer matrix significantly enhances the tribological properties of the polymer composite.
- The PLA-Al composite performs better than PLA with respect to wear resistance, according to the wear test results. The composite material's improved wear-resistant properties are a result of the inclusion of Al fillers as reinforcement within the PLA matrix.
- The results of the wear test demonstrated that the PLA has a poor wear rate of 1,188 μm and a COF of 1.8, whereas the PLA-Al composite with 0.15 mm layer thickness samples exhibits an excellent wear rate of 501 μm and a COF of 2.6.
- Even though this study concentrated on PLA-Al composites with particular parameters, additional research into various combinations of polymer matrices and reinforcement materials is required. Additionally, investigating the impact of other 3D printing settings on the tribological properties of polymer composites can offer useful insights for optimizing wear resistance in additive manufacturing applications.

ACKNOWLEDGMENT

The authors are obliged to University Centre for Research and Development, Chandigarh University for technical support.

REFERENCES

Al Abir A, Chakrabarti D, Trindade B. Fused filament fabricated poly (lactic acid) parts reinforced with short carbon fiber and graphene nanoparticles with improved tribological properties. *Polymers*. 2023;15(11):2451.

Arif MF, Alhashmi H, Varadarajan KM, Koo JH, Hart AJ, Kumar S. Multifunctional performance of carbon nanotubes and graphene nanoplatelets reinforced PEEK composites enabled via FFF additive manufacturing. *Composites Part B: Engineering.* 2020;184:107625.

Bahadur S. The development of transfer layers and their role in polymer tribology. *Wear.* 2000;245(1–2):92–9.

Bahadur S, Gong D. The action of fillers in the modification of the tribological behavior of polymers. *Wear.* 1992;158(1–2):41–59.

Batakliev TO. Tribological investigation of PLA-based nanocomposites by scratch and wear experiments. *Journal of Theoretical and Applied Mechanics.* 2020;50:105–13.

Boparai K, Singh R, Singh H. Comparison of tribological behaviour for Nylon6-Al-Al2O3 and ABS parts fabricated by fused deposition modelling: This paper reports a low cost composite material that is more wear-resistant than conventional ABS. *Virtual and Physical Prototyping.* 2015;10(2):59–66.

Bustillos J, Montero D, Nautiyal P, Loganathan A, Boesl B, Agarwal A. Integration of graphene in poly (lactic) acid by 3D printing to develop creep and wear-resistant hierarchical nanocomposites. *Polymer Composites.* 2018;39(11):3877–88.

Cardoso PH, de Oliveira MF, de Oliveira MG, da Silva Moreira Thiré RM. 3D printed parts of polylactic acid reinforced with carbon black and alumina nanofillers for tribological applications. *Macromolecular Symposia.* 2020;394(1):2000155.

Chan JX, Wong JF, Petrů M, Hassan A, Nirmal U, Othman N, Ilyas RA. Effect of nanofillers on tribological properties of polymer nanocomposites: A review on recent development. *Polymers.* 2021;13(17):2867.

Friedrich K. Polymer composites for tribological applications. *Advanced Industrial and Engineering Polymer Research.* 2018;1(1):3–9.

Garg H, Singh R. Tribological properties of Fe-Nylon6 composite parts prepared using fused deposition modelling. *Transactions of the Indian Institute of Metals.* 2017;70(5):1241–4.

Hanon MM, Alshammas Y, Zsidai L. Effect of print orientation and bronze existence on tribological and mechanical properties of 3D-printed bronze/PLA composite. *The International Journal of Advanced Manufacturing Technology.* 2020;108:553–70.

Karabeyoğlu SS, Olcay EK, Feratoğlu K. Wear characteristics of PLA-Cu composites manufactured by fused deposition modelling under different temperature conditions. *BalıkesirÜniversitesi Fen BilimleriEnstitüsüDergisi.* 2021;23(1):358–65.

Kumar V, Singh R, Ahuja IS. On wear properties of mechanical blended and chemical assisted mechanical blended ABS-graphene reinforced composites. In: Hashmi MSJ (ed.), *Encyclopedia of Materials: Plastics and Polymers*, 2022, Vol. 1, pp. 434–441, Elsevier.

Li Z, Chen G, Lyu H, Ko F. Experimental investigation of compression properties of composites with printed braiding structure. *Materials.* 2018;11(9):1767.

Myshkin N, Kovalev A. Adhesion and surface forces in polymer tribology-A review. *Friction.* 2018;6:143–55.

Omrani E, Menezes PL, Rohatgi PK. State of the art on tribological behavior of polymer matrix composites reinforced with natural fibers in the green materials world. *Engineering Science and Technology.* 2016;19(2):717–36.

Pawlak W, Kowalewski P, Przekop R. The influence of MoS2 on the tribological properties of polylactide (PLA) applied in 3D printing technology. *Tribologia.* 2020;289(1):57–62.

Przekop RE, Kujawa M, Pawlak W, Dobrosielska M, Sztorch B, Wieleba W. Graphite modified polylactide (PLA) for 3D printed (FDM/FFF) sliding elements. *Polymers.* 2020;12(6):1250.

Ramachandran MG, Rajeswari N. Influence of nano silica on mechanical and tribological properties of additive manufactured PLA bio nanocomposite. *Silicon.* 2022;14:703–709. https://doi.org/10.1007/s12633-020-00878-4

Tambrallimath V, Keshavamurthy R, Patil A, Adarsha H. Mechanical and tribological characteristics of polymer composites developed by fused filament fabrication. In: Dave HK, Davim JP (eds.), *Fused Deposition Modeling Based 3D Printing*, 2021, pp. 151–66. Cham: Springer.

Valente CA, Boutin FF, Rocha LP, do Vale JL, da Silva CH. Effect of graphite and bronze fillers on PTFE tribological behavior: A commercial materials evaluation. *Tribology Transactions*. 2020;63(2):356–70.

Vishal K, Rajkumar K, Sabarinathan P, Dhinakaran V. Mechanical and wear characteristics investigation on 3D printed silicon filled poly (lactic acid) biopolymer composite fabricated by fused deposition modeling. *Silicon*. 2022;14(15):9379–91.

Zhou Y, Lei L, Yang B, Li J, Ren J. Preparation of PLA-based nanocomposites modified by nano-attapulgite with good toughness-strength balance. *Polymer Testing*. 2017;60:78–83.

4 Tribocorrosion Properties of Additively Manufactured Parts

Gaurav Parmar, Mukul Anand, Harish Bishwakarma, and Nitesh Kumar
Indian Institute of Technology (Indian School of Mines, Dhanbad)

Rashi Tyagi
Chandigarh University

Alok Kumar Das
Indian Institute of Technology (Indian School of Mines, Dhanbad)

4.1 INTRODUCTION

Additive manufacturing (AM) has become a notable method within the last decade with promising potential (Tyagi et al., 2023; Ranjan et al., 2023a). Its applications have extended to nearly every sector of manufacturing. Industries like automobiles, aerospace, and medicine have greatly been aided by the introduction of this type of technology (Anand et al., 2022).

The economic importance of tribocorrosion investigations in terms of the longevity and durability of engineered materials has garnered more attention in recent years. Nearly all metals react chemically with the environment and develop some type of corrosion, which degrades the metal. The majority of systems work in environments with frequent mechanical engagement and corrosive environments. The mechanical qualities can also be harmed by mechanical interactions such as loading, tension, and friction. The material can be severely damaged by wear that results from the rubbing of two surfaces or the collision of solid particles or liquids. The corrosive environment frequently affects the material loss brought on by wear (Cao & Mischler, 2018). A corrosive environment can lower the service life of the materials by accelerating the rate of wear and, at the same time, enhancing or complementing corrosion damage.

Frictional interactions contribute to the overall loss of 23% of the world's energy. Furthermore, around a fifth of this energy is utilized to reduce friction, and about 3%

is used to remanufacture or replace damaged components (Ahmed et al., 2020). Good tribological practices can reduce energy losses from friction and wear by around 40%, and these savings might add up to about 1.4% of the GDP of any wealthy country (Holmberg & Erdemir, 2017). Similar to this, corrosion losses represent between 1% and 5% of a country's GDP. This cost also includes indirect losses from corrosion such as production, maintenance, and breakdown expenses (Javaherdashti, 2000). The collaboration between corrosion and wear can cause a large increase in material loss. Tribocorrosion is the name for this cooperative relationship between wear and corrosion. This chapter introduces the concept of AM, the concept of tribocorrosion, and the tribological behavior of various parts made using AM.

4.2 ADDITIVE MANUFACTURING

According to the ASTM committee, AM has been defined as a process of joining materials from 3D data usually layer by layer to make objects contrary to subtractive manufacturing (Tyagi and Tripathi, 2023; Ranjan et al., 2023b; Anand and Das, 2022). It has several nomenclatures such as rapid fabrication, rapid prototyping, and additive fabrication (Huang et al., 2015). In the initial days, the concept of AM was only used to create prototypes (Anand & Das, 2021). The first AM technique to employ a laser to cure photosensitive polymer was stereolithography (SLA), which was introduced in 1987 (Abdulhameed et al., 2019). The first technologies to be patented were selective laser sintering (SLS) and fused deposition modeling (FDM), which occurred in 1988 and 1989, respectively. Since then, this technology has become increasingly prevalent in industrial settings all around the world (Rouf et al., 2022).

The process of adding layers of materials on top of one another to create the final product is known as AM. AM uses a far less material than traditional subtractive manufacturing techniques. By combining succeeding layers, it creates a part using the layer-by-layer manufacturing technique. AM has steadily transformed from being used only for prototyping into a fully functional manufacturing method in a variety of sectors (Baumers et al., 2020; Saade et al., 2020). Initially, quick prototyping and analysis were all that AM was employed for, but it has since emerged as one of the highly capable technologies for the future (Egger & Masood, 2020). In AM, the final object is developed in 3D CAD software before being manufactured. The Standard Tessellation Language (STL) file, which contains all the details about the geometry of the component to be made, is created from this 3D computer data (Borgianni et al., 2019). The final result is then produced once the item is printed using an appropriate AM technique. The AM process is shown in Figure 4.1.

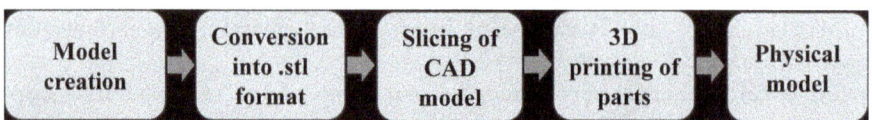

FIGURE 4.1 Additive manufacturing process.

4.2.1 INDUSTRIAL APPLICATIONS OF AM

Globally, AM is transforming the landscape of product design and on-site production, assisting radical product redesigns that hasten the discovery of novel material characteristics. Table 4.1 describes the industrial applications and advantages of AM.

According to the ASTM committee, AM has been classified into seven categories as described in Table 4.2.

TABLE 4.1
Applications and Advantages of Additive Manufacturing

Industry	Applications	Advantages
Automotive	An engine control unit, air inlet, intake valves, engine parts, gearboxes, and bottom fairing baffle	Cost-effective luxury car customization method Testing and production of highly strong, lightweight components Management of undesirability for faulty parts
Aerospace and defense	Landing gears, push reverser doors, miniaturized drones for reconnaissance, gimbal eyes, projectile launcher for grenades, intricate struts, and set engine constituents Repair of high-value parts and turbine blades	Low-volume manufacture of intricately shaped goods with great value Fuel economy through component weight reduction Increased usefulness of the product by the creation of replacement components as needed
Healthcare	Medical instruments used for grasping and manipulating objects or tissues, retractors, medical immobilizers, and dagger levers are examples of surgical instruments Implants: Casts, Stents, Craniofacial Implants, and Limbs	Manufacturing of specialized implants, equipment, dental tops, etc. Cost savings in medical care due to fewer re-interventions made possible by precise diagnosis Rapid production scaling to provide quick reaction during emergencies
Electronics	Wearable technology, soft robotics, structural monitoring, and construction components, as well as RFID (Radio Frequency Identification) technology integrated with robust bases	High resolution, multi-material. The manufacture of electronic devices on a significant scale without the use of printed circuit boards (PCBs) in their construction Creation of intricate, lightweight structures with impact resistance and various uses. Designing intricately shaped pieces with embedded electronics, sensors, and antennas that are impossible to manufacture using traditional production methods Internal production of circuits and circuit boards shortens procurement stretch and removes difficulties with intellectual assets

(Continued)

TABLE 4.1 (*Continued*)
Applications and Advantages of Additive Manufacturing

Industry	Applications	Advantages
Consumer goods	Electronics for the home, jewelry, footwear, apparel, cosmetics, toys, breakables, cabinets, office supplies, musical appliances, bicycles, and food goods	Innovative product design is required due to the fabrication of complicated interior and exterior structures Customizing a product that is focused on the needs of the consumer more quickly and affordably Decentralized production lowers costs that are passed on to consumers

TABLE 4.2
Classification of Additive Manufacturing

Process	Technology	Materials	Applications
Vat photopolymerization: A method to selectively use light-activated polymerization to cure liquid photopolymers in a vat	Stereolithography (SLA), digital light processing (DLP)	Photopolymers	Purchaser Toys, electronics, guides, and stuffs
Binder jetting: A method for selectively applying liquid adhesive to connect powder materials	Powder bed and inkjet head, plaster-based 3D printing	Polymers, waxes	Prototyping and tooling
Directed energy deposition: A technique in which materials are fused by melting as they are being deposited using focused thermal energy (such as a laser, electron beam, or plasma arc)	Laser metal deposition (LMD)	Metals	Fixing or assembling high-volume parts
Material extrusion: A technique for selectively dispensing material through an aperture or nozzle	Fused deposition modeling (FDM)	Polymers	Prototyping, tooling, and office manufacturing
Material jetting: Droplets of built material are dropped in specific locations by this procedure	Multijet modeling (MMM)	Polymers, waxes	High-resolution prototypes, circuit boards and other electronics, purchaser merchandise, and tooling
Powder bed fusion: A method wherein a powder bed's various sections are fused using heat energy	Electron beam melting (EBM), selective laser sintering (SLS), selective heat sintering (SHS)	Metals, polymers	Aerospace, automotive, medical products, tooling and dental implants
Sheet lamination: A procedure for adhering sheets of material together to create an object	Laminated object manufacturing (LOM), ultrasonic consolidation (UC)	Paper, metals	Large portions and tooling

4.3 TRIBOCORROSION

In the 1990s, tribocorrosion was initially examined, although there were only a few studies done at that time (Malik et al., 2022b). Today, tribocorrosion is a significant study area with several applications in industry (Shahini et al., 2022). In essence, it is the fusion of two key disciplines, explicitly tribology and corrosion, which have considerable and useful submissions in mechanical frameworks. Figure 4.2 depicts a tribo-mechanical system. Tribocorrosion systems are extremely complicated and depend on a multitude of factors, including mechanical, material, physiochemical, and electrochemical ones (Mathew et al., 2009) as depicted in Figure 4.3.

While corrosion pertains to chemical deterioration of materials, tribology addresses friction, wear, and lubrication (Vieira et al., 2006).

"A degradation phenomenon of material surfaces (wear, cracking, corrosion, etc.) subjected to the combined action of mechanical loading (friction, abrasion, erosion, etc.) and corrosion attack caused by the environment (chemical and/or electrochemical interaction)"

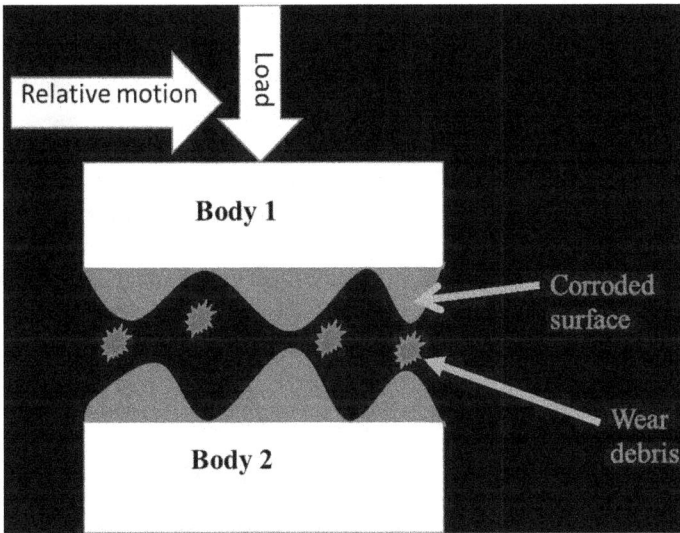

FIGURE 4.2 Tribocorrosion between two surfaces.

Mechanical	Material	Electrochemical	Physiochemical
• Load	• Hardness	• Resistance	• Temperature
• Friction	• Elasticity	• Potential	• Viscosity
• Vibration	• Roughness	• Kinetics	• Conductivity
	• Microstructure	• Mass transfer	• Oxidants
			• Ions

FIGURE 4.3 Factors affecting tribocorrosion.

is the definition of tribocorrosion (Celis et al., 2006). Tribocorrosion, in other terms, refers to the combined effects of wear and corrosion. Tribocorrosion is largely concerned with the permanent alteration of materials or their function brought on by concurrent mechanical, chemical, and electrochemical contact amid faces undergoing relative motion (Mischler, 2008).

Tribocorrosion is recognized by the combined effect of mechanical and environmental consequences it causes. Compared to the separate effects of these procedures, the united influence of these procedures produces higher material loss and material deterioration. The reaction products can either protect the exterior (self-oiling and self-restorative layers) or worsen material deterioration and accelerate material loss, depending on the reactions occurring at the surface (Basak et al., 2006).

Tribocorrosion primarily focuses on the interaction of two surfaces in a corrosive media by mechanical, chemical, and electrochemical means. It is undesirable in many devices and installations because of its negative consequences. Nevertheless, it may also be utilized effectively, for example in electrochemical machining, when chemical processes play a relatively little role. Metal implants used in biological systems, such as dental implants, prosthetic joints, screws, and plates, are similarly subject to tribocorrosion. Tribocorrosion is therefore divided into two categories: tribocorrosion in industrial systems and tribocorrosion in biological systems (Wood, 2007).

Tribocorrosion is a phenomenon that occurs in a variety of real-world settings and causes material loss while also impairing a manufacturing process's robustness, performance, efficiency, and dependability. The nuclear, chemical, petrochemical, material handling, maritime, aerospace, automotive, mining, and biomedical sectors are the industries where tribocorrosion occurs most frequently (Landolt et al., 2001). Tribocorrosion is also found in biological organisms, where it may harm joints, prosthetics, and dental restoration, in addition to the previously stated industrial uses. Different situations and forms of interaction can cause tribocorrosion. Typically, it combines corrosion with biological solutions, sliding wear, fretting, solid particle erosion, cavitation erosion, and abrasion. Tribocorrosion is mostly caused by two- and three-body interactions between sliding exteriors. Reciprocating or unidirectional relative motion between sliding surfaces is also possible. Another tribological contact is fretting, which involves a slight reciprocating motion. Ball bearings include rolling contacts that also come into touch with tribocorrosion. Impingement of particles occurs during erosion corrosion as a result of mechanical and chemical acts, and it is especially prevalent in pipelines and pumps that transport slurries. Thus, it can be stated that tribocorrosion is made up of a variety of mechanical and chemical phenomena, including corrosion fatigue, oxidative wear, fretting corrosion, erosion-corrosion, and wear due to corrosion (Ponthiaux et al., 2004).

Many of the products for tribocorrosion applications manufactured through AM have their uses in medical fields for orthopedic applications. The number of orthopedic patients has increased as a result of the expanding population and rising accident rates each year which has rapidly increased the demand for medical implants. AM is a kind of technology that can solve this problem in a time-efficient way. Once these medical implants are installed, they must be able to withstand the effects of crucial surface phenomena similar to corrosion, friction, and wear.

4.4 TRIBOLOGY OF ADDITIVE MANUFACTURED ORTHOPEDIC IMPLANTS

In orthopedics, AM is widely employed in the creation of different implants and the cure of numerous ruptures. These implants are subjected to a variety of stresses and pressures when they are placed in a patient's body. Implants must endure these loads throughout their service life while retaining functionality. As a result of a person's constant mobility, friction is a crucial factor that has to be researched in these implants. Wear also comes into play as a result of friction and might harm or shorten the life of the implant. In these implants, lubrication is frequently used to lower friction and wear. The crucial tribological features of these implants, such as friction, wear, and lubrication, are covered below.

4.4.1 Friction and Wear

Strength, hardness, and wear resistance are among the mechanical and tribological qualities that many orthopedic implants and implant materials must possess. A joint prosthesis has at least two pieces moving continuously with one another (Landolt, 2006). Because of this, it is crucial in these situations to analyze various tribological variables including friction and wear. To ensure that orthopedic implants work and last when exposed to prolonged friction conditions, it is critical to analyze these factors. Another issue with these implants is the build-up of wear debris because of the ongoing presence of friction. The best materials to use are ones that don't produce and disseminate wear debris in the surroundings (Paterlini et al., 2021). These substances typically have a low coefficient of friction and good wear resistance. Sophisticated ceramics such as alumina and zirconia, metallic elements like titanium and chromium alloys, or polymer materials, particularly ultra-high molecular weight polyethylene (UHMWPE), are the materials that display the aforementioned characteristics most frequently (Affatato et al., 2017; Bal et al., 2007).

Metal-on-metal (MoM) configurations, ceramics or metal paired with UHMWPE (CoP or MoP), and ceramic-on-ceramic (CoC) are the three most significant bearing pairs that have been employed more often. Pairings of ceramic with ceramic, metal with metal, and metal combined with UHMWPE all have coefficients of friction that, in a dry environment, fall in the ranges of 0.40–0.50 (Malik et al., 2022a), 0.40–0.4592 (Heshmat & Jahanmir, 2004), and 0.2–0.2593 (Blau, 1998), respectively. These constituents were evaluated in simulated body settings with different bodily fluids, such as bovine and calf serum, for orthopedic applications, and it was discovered that the coefficient of friction significantly declined (Guezmil et al., 2016; Baykal et al., 2014). Coefficients of friction offered by CoP and MoP are both significantly lower than those offered by UHMWPE systems, which frequently create more wear debris and pose considerable health risks (Pawlak et al., 2013; Nine et al., 2014). Thus, the production of load-bearing components using ceramic materials is a viable alternative, and a variety of arthroplasty implants may be effectively produced via AM. The tribological examination of items crafted from alumina, zirconia stabilized with yttria, and alumina reinforced with zirconia (ZTA) produced using SLA was done by researchers. They found that printing orientation had the least impact on micro-hardness, wettability, and micro-porosity. Zirconia toughened alumina showed the slightest wear rate and provided the best surface characteristics among these materials. As opposed to alumina, which had stable and lower frictional values, yttria-stabilized zirconia had greater wear rates and coefficients of friction. Additionally, a self-mated tribe-sintered layer was seen to develop in alumina-based components, expanding its applications to joint implants and other tribological ones.

Ti6Al4V, a titanium alloy, is one of the most often utilized materials for creating different orthopedic implants utilizing AM (Plumlee & Schwartz, 2013; Bartolomeu et al., 2017). This material has good corrosion resistance, lower density when compared to metal, and excellent mechanical properties (Bruschi et al., 2017; Dantas et al., 2017). The biggest issue with this material, though, is its resistance to wear. Ti4Al6V has low wear resistance, which over time may cause these wear particles to become loose and could even cause the implant to fail (Sampaio et al., 2016a). To create a variety of Ti6Al4V alloys, the authors employed several AM processes. Studies showed that AM has no appreciable impact on it. Despite this, the specific wear rate (SWR) somewhat lowered as a result of the improved microstructure and residual stresses discovered by AM When compared to other AM materials, the samples made utilizing electron beam melting (EBM) were very hard yet had the greatest SWR, despite having a martensitic microstructure and being the hardest. Different techniques, including surface modifications, coatings, functionally graded materials, and composite materials, have been used to extend the lifespan of these implants (Gorsse et al., 2017). Poly-ether-ether-ketone (PEEK) was one of these materials that were utilized to create a composite. Due to its weak mechanical characteristics and low elastic modulus (3 GPa) compared to cortical human bone (20 GPa), PEEK cannot be utilized alone as an implant material. For usage primarily in hip implants, researchers created and manufactured a cellular-structured component with the characteristics of Ti6Al4V and PEEK (Sampaio et al., 2016b). Ti6AlV-PEEK multi-material structures were obtained using SLM and pressure-assisted injection techniques. Tribological experiments were conducted to determine how this multi-material behaved. The addition of PEEK in Ti6Al4V was shown to greatly boost wear resistance based on wear data. Due to the self-lubricating qualities of PEEK, mass loss was reduced by over 62% when compared to cellular Ti6Al4V structures and by 40% when compared to traditionally manufactured implant materials.

4.4.2 LUBRICATING BEHAVIOR

As was mentioned in the preceding section, friction and wear lead to material detachment from implants, which accumulates as debris and causes implant failure. Different composite materials with great wear resistance and low coefficients of friction are being developed to address this issue. The idea of lubrication in these implants is another workable answer to these issues. One of the main foundations of tribology is lubrication, which has been successfully applied for many years to lessen friction between two contacting surfaces.

According to reports, globally more than 1 million hip replacement procedures are carried out each year, and this number is anticipated to rise as people live longer due to an increase in arthritis concerns among the elderly (Cowie et al., 2019). Wear between the articulating surfaces is the primary cause of hip implants' current failure rates. Thus, it is crucial to evaluate these implants in a setting that closely resembles the movements that the patient's body would subject them to. In this context, dedicated joint simulators have been developed by researchers that offer an authentic background for testing these implants. According to ISO standard requirements, which allow one or the other bovine or calf serum to be used in wear testing, a diluted bovine serum is the lubricant utilized in these simulators (Bauer & Schils, 1999). Although bovine serum is frequently employed as a lubricant in wear testing, its principal drawbacks are deterioration, batch fluctuation, and precipitation (ISO, 2002). The bovine serum also contains lipids (cholesterol and triglycerides) and proteins (albumin and globulin), whereas synovial fluid, the body's natural lubricant, is made up of albumin, globulin, phospholipids, and hyaluronan. These variations cause these fluids' rheological behavior to differ. This is why different lubricants have been created and employed in these implants, always retaining biocompatibility as a key consideration. When compared to bovine serum albumin, the usage of this biopolymer modestly decreased the COF at both low and high sliding velocities. Other methods are used to enhance the surface and tribological performance of implants without the use of lubricants. These entail surface alterations, and one such method is surface texturing, which is covered in the subsequent section.

4.5 ORTHOPEDIC USES FOR TEXTURING AM PARTS

To give materials and products particular properties that can improve their functioning and effectiveness within particular and focused operational conditions, surface modifications have been applied. Numerous surface modification techniques have also been applied in engineering to increase the materials' resistance to corrosion and minimize wear and friction (Ching et al., 2014). Selective laser melting (SLM), an AM technique, offers a lot of perspective for producing complex metallic components, which has piqued the curiosity of many academics. SLM enables several crucial industrial aspects, including precise and exact control over geometrical dimensions, and forms down to the microscale (Weng et al., 2014). Since the majority of implant surfaces are constantly rubbing against one another, surface features have a significant influence on the functionality and behavior of these implants. Due to its excellent mechanical and anti-corrosive qualities, titanium and its alloys are the most often utilized materials for implants for dental and bone tissue restoration (Li et al., 2020). The majority of investigations on the alteration of the surface of these titanium alloys have focused on analyzing the impacts of functional coatings to enhance the bio-functionality of these alloys. In investigations on the surface alteration of these titanium alloys, the focus has mostly been on analyzing the impacts of functional coatings to enhance the bio-functionality of these alloys (Chen et al., 2020).

Surface texturing is a useful technique for enhancing surface behavior and lengthening product lifespan. It has also been demonstrated to lower the coefficient of friction, which results in less wear detritus and extends the lifespan of items (Brown & Badylak, 2013). Surface texturing has drawn a lot of attention in orthopedic implants and synthetic joints since it has been shown to both enhance hydrodynamic pressure and store wear debris. To enhance Ti-6Al-4V's tribological characteristics, cyto-biocompatibility, and anti-inflammatory effects, researchers focused on surface texturing. They produced Ti-6Al-4V components with specified rhombus surface textures using SLM and then tracked it up with post-heating in N_2 (nitriding), which caused thick coatings to develop on the patterned surfaces. The results of the investigation showed that surface texturing improved the material's wear resistance while assisting in lowering the coefficient of friction from 0.343 to 0.166.

For example, surface texturing in different orthopedic applications uses a technology called laser surface texturing (LST) to modify surfaces (Guo et al., 2018). Researchers carefully investigated LST for functionalizing surfaces. LST improved implant performance in the area of orthopedic applications. In addition to strengthening their mechanical interaction with bones, it has increased the materials' biocompatibility (Wahab et al., 2016). Laser induced periodic surface texture (LIPSS) has also assisted in enhancing the performance of biomaterials. Given that the laser intensity is more or less equivalent to the material's damage threshold, these structures, which take the shape of consistent ripple patterns, are simple to recreate on any material. The majority of laser surface texturing research has been conducted on flat substrates. Several processing disturbances are encountered when processing complicated forms, for instance, the acetabular shells used in complete hip replacements. This problem is resolved by creating these spherical components using the LPBF approach, which can create these parts in a form that is very close to a net.

4.6   AM ORTHOPEDIC COMPONENTS' CORROSION BEHAVIOR

As covered in earlier sections, orthopedic applications frequently use titanium and its alloys produced using AM. Considering the characteristics and functioning circumstances of implants, their success is heavily dependent on the final product's tribological performance, which is allegedly poor for titanium alloys, particularly Ti-6Al-4V (Mariscal-Muñoz et al., 2016). These materials are vulnerable to corrosion because the human body is exposed to a corrosive environment and this alloy has poor wear resistance. Once corrosion begins, the passive layers on the material's surface are removed, causing the growth of new oxide layers. The development and build-up of wear debris as a result of the full chain of events has an adverse effect on the patient (Buciumeanu et al., 2018). Biomaterials created from 316L stainless steel are also employed in the creation of implants and prostheses besides Ti-6Al-4V. They are primarily employed in bone screws, different dental implants, and hip-knee prostheses. Ti-6Al-4V is employed in high-load applications in addition to 316L stainless steel due to its low weight, substantial strength, biocompatible nature, elevated stability, and superior corrosion resistance (Sun et al., 2016). Titanium alloys are demonstrated to have superior surface characteristics like corrosion and biocompatibility in comparison to stainless steel.

Anodic oxidation, one of the strategies listed above, is quite helpful in reducing the rate of corrosion in metallic components. It has also improved the electrochemical and tribological characteristics of certain titanium alloys. Additionally, it saves time and money, and it is incredibly simple to use (Revilla et al., 2018). Researchers discovered that anodizing titanium alloys increased their corrosion and wear resistance (Luz et al., 2018). Researchers created multilayer Ti-6Al-4V/316L substrates using laser powder bed fusion (Turalıoğlu et al., 2021). TiO_2 was applied as a layer on Ti-6Al-4V/316L using the anodic oxidation technique. The study found that compared to non-anodized specimens, anodized specimens had much superior wear and corrosion resistance.

4.7 OBSTACLES WITH AM ORTHOPEDIC IMPLANTS

The majority of industrial sectors worldwide have already started implementing intelligent manufacturing techniques like machine learning (ML), computing, artificial intelligence (AI), and the Internet of Things (IoT) in their networks. Such methods constitute a barrier to advancing AM since they have not yet been commonly incorporated with AM technology in orthopedic applications. Process variable factors such as layer thickness, internal structure, printing direction, size of metal particles, printing speed, supportive material, and after-printing procedures have a significant impact on the majority of additively manufactured orthopedic items, such as bone screws. Before printing, all of these variables must be optimized to get the greatest quality. Many implant qualities might be impacted if one parameter is not adequately optimized. By properly optimizing these parameters, using different ML algorithms may assist in increasing the implant quality. Another significant issue with AM that precludes it from being utilized for large manufacturing is scalability. Only a relatively small number of medical practitioners and engineers have the acquaintance and expertise to scale up the technology economically, and AM requires manual inputs for running machines. Additionally, AM must be extremely dependable in creating consistent high-quality components at greater rates (Agarwal et al., 2022).

The absence of manufacturing in real-time is another major issue in AM orthopedic parts. This is especially important for operations where a lack of communication between the medical doctor and the design engineer can lead to delays in part manufacturing and can have serious consequences for the patient's health. This is mainly because several software are required for the design, slicing, and correction of mistakes, which the medical practitioner is not familiar with and therefore has to rely on someone else (Putra et al., 2020). Another current issue with orthopedic parts is the use of identical products with standardized geometry, such as bone screws, for fractured bones in all age groups. Due to their less thick bones, older individuals may experience bone screws loosening if bone screws are used with the same geometry for all age groups. Making personalized, age-specific items using AM is a simple solution to this issue (Chandra & Pandey, 2020).

4.8 CONCLUSIONS

The main topic of this chapter is the tribological behavior of components made via AM. Tribocorrosion is a manifold phenomenon that includes mechanical, physical, and electrochemical elements. Tribocorrosion is reliant on several conditions since it combines tribology with corrosion. The capacity of AM to work with complicated geometries may be used to create orthopedic implants with intricate shapes and textures. AM affects a few factors that are crucial for maintaining the durability and effectiveness of implants, particularly when they are subjected to prolonged frictional conditions. When it comes to orthopedic implants, the performance of various materials in corrosive conditions is harmful, and the usage of titanium alloys has produced improved surface qualities in contrast to other metals. Anodic oxidation is particularly effective at reducing the rate of corrosion in metallic components.

ACKNOWLEDGMENTS

The authors are grateful to IIT(ISM) Dhanbad and University Centre for Research and Development, Chandigarh University for their support.

REFERENCES

Abdulhameed, O., Al-Ahmari, A., Ameen, W., & Mian, S. H. (2019). Additive manufacturing: Challenges, trends, and applications. *Advances in Mechanical Engineering*, 11(2), 1687814018822880.

Affatato, S., Ruggiero, A., Merola, M., & Logozzo, S. (2017). Does metal transfer differ on retrieved Biolox(r) Delta composites femoral heads? Surface investigation on three Biolox(r) generations from a biotribological point of view. *Composites Part B: Engineering*, 113, 164–173.

Agarwal, R., Gupta, V., & Singh, J. (2022). Additive manufacturing-based design approaches and challenges for orthopaedic bone screws: A state-of-the-art review. *Journal of the Brazilian Society of Mechanical Sciences and Engineering*, 44(1), 37.

Ahmad, N. A., Samion, S., Rahim, E. A., & Jamir, M. R. M. (2020). Environmentally approach for enhancing tribological characteristics in metal forming: A review. *Jurnal Tribologi*, 26(1), 37–59.

Anand, M., Bishwakarma, H., Kumar, N., Ujjwal, K., & Das, A. K. (2022). Fabrication of multilayer thin wall by WAAM technique and investigation of its microstructure and mechanical properties. *Materials Today: Proceedings*, 56, 927–930.

Anand, M., & Das, A. K. (2021). Issues in fabrication of 3D components through DMLS technique: A review. *Optics & Laser Technology*, 139, 106914.

Anand, M., & Das, A. K. (2022). Grain refinement in wire-arc additive manufactured Inconel 82 alloy through controlled heat input. *Journal of Alloys and Compounds*, 929, 166949.

Bal, B. S., Garino, J., Ries, M., & Rahaman, M. N. (2007). A review of ceramic bearing materials in total joint arthroplasty. *Hip International*, 17(1), 21–30.

Bartolomeu, F., Sampaio, M., Carvalho, O., Pinto, E., Alves, N., Gomes, J. R., & Miranda, G. (2017). Tribological behavior of Ti6Al4V cellular structures produced by selective laser melting. *Journal of the Mechanical Behavior of Biomedical Materials*, 69, 128–134.

Basak, A. K., Matteazzi, P., Vardavoulias, M., & Celis, J. P. (2006). Corrosion-wear behaviour of thermal sprayed nanostructured FeCu/WC-Co coatings. *Wear*, 261(9), 1042–1050.

Bauer, T. W., & Schils, J. (1999). The pathology of total joint arthroplasty: II. Mechanisms of implant failure. *Skeletal Radiology*, 28, 483–497.

Baumers, M., Carmignato, S., & Leach, R. (2020). Introduction to precision metal additive manufacturing. In: *Precision Metal Additive Manufacturing*, Edited By Richard Leach & Simone Carmignato, pp. 1–10. CRC Press.

Baykal, D., Siskey, R. S., Haider, H., Saikko, V., Ahlroos, T., & Kurtz, S. M. (2014). Advances in tribological testing of artificial joint biomaterials using multidirectional pin-on-disk testers. *Journal of the Mechanical Behavior of Biomedical Materials*, 31, 117–134.

Blau, P. J. (1998). Four great challenges confronting our understanding and modeling of sliding friction. In: *Tribology Series*, Edited by D. Dowson, C.M. Taylor, T.H.C. Childs, G. Dalmaz, Y. Berthier, L. Flamand, J.-M. Georges, & A.A. Lubrecht. Vol. 34, pp. 117–128. Elsevier.

Borgianni, Y., Maccioni, L., & Basso, D. (2019). Exploratory study on the perception of additively manufactured end-use products with specific questionnaires and eye-tracking. *International Journal on Interactive Design and Manufacturing (IJIDeM)*, 13(2), 743–759.

Brown, B. N., & Badylak, S. F. (2013). Expanded applications, shifting paradigms and an improved understanding of host-biomaterial interactions. *Acta Biomaterialia*, 9(2), 4948–4955.

Bruschi, S., Bertolini, R., & Ghiotti, A. (2017). Coupling machining and heat treatment to enhance the wear behaviour of an additive manufactured Ti6Al4V titanium alloy. *Tribology International*, 116, 58–68.

Buciumeanu, M., Almeida, S., Bartolomeu, F., Costa, M. M., Alves, N., Silva, F. S., & Miranda, G. (2018). Ti6Al4V cellular structures impregnated with biomedical PEEK-new material design for improved tribological behavior. *Tribology International*, 119, 157–164.

Cao, S., & Mischler, S. (2018). Modeling tribocorrosion of passive metals - A review. *Current Opinion in Solid State and Materials Science*, 22(4), 127–141.

Celis, J. P., Ponthiaux, P., & Wenger, F. (2006). Tribo-corrosion of materials: Interplay between chemical, electrochemical, and mechanical reactivity of surfaces. *Wear*, 261(9), 939–946.

Chandra, G., & Pandey, A. (2020). Biodegradable bone implants in orthopedic applications: A review. *Biocybernetics and Biomedical Engineering*, 40(2), 596–610.

Chen, X., Shah, K., Dong, S., Peterson, L., La Plante, E. C., & Sant, G. (2020). Elucidating the corrosion-related degradation mechanisms of a Ti-6Al-4V dental implant. *Dental Materials*, 36(3), 431–441.

Ching, H. A., Choudhury, D., Nine, M. J., & Osman, N. A. A. (2014). Effects of surface coating on reducing friction and wear of orthopaedic implants. *Science and Technology of Advanced Materials*, 15(1), 014402.

Cowie, R. M., Briscoe, A., Fisher, J., & Jennings, L. M. (2019). Wear and friction of UHMWPE-on-PEEK OPTIMA(tm). *Journal of the Mechanical Behavior of Biomedical Materials*, 89, 65–71.

Dantas, T. A., Abreu, C. S., Costa, M. M., Miranda, G., Silva, F. S., Dourado, N., & Gomes, J. R. (2017). Bioactive materials driven primary stability on titanium biocomposites. *Materials Science and Engineering: C*, 77, 1104–1110.

Egger, J., & Masood, T. (2020). Augmented reality in support of intelligent manufacturing-a systematic literature review. *Computers & Industrial Engineering*, 140, 106195.

Gorsse, S., Hutchinson, C., Gouné, M., & Banerjee, R. (2017). Additive manufacturing of metals: A brief review of the characteristic microstructures and properties of steels, Ti-6Al-4V and high-entropy alloys. *Science and Technology of Advanced Materials*, 18(1), 584–610.

Guezmil, M., Bensalah, W., & Mezlini, S. (2016). Tribological behavior of UHMWPE against TiAl6V4 and CoCr28Mo alloys under dry and lubricated conditions. *Journal of the Mechanical Behavior of Biomedical Materials.* doi:10.1016/j.jmbbm.2016.07.002.

Guo, J., Mei, T., Li, Y., Ren, S., Hafezi, M., Lu, H., … & Dong, G. (2018). Sustained-release application of PCEC hydrogel on laser-textured surface lubrication. *Materials Research Express*, 5(6), 065315.

Heshmat, H., & Jahanmir, S. (2004). Tribological behavior of ceramics at high sliding speeds in steam. *Tribology Letters*, 17(3), 359–366.

Holmberg, K., & Erdemir, A. (2017). Influence of tribology on global energy consumption, costs and emissions. *Friction*, 5, 263–284.

Huang, Y., Leu, M. C., Mazumder, J., & Donmez, A. (2015). Additive manufacturing: Current state, future potential, gaps and needs, and recommendations. *Journal of Manufacturing Science and Engineering*, 137(1), 014001.

ISO, B. (2002). 14242-1, *Implants for surgery. Wear of total hip joint prostheses. Loading and displacement parameters for wear-testing machines and corresponding environmental conditions for test.* London: British Standards Institute.

Javaherdashti, R. (2000). How corrosion affects industry and life. *Anti-Corrosion Methods and Materials*, 47(1), 30–34.

Landolt, D. (2006). Electrochemical and materials aspects of tribocorrosion systems. *Journal of Physics D: Applied Physics*, 39(15), 3121.

Landolt, D., Mischler, S., & Stemp, M. (2001). Electrochemical methods in tribocorrosion: A critical appraisal. *Electrochimica Acta*, 46(24–25), 3913–3929.

Li, Y., Chen, D., Sheng, Y., Li, W., & Wang, X. (2020). In situ preparation of antibacterial Ag particles on Ti6Al4V surfaces by spray deposition. *Surface Innovations*, 9(2-3), 166–173.

Luz, A. R., De Souza, G. B., Lepienski, C. M., Siqueira, C. J., & Kuromoto, N. K. (2018). Tribological properties of nanotubes grown on Ti-35Nb alloy by anodization. *Thin Solid Films*, 660, 529–537.

Malik, A., Haq, M. I. U., Raina, A., & Gupta, K. (2022a). 3D printing towards implementing Industry 4.0: Sustainability aspects, barriers and challenges. *Industrial Robot: The International Journal of Robotics Research and Application*, 49(3), 491–511.

Malik, A., Rouf, S., Haq, M. I. U., Raina, A., Puerta, A. P. V., Sagbas, B., & Ruggiero, A. (2022b). Tribo-corrosive behavior of additive manufactured parts for orthopaedic applications. *Journal of Orthopaedics*, 34, 49–60.

Mariscal-Muñoz, E., Costa, C. A., Tavares, H. S., Bianchi, J., Hebling, J., Machado, J. P., … & Souza, P. P. (2016). Osteoblast differentiation is enhanced by a nano-to-micro hybrid titanium surface created by Yb: YAG laser irradiation. *Clinical Oral Investigations*, 20, 503–511.

Mathew, M. T., Srinivasa Pai, P., Pourzal, R., Fischer, A., & Wimmer, M. A. (2009). Significance of tribocorrosion in biomedical applications: Overview and current status. *Advances in Tribology*, 2009, 1–12.

Mischler, S. (2008). Triboelectrochemical techniques and interpretation methods in tribocorrosion: A comparative evaluation. *Tribology International*, 41(7), 573–583.

Nine, M. J., Choudhury, D., Hee, A. C., Mootanah, R., & Osman, N. A. A. (2014). Wear debris characterization and corresponding biological response: Artificial hip and knee joints. *Materials*, 7(2), 980–1016.

Paterlini, A., Stamboulis, A., Turq, V., Laloo, R., Schwentenwein, M., Brouczek, D., … & Bertrand, G. (2021). Lithography-based manufacturing of advanced ceramics for orthopaedic applications: A comparative tribological study. *Open Ceramics*, 8, 100170.

Pawlak, Z., Petelska, A. D., Urbaniak, W., Yusuf, K. Q., & Oloyede, A. (2013). Relationship between wettability and lubrication characteristics of the surfaces of contacting phospholipid-based membranes. *Cell Biochemistry and Biophysics*, 65, 335–345.

Plumlee, K. G., & Schwartz, C. J. (2013). Surface layer plastic deformation as a mechanism for UHMWPE wear, and its role in debris size. *Wear*, 301(1–2), 257–263.

Ponthiaux, P., Wenger, F., Drees, D., & Celis, J. P. (2004). Electrochemical techniques for studying tribocorrosion processes. *Wear*, 256(5), 459–468.

Putra, N. E., Mirzaali, M. J., Apachitei, I., Zhou, J., & Zadpoor, A. A. (2020). Multi-material additive manufacturing technologies for Ti-, Mg-, and Fe-based biomaterials for bone substitution. *Acta Biomaterialia*, 109, 1–20.

Ranjan, N., Tyagi, R., Kumar, R., & Babbar, A. (2023a). 3D printing applications of thermo-responsive functional materials: A review. *Advances in Materials and Processing Technologies*. https://www.tandfonline.com/doi/abs/10.1080/2374068X.2023.2205669.

Ranjan, N., Tyagi, R., Kumar, R., & Kumar, V. (2023b). On fabrication of acrylonitrile butadiene styrene-zirconium oxide composite feedstock for 3D printing-based rapid tooling applications. *Journal of Thermoplastic Composite Materials*, 0(0), 08927057231186310. https://doi.org/10.1177/08927057231186310

Revilla, R. I., Terryn, H., & De Graeve, I. (2018). Role of Si in the anodizing behavior of Al-Si alloys: Additive manufactured and cast Al-Si10-Mg. *Journal of the Electrochemical Society*, 165(9), C532.

Rouf, S., Raina, A., Haq, M. I. U., Naveed, N., Jeganmohan, S., & Kichloo, A. F. (2022). 3D printed parts and mechanical properties: Influencing parameters, sustainability aspects, global market scenario, challenges and applications. *Advanced Industrial and Engineering Polymer Research*, 5(3), 143–158.

Saade, M. R. M., Yahia, A., & Amor, B. (2020). How has LCA been applied to 3D printing? A systematic literature review and recommendations for future studies. *Journal of Cleaner Production*, 244, 118803.

Sampaio, M., Buciumeanu, M., Askari, E., Flores, P., Souza, J. C. M., Gomes, J. R., ... & Henriques, B. (2016a). Effects of poly-ether-ether ketone (PEEK) veneer thickness on the reciprocating friction and wear behavior of PEEK/Ti6Al4V structures in artificial saliva. *Wear*, 368, 84–91.

Sampaio, M., Buciumeanu, M., Henriques, B., Silva, F. S., Souza, J. C., & Gomes, J. R. (2016b). Tribocorrosion behavior of veneering biomedical PEEK to Ti6Al4V structures. *Journal of the Mechanical Behavior of Biomedical Materials*, 54, 123–130.

Shahini, M. H., Mohammadloo, H. E., & Ramezanzadeh, B. (2022). Recent approaches to limit the tribocorrosion of biomaterials: A review. *Biomass Conversion and Biorefinery*, 1–21.

Sun, Z., Tan, X., Tor, S. B., & Yeong, W. Y. (2016). Selective laser melting of stainless steel 316L with low porosity and high build rates. *Materials & Design*, 104, 197–204.

Turalıoğlu, K., Taftalı, M., Tekdir, H., Çomaklı, O., Yazıcı, M., Yetim, T., & Yetim, A. F. (2021). The tribological and corrosion properties of anodized Ti6Al4V/316L bimetallic structures manufactured by additive manufacturing. *Surface and Coatings Technology*, 405, 126635.

Tyagi, R., Singh. G., Kuma, R., Kumar, V., & Singh, S. (2023). 3D-printed sandwiched acrylonitrile butadiene styrene/carbon fiber composites: Investigating mechanical, morphological, and fractural properties. *Journal of Materials Engineering and Performance*, 1–14. https://link.springer.com/article/10.1007/s11665-023-08292-8.

Tyagi, R., & Tripathi, A. (2023). Coating/cladding based post-processing in additive manufacturing. In: *Handbook of Post-Processing in Additive Manufacturing: Requirements, Theories, and Methods*. Edited by Gurminder Singh, Ranvijay Kumar, Kamalpreet Sandhu, Eujin Pei, & Sunpreet Singh, p. 127. CRC Press.

Vieira, A. C., Ribeiro, A. R., Rocha, L. A., & Celis, J. P. (2006). Influence of pH and corrosion inhibitors on the tribocorrosion of titanium in artificial saliva. *Wear*, 261(9), 994–1001.

Wahab, J. A., Ghazali, M. J., Yusoff, W. M. W., & Sajuri, Z. (2016). Enhancing material performance through laser surface texturing: A review. *Transactions of the IMF*, 94(4), 193–198.

Weng, F., Chen, C., & Yu, H. (2014). Research status of laser cladding on titanium and its alloys: A review. *Materials & Design*, 58, 412–425.

Wood, R. J. (2007). Tribo-corrosion of coatings: A review. *Journal of Physics D: Applied Physics*, 40(18), 5502.

5 Future Trends in Laser Powder Bed Fusion Process for Tribological Applications

Nishant Ranjan
Chandigarh University

5.1 INTRODUCTION

A significant transition is now taking place in the field of materials science and production as a result of the contemporary era's fast technological breakthroughs. The Laser Powder Bed Fusion (LPBF) method, a additive manufacturing (AM) approach that has ushered in a new era of accuracy, customization, and efficiency in the creation of complex components, is one of the most promising breakthroughs to emerge in recent years (Yuan et al., 2018; Ranjan et al., 2023; Tyagi et al., 2023). While LPBF has found uses in a range of sectors, including healthcare and aerospace, one in particular bears great promise: tribological applications. Tribology, the study of friction, wear, and lubrication, is essential to guaranteeing the dependable and effective operation of mechanical systems and machinery (Rosenkranz et al., 2019). Since components that interact under sliding, rolling, or abrasive circumstances are designed and manufactured differently now than they once did, the incorporation of LPBF into tribological applications constitutes a substantial change in that regard (Renner et al., 2021).

This study intends to offer a thorough examination of potential developments in the LPBF process for tribological applications (Nazir et al., 2023; Narasimharaju et al., 2022). Tribology is a crucial component of contemporary engineering, affecting everything from the effectiveness of energy conversion systems to the performance and lifetime of automobile components (Tung and McMillan, 2004). It is of utmost importance in both academic and industry circles to create cutting-edge materials and production techniques that can improve tribological performance. Because of its potential to fundamentally alter the way tribological components are designed, made, and optimized, LPBF has drawn a lot of interest in this area. A high-powered laser is used in the LPBF method, also known as selective laser melting or direct metal laser sintering, to produce three-dimensional structures layer by layer from powdered materials, usually metals (Khan et al., 2020; Mahale et al., 2022). It has attracted attention as a possibility for tribological applications, where the optimization of material composition and microstructure is crucial, due to its capacity to make complex and customized components with outstanding material characteristics. In comparison to conventional subtractive

DOI: 10.1201/9781003400523-5

manufacturing techniques like machining, this technology has several advantages, including less material waste, shorter lead times, and the capacity to create extremely complex geometries that were previously impractical or prohibitively expensive to make (Wang et al., 2022; Lozano et al., 2022; Vaezi et al., 2013).

It is crucial to consider the present state of the art and the main issues that researchers and business experts are now trying to solve to comprehend the future developments in LPBF for tribological applications. The study of tribology has always relied significantly on empirical research and experimentation. Based on prior knowledge and readily accessible materials, components were developed and produced, and rigorous testing was carried out to evaluate their tribological performance. Even while this strategy has resulted in substantial advancement, it is intrinsically unable to tap into the full potential of novel materials and production methods. The LPBF technique promises a paradigm change in tribological research and development because of its fine control over material deposition and the possibility to customize microstructures at the microscopic level (Fu et al., 2022; Pandiyan et al., 2022). Utilizing the special qualities of LPBF enables engineers and scientists to develop and fabricate components particularly optimized for tribological applications. One of the main drivers influencing future advancements in LPBF for tribological applications is this transition from empirical to knowledge-driven design. The capacity to produce customized material compositions and microstructures is the most significant component of LPBF's influence on tribology (Cunha et al., 2022). Traditional production procedures sometimes restrict material possibilities to those that are easily available in quantity (Crini, 2006). LPBF allows for the use of powder mixes with customized compositions, enabling to production of innovative materials optimized for specific tribological circumstances (Nazir et al., 2023). Depending on the application, researchers can create alloys with increased hardness, wear resistance, or lubricity. LPBF provides fine control over the microstructure of the material, allowing for the adjustment of grain size, orientation, and phase distribution to optimize tribological characteristics (Knezevic et al., 2021; Kong et al., 2021). Another important component of LPBF's tribological potential is its capacity to construct complicated and elaborate geometries that were previously difficult or impossible to obtain. The structure and surface characteristics of components can significantly affect friction, wear, and lubrication in tribological applications (Rosenkranz et al., 2021). The flexibility of LPBF enables the construction of components with carefully regulated surface patterns, textures, and characteristics that might affect tribological performance (Elambasseril et al., 2023). With this skill, designers may create high-performance bearings, seals, and other parts that can function more effectively and last longer. When it comes to quality assurance and process control, LPBF has a distinct edge. Layer-by-layer building techniques (LPBF) provide real-time build process monitoring, enabling early discovery of flaws and deviations from the original design. In tribological applications, where the dependability and consistency of components are critical, this skill is essential (Imani et al., 2018; Chadha et al., 2023). Researchers and manufacturers may make sure that tribological components satisfy the highest requirements of performance and reliability by incorporating cutting-edge monitoring and quality assurance tools into the LPBF process. The potential of LPBF in tribology goes beyond making specific parts. It creates possibilities for the creation of integrated systems and assemblies with improved tribological characteristics. Researchers may create intricate mechanical systems with embedded sensors for real-time monitoring of wear and friction, internal lubrication channels, and self-lubricating surfaces (Peng et al., 2022). By lesser maintenance requirements, increasing efficiency, and prolonging the operating life of crucial components, these breakthroughs have the potential to revolutionize sectors including automotive, aerospace, and energy (Basheer et al., 2020).

While the potential benefits of LPBF in tribological applications are significant, various problems and research pathways must be addressed before this potential can be completely realized. These difficulties include optimizing process parameters for individual materials and applications, building trustworthy prediction models for tribological performance, and assuring LPBF scalability and cost-effectiveness for mass production. The incorporation of LPBF into current supply chains and quality assurance standards is a challenging undertaking that must be approached with caution (Ravalji and Raval, 2023). The use of the LPBF technique in tribological applications is an exciting new area in materials research and industry (Rahmani et al., 2023). The capacity of LPBF to develop bespoke materials, complicated geometries, and integrated systems has the potential to revolutionize the way tribological components are designed and manufactured (Gu et al., 2021; Jalalahmadi et al., 2021). This research paper will go further into the present status of LPBF in tribological applications, investigating recent advances, continuing research efforts, and future trends impacting the landscape of this intriguing topic. We want to contribute significant insights to researchers, engineers, and industry experts by analysing the problems and possibilities presented by LPBF.

5.2 FUNDAMENTALS OF LPBF FOR TRIBOLOGICAL APPLICATIONS

LPBF, also known as Selective Laser Melting (SLM), is a cutting-edge AM method that has received a lot of interest in recent years because of its promise in a variety of technical domains, including tribological applications. Figure 5.1 shows the schematic diagram of the LPBF working process. This section delves into the principles of LPBF and how it may be used to solve tribological problems.

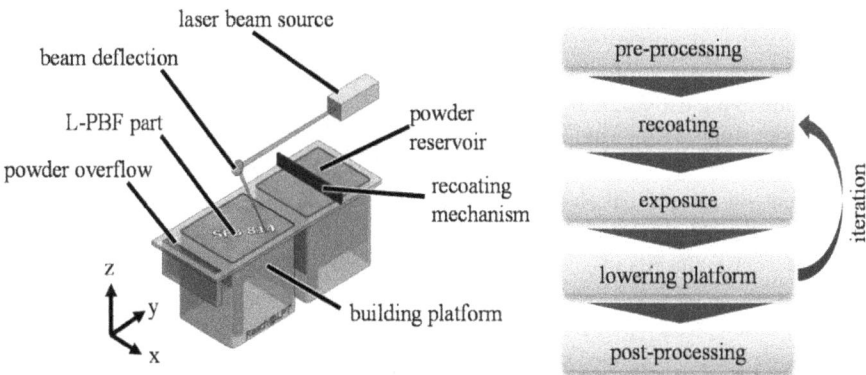

FIGURE 5.1 Schematic of the Laser Powder Bed Fusion working process (Huber et al., 2018).

5.2.1 Introduction to LPBF

LPBF is an AM method of the powder bed fusion family. It entails selectively melting a powdered substance with a high-energy laser beam to layer by layer construct a three-dimensional object. LPBF is very accurate and provides good material placement control, making it ideal for constructing delicate and complicated geometries. When developing components for tribological applications, this level of control is critical since specific surface characteristics and material composition may have a considerable influence on friction, wear, and lubrication qualities.

5.2.2 Material Selection and Characteristics for Tribology

Material selection is an important factor in LPBF for tribological applications. The material used has a significant impact on the performance and lifetime of components in tribological systems. Several things must be considered:

The material used in LPBF must be compatible with the tribological conditions it will be subjected to. Materials having great temperature resistance, such as nickel-based superalloys, may be favoured in high-temperature applications, for example. High-wear resistance materials are critical for lowering the wear rate of components in tribological systems. For this reason, hardened steels, ceramics, and carbides are frequently employed. Some tribological applications necessitate materials with inherent lubricity or the capacity to efficiently retain lubricants. In such instances, self-lubricating materials, such as certain polymers or composites, are useful. LPBF enables fine control over the microstructure of materials, allowing them to be customized to improve their tribological characteristics. For example, the distribution of phases, grain size, and orientation may all be optimized to enhance wear resistance and minimize friction.

5.2.3 Tribological Challenges Addressed by LPBF

Tribological systems are concerned with the investigation of friction, wear, and lubrication phenomena, which are all critical to the performance and dependability of mechanical components. LPBF can address several tribological issues:

Parts with exceptional surface quality may be produced with LPBF, minimizing the need for extra post-processing. A smoother surface can reduce friction and wear, hence improving tribological performance. LPBF enables the design of complicated geometries, such as internal channels and textured surfaces, to increase lubricant distribution and minimize friction. LPBF allows for component customization depending on unique tribological criteria. Bearing surfaces, for example, can be constructed with specific roughness profiles to optimize lubrication. LPBF may mix different materials in a single component, enabling the development of hybrid constructions with a balance of qualities such as wear resistance and toughness. Because it selectively melts the powder, LPBF is noted for its material efficiency. When compared to standard subtractive manufacturing processes, this lowers material waste. The capacity of LPBF to rapidly create prototypes and small batches of

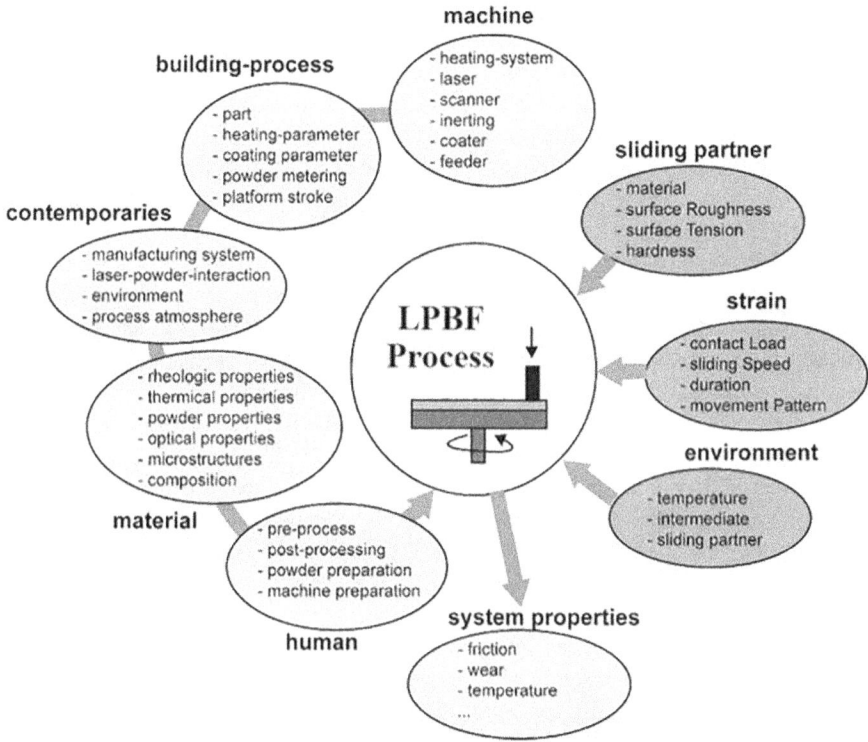

FIGURE 5.2 Internal and external factors that affect the LPBF process.

parts is useful for testing and optimizing tribological designs before large-scale manufacturing. Heat treatments and post-processing processes can be used to improve the tribological qualities of LPBF components. Stress relief, surface hardening, and coating deposition are all examples of this.

According to the findings of this study, LPBF is critical for realizing its promise in tribological applications. LPBF's precision control over material properties and component shape offers new avenues for tackling tribological difficulties and creating components with enhanced friction, wear, and lubrication attributes. As technology advances, LPBF is set to play an increasingly important part in tribology's future, allowing novel solutions for a wide range of industries and applications. Figure 5.2 shows that internal and external factors affect the LPBF process.

5.3 STATE OF THE ART IN LPBF FOR TRIBOLOGICAL APPLICATIONS

In the area of tribological applications, LPBF technology has advanced significantly, providing creative answers to problems including friction, wear, and lubrication. We go into the most recent advancements and significant accomplishments in LPBF for tribological applications in this part.

5.3.1 CASE STUDIES

Numerous case studies and success stories over the previous ten years have demonstrated the effectiveness of LPBF in tribological applications. One prominent instance is the aircraft sector, where components made by LPBF have been used to improve fuel economy and lower maintenance costs. For example, LPBF has been used to create components made of lightweight, high-strength titanium alloy with customized surface qualities, decreasing wear and friction in crucial aviation systems. In the automobile industry, LPBF has proven its skill in creating specialized parts with complex internal structures that are enhanced for tribological performance, such as gears and bearings. The fuel efficiency and durability of the vehicles have significantly increased as a result of this customization. The application of LPBF in the medical industry allows for the creation of joint implants from biocompatible metals like titanium and cobalt-chromium alloys. The lifetime and functionality of these implants in the human body depend on their outstanding tribological qualities, which lower friction and minimize wear. The tribological capabilities of components manufactured by LPBF have been improved because of advancements in material development. New materials and alloys being investigated by researchers aim to provide better thermal stability, decreased friction, and enhanced wear resistance. For instance, progress has been made in the creation of self-lubricating materials that embed solid lubricant particles into the metal matrix. Under high pressure, these materials release lubricants that reduce wear and increase component longevity. Many scientists have already looked at the addition of nanoparticles like graphene and molybdenum disulfide to LPBF feedstock powders. Due to the outstanding lubrication and strengthening that these nanoparticles can offer, LPBF components are now even more wear- and temperature-resistant. In tribological applications, achieving the desired surface finish is essential. Due to the AM technique's layer-by-layer construction, LPBF items often have a rough surface. Advanced post-processing methods have been created, nevertheless, to enhance tribological performance and surface quality. Surface finishing via abrasive techniques like grinding and polishing is one such method.

These techniques can smooth the surface and lessen its roughness, reducing wear and friction. To alter the surface chemistry, improve lubrication, and decrease adhesion, chemical and electrochemical treatments are being investigated. The application of laser-based surface texturing in LPBF is a further potential direction. Component surfaces can be given micro- and nanoscale textures to attract and hold lubricants, produce hydrodynamic effects, and lessen friction. This cutting-edge method can significantly improve the tribological characteristics of LPBF components. Several obstacles still exist in LPBF for tribological applications despite the substantial progress that has been made. Controlling residual stresses and microstructural flaws in components produced by LPBF is a major problem. These variables may have an impact on a component's mechanical and tribological characteristics. To reduce residual stresses and improve material qualities, researchers are actively striving to optimize process variables and heat treatments. The requirement for standardized testing and characterization techniques especially designed for LPBF-produced tribological components presents another difficulty. For correct evaluation of the wear resistance, friction, and lubrication qualities of these parts, trustworthy testing procedures must be established. Exciting opportunities exist for the use of LPBF in tribological applications in the future. Multi-material printing, which enables the integration of many materials with diverse tribological characteristics into a single component, is a research area that is receiving more attention. This creates possibilities for the development of hybrid constructions that display customized tribological behaviour in different parts of a component.

It is projected that LPBF will increasingly be used with other production techniques including post-machining and surface treatments. Components with improved surfaces and improved tribological properties can be produced using LPBF in combination with conventional production methods. Based on the information in this section, it can be concluded that LPBF has become a game-changing technology in tribological applications. LPBF is well-positioned to play a key role in tackling friction, wear, and lubrication concerns in a variety of sectors thanks to success stories in the aerospace, automotive, and medical domains as well as continuous material breakthroughs and surface finish improvements.

5.4 DESIGN CONSIDERATIONS FOR TRIBOLOGICAL PERFORMANCE

In LPBF, it is crucial to design components with optimized tribological performance. Tribology is the study of friction, wear, and lubrication, and it is essential for figuring out how effectively and long-lasting engineering systems operate. The following design factors can have a major influence on tribological performance in LPBF:

5.4.1 TOPOLOGY OPTIMIZATION FOR LPBF COMPONENTS

A computational design strategy called topology optimization uses algorithms to identify the most effective material distribution within a particular design space. Engineers may design buildings using this method that are both lightweight and tribologically very effective. Manufacturing complicated, optimized geometries is where LPBF excels. With the capacity to print complex lattice structures or internal channels, component weight may be decreased while tribological qualities are maintained or even improved.

5.4.2 LATTICE STRUCTURES AND THEIR TRIBOLOGICAL BENEFITS

The recurring, open-cell geometries of lattice structures distinguish them. They provide various tribological benefits, such as reduced weight, increased lubricant retention, and improved heat dissipation. Lattice structures can be designed to meet certain tribological needs. Some lattice designs, for example, are better at trapping and spreading lubricants than others at dissipating heat created during frictional contact.

5.4.3 SURFACE TEXTURING AND ITS IMPACT ON FRICTION AND WEAR

Surface texturing is the process of incorporating microscale or nanoscale characteristics into component surfaces. These textures can influence tribological behaviour by retaining lubricants, lowering adhesion, and changing frictional contact. LPBF may be used to generate highly precise textured surfaces, allowing for customized patterns and geometries to meet specific tribological requirements.

5.5 CHALLENGES AND FUTURE DIRECTIONS

While LPBF has a lot of potential for tribological applications, there are some barriers and future approaches to consider:

5.5.1 Residual Stresses and Microstructural Defects

Because of the fast heating and cooling, LPBF naturally produces residual stresses and microstructural flaws in manufactured items. These variables can have an impact on tribological performance. Researchers are attempting to minimize residual stresses and flaws by optimizing LPBF process parameters such as laser power, scanning speed, and construction orientation. Advanced heat treatments are also being investigated as a means of mitigating their effects.

5.5.2 Standardization and Testing Protocols

It is critical to develop standardized testing and characterization techniques for LPBF-produced tribological components. These techniques guarantee that wear resistance, friction, and lubrication qualities are consistently and reliably assessed. To produce comprehensive testing standards and recommendations, collaboration between academics, industry, and regulatory agencies is required.

5.5.3 Multi-Material Printing and Integration

Multi-material printing, or the integration of several materials with varied tribological characteristics into a single component, is an interesting future path. This method permits the development of hybrid constructions with specific tribological behaviour in different parts of a component. Researchers are investigating strategies for blending multiple materials smoothly throughout the LPBF process, allowing for fine control over material transitions and characteristics within a single component.

5.6 INDUSTRY APPLICATIONS AND CASE STUDIES

Due to LPBF's adaptability for tribological applications, it has been adopted by several industries, each with its own set of difficulties and successes:

In the automobile sector, LPBF is increasingly used to produce vital parts including gears, bearings, and lightweight structural elements. These parts gain from LPBF's capacity to design complex geometries and perfect material placement for enhanced tribological performance. LPBF has been used in aerospace applications to provide lightweight, very durable components with less wear and friction. Because of LPBF's ability to deal with cutting-edge materials like titanium alloys and superalloys, aviation systems have changed and maintenance costs have gone down. By making it possible to create specialized orthopaedic and joint implants with exceptional biocompatibility and tribological qualities, LPBF has revolutionized the biomedical industry. Patients gain from implants that last longer and have fewer issues from use. To create components that are wear-resistant and utilized in drilling and exploration equipment, the oil and gas industry has adopted LPBF. With the use of LPBF, the energy sector has also produced effective turbine parts with improved tribological characteristics.

5.7 CONCLUSIONS

Based on review work, it has been observed that LPBF has become a fast growing technology in the field of tribological applications. It is clear from case studies and success stories from a variety of sectors that LPBF can handle significant problems with friction, wear, and lubrication. The creation of new materials, improvements to surface finishes, and design factors like topology optimization and lattice structures have all increased the possibilities of LPBF in tribology.

There are also difficulties in establishing standardized testing procedures and controlling residual strains. These difficulties, together with upcoming trends like multi-material printing and seamless integration, give exciting chances for LPBF for tribological applications to continue growing and innovating. The future of engineering systems across a variety of sectors is expected to be significantly shaped by LPBF as it develops and gains wider adoption. LPBF is poised to make a substantial contribution to the creation of more effective, long-lasting, and dependable components in the field of tribology through continued research and collaboration.

ACKNOWLEDGEMENTS

The authors are thankful to University Centre for Research and Development, Chandigarh University for technical support.

REFERENCES

Basheer AA. Advances in the smart materials applications in the aerospace industries. *Aircraft Engineering and Aerospace Technology*. 2020;92(7):1027–35.

Chadha U, Selvaraj SK, Abraham AS, Khanna M, Mishra A, Sachdeva I, Kashyap S, Dev SJ, Swatish RS, Joshi A, Anand SK. Powder bed fusion via machine learning-enabled approaches. *Complexity*. 2023; in press, https://doi.org/10.1155/2023/9481790

Crini G. Non-conventional low-cost adsorbents for dye removal: A review. *Bioresource Technology*. 2006;97(9):1061–85.

Cunha A, Marques A, Silva MR, Bartolomeu F, Silva FS, Gasik M, Trindade B, Carvalho Ó. Laser powder bed fusion of the steels used in the plastic injection mould industry: A review of the influence of processing parameters on the final properties. *The International Journal of Advanced Manufacturing Technology*. 2022;121(7–8):4255–87.

Elambasseril J, Rogers J, Wallbrink C, Munk D, Leary M, Qian M. Laser powder bed fusion additive manufacturing (LPBF-AM): The influence of design features and LPBF variables on surface topography and effect on fatigue properties. *Critical Reviews in Solid State and Materials Sciences*. 2023;48(1):132–68.

Fu Y, Downey AR, Yuan L, Zhang T, Pratt A, Balogun Y. Machine learning algorithms for defect detection in metal laser-based additive manufacturing: A review. *Journal of Manufacturing Processes*. 2022;75:693–710.

Gu D, Shi X, Poprawe R, Bourell DL, Setchi R, Zhu J. Material-structure-performance integrated laser-metal additive manufacturing. *Science*. 2021;372(6545):eabg1487.

Huber F, Papke T, Scheitler C, Hanrieder L, Merklein M, Schmidt M. In situ formation of a metastable β-Ti alloy by laser powder bed fusion (L-PBF) of vanadium and iron modified Ti-6Al-4V. *Metals*. 2018;8(12):1067.

Imani F, Gaikwad A, Montazeri M, Rao P, Yang H, Reutzel E. Process mapping and in-process monitoring of porosity in laser powder bed fusion using layerwise optical imaging. *Journal of Manufacturing Science and Engineering*. 2018;140(10):101009.

Jalalahmadi B, Liu J, Liu Z, Vechart A, Weinzapfel N. An integrated computational material engineering predictive platform for fatigue prediction and qualification of metallic parts built with additive manufacturing. *Journal of Tribology*. 2021;143(5):051112.

Khan HM, Karabulut Y, Kitay O, Kaynak Y, Jawahir IS. Influence of the post-processing operations on surface integrity of metal components produced by laser powder bed fusion additive manufacturing: A review. *Machining Science and Technology*. 2020;25(1):118–76.

Knezevic M, Ghorbanpour S, Ferreri NC, Riyad IA, Kudzal AD, Paramore JD, Vogel SC, McWilliams BA. Thermo-hydrogen refinement of microstructure to improve mechanical properties of Ti-6Al-4V fabricated via laser powder bed fusion. *Materials Science and Engineering: A*. 2021;809:140980.

Kong D, Dong C, Wei S, Ni X, Zhang L, Li R, Wang L, Man C, Li X. About metastable cellular structure in additively manufactured austenitic stainless steels. *Additive Manufacturing*. 2021;38:101804.

Lozano AB, Álvarez SH, Isaza CV, Montealegre-Rubio W. Analysis and advances in additive manufacturing as a new technology to make polymer injection molds for world-class production systems. *Polymers*. 2022;14(9):1646.

Mahale RS, Shamanth V, Hemanth K, Nithin SK, Sharath PC, Shashanka R, Patil A, Shetty D. Processes and applications of metal additive manufacturing. *Materials Today: Proceedings*. 2022;54:228–33.

Narasimharaju SR, Zeng W, See TL, Zhu Z, Scott P, Jiang X, Lou S. A comprehensive review on laser powder bed fusion of steels: Processing, microstructure, defects and control methods, mechanical properties, current challenges and future trends. *Journal of Manufacturing Processes*. 2022;75:375–414.

Nazir A, Gokcekaya O, Billah KM, Ertugrul O, Jiang J, Sun J, Hussain S. Multi-material additive manufacturing: A systematic review of design, properties, applications, challenges, and 3D Printing of materials and cellular metamaterials. *Materials & Design*. 2023;226:111661.

Pandiyan V, Cui D, Le-Quang T, Deshpande P, Wasmer K, Shevchik S. In situ quality monitoring in direct energy deposition process using co-axial process zone imaging and deep contrastive learning. *Journal of Manufacturing Processes*. 2022;81:1064–75.

Peng H, Zhang H, Shangguan L, Fan Y. Review of tribological failure analysis and lubrication technology research of wind power bearings. *Polymers*. 2022;14(15):3041.

Rahmani R, Karimi J, Kamboj N, Kumar R, Brojan M, Tchórz A, Skrabalak G, Lopes SI. Fabrication of localized diamond-filled copper structures via selective laser melting and spark plasma sintering. *Diamond and Related Materials*. 2023;136:109916.

Ranjan N, Tyagi R, Kumar R, Kumar V. On fabrication of acrylonitrile butadiene styrene-zirconium oxide composite feedstock for 3D printing-based rapid tooling applications. *Journal of Thermoplastic Composite Materials*. 2023:in press. https://doi.org/10.1177/08927057231186310.

Ravalji JM, Raval SJ. Review of quality issues and mitigation strategies for metal powder bed fusion. *Rapid Prototyping Journal*. 2023;29(4):792–817.

Renner P, Jha S, Chen Y, Raut A, Mehta SG, Liang H. A review on corrosion and wear of additively manufactured alloys. *Journal of Tribology*. 2021;143(5):050802.

Rosenkranz A, Costa HL, Baykara MZ, Martini A. Synergetic effects of surface texturing and solid lubricants to tailor friction and wear-a review. *Tribology International*. 2021;155:106792.

Rosenkranz A, Grützmacher PG, Gachot C, Costa HL. Surface texturing in machine elements – A critical discussion for rolling and sliding contacts. *Advanced Engineering Materials*. 2019;21(8):1900194.

Tung SC, McMillan ML. Automotive tribology overview of current advances and challenges for the future. *Tribology International*. 2004;37(7):517–36.

Tyagi R, Singh G, Kumar R, Kumar V, Singh S. 3D-printed sandwiched acrylonitrile butadiene styrene/carbon fiber composites: Investigating mechanical, morphological, and fractural properties. *Journal of Materials Engineering and Performance*. 2023:1–4: in press, doi. org/10.1007/s11665-023-08292-8.

Vaezi M, Chianrabutra S, Mellor B, Yang S. Multiple material additive manufacturing-Part 1: A review: This review paper covers a decade of research on multiple material additive manufacturing technologies which can produce complex geometry parts with different materials. *Virtual and Physical Prototyping*. 2013;8(1):19–50.

Wang H, Lamichhane TN, Paranthaman MP. Review of additive manufacturing of permanent magnets for electrical machines: A prospective on wind turbine. *Materials Today Physics*. 2022;24:100675.

Yuan B, Guss GM, Wilson AC, Hau-Riege SP, DePond PJ, McMains S, Matthews MJ, Giera B. Machine-learning-based monitoring of laser powder bed fusion. *Advanced Materials Technologies*. 2018;3(12):1800136.

6 Role of Natural Fiber-Based Composite on Wear and Friction Resistance

Rajnish P. Modanwal and Dan Sathiaraj
Indian Institute of Technology Indore

Pradeep K. Singh
Sant Longowal Institute of Engineering and Technology

Rashi Tyagi
Chandigarh University

Ashwath Pazhani
Coventry University

6.1 INTRODUCTION

The term "composite material" is narrated as a combination of at least two different materials with different physical, chemical, electrical, and thermo-mechanical properties that produces a new material with better properties than that of the individual material. From ancient times, one material was combined with other materials to produce a new material with improved properties, such as the mixing of mud into the hay for fabricating powerful mud walls, rubber in carbon black, steel rod in concrete, fiberglass in resin, and sand in the cement (Nirmal et al., 2011; Singh et al., 2021). Recently, polymer composites (PCs) have attracted researchers' attention because of their excellent thermo-mechanical characteristics. In order to address the rising need for high wear and frictional resistance in materials, various researchers have fabricated natural fiber (NF)-based PCs. NF PCs find extensive utilization across various engineering disciplines due to their wide range of applications. NFs possess significant potential as reinforcement materials in polymers for a wide range of industrial applications, thereby exerting a substantial influence on socioeconomic progress. Due to their cost-effectiveness, these materials are particularly well-suited for applications in storage devices, low-cost housing, automotive interiors, packaging, and the construction sector (Shuhimi et al., 2017).

DOI: 10.1201/9781003400523-6

Numerous varieties of NFs can be identified, including jute, flax, banana, oil palm, kenaf, hemp, bamboo, and sisal fiber. The utilization of NFs as reinforcement for PCs has experienced a surge in interest in the current era. This heightened attention can be attributed to the advantageous characteristics exhibited by these fibers, including but not limited to their low density, high specific strength, lightweight nature, cost-effectiveness in processing, renewability, non-toxicity, and biodegradability (Nishino et al., 2003; Shehu et al., 2014). Composite science presents substantial opportunities for the development of enhanced materials derived from renewable resources, thereby contributing to the advancement of global sustainability (Bongarde et al., 2014). The field of tribology plays a crucial role in enhancing the dependability of various machine components and systems. The subject matter pertains to the examination of lubrication, wear, friction, and other phenomena that occur between surfaces that are in relative motion with each other. Every individual element within a mechanical system experiences degradation and energy losses due to wear and friction. According to research findings, it has been predicted that the friction and wear losses pertaining to significant mechanical system components can contribute to as much as 63% of the overall expenses (Medalia, 1980). The increasing focus on the optimization of systems and energy efficiency, along with the consideration of cost-effectiveness, has generated a growing fascination in the domain of tribology.

The preceding decade has observed significant advancements in the field of tribology pertaining to PCs that are exposed to NFs. Extensive research was undertaken on both synthetics and NFs. Figure 6.1 depicts the quantity of scholarly articles that have been published within the last 20 years regarding research endeavors in the domain of tribology, specifically focusing on both synthetics and NFs. This observation signifies the growing global research interest in the subject matter. The latter half of the 20th century was primarily characterized by a significant emphasis on tribology research. However, the initial period of the 21st century witnessed a shift in attention towards renewable sources of NFs, primarily for the purpose of tribo-composite applications, driven by concerns surrounding environmental matters. During the current era, a multitude of scientific investigations have been conducted pertaining to the utilization of plants as alternative resources in various industrial applications. The subject of interest regarding the utilization of plant-based alternatives for industrial purposes has led (Faruk et al., 2012) to refer to the present era as the "cellulose era." Consequently, the researchers deem it suitable for compilation of research endeavors associated with the tribological performance of NFs in PCs. This presentation will provide information on the historical and scientific advancements in tribology, specifically focusing on NFs and PCs. It is important to acknowledge that this book chapter will solely focus on the frictional and wear properties of synthetic PCs reinforced with plant-derived NFs, as the science of tribology includes various testing modes of considerable complexity.

Therefore, the above discussion suggests that PCs are widely utilized in a range of tribological applications, including dry sliding bearings, gears, and rollers. This is primarily due to their inherent benefits, such as their self-lubricating properties, resistance to corrosion, and high specific strength (Nishino et al., 2003; Shehu et al., 2014). The incorporation of different fillers or reinforcing fibers has been widely

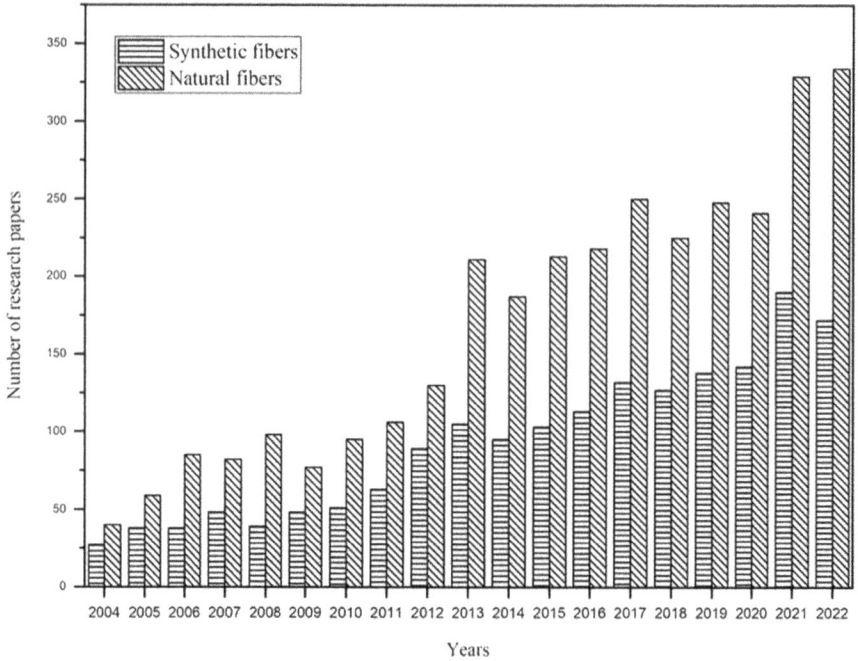

FIGURE 6.1 Number of research paper published related to wear and friction of NFs and synthetic fibers.

Source: https://www.sciencedirect.com/ Keywords used: Friction resistance, wear, tribology, NFs composites, and synthetic fibers composites.

acknowledged as a means to enhance the tribological performance of polymers (Shuhimi et al., 2017; Nishino et al., 2003; Shehu et al., 2014). When considering the comparison between short fibers and continuous fiber reinforcements, it becomes evident that continuous fibers provide certain advantages. These advantages include a higher fiber volume percentage and more efficient load transmission between the matrix and fibers. Consequently, these reinforcements possess the capability to substantially enhance the load-bearing capacity and WRE of PCs (Singh et al., 2021). Nevertheless, the conventional manufacturing methods used for producing continuous fiber-reinforced composites (FRCs), such as pultrusion, filament winding, automated tape laying, automated fiber placement, and resin transfer molding, have certain limitations. These limitations include the high cost of molds and metal tools and limited structural flexibility (Huimin et al., 2017; Liu et al., 2020). In the past few decades, research has focused more on the production of additively manufactured continuous FRCs, owing to the ongoing advancements and larger acceptance of 3DP, also known as AM technology. This phenomenon can primarily be attributed to its inherent capability to convert complex geometries into their ultimate form without necessitating the use of molds or specialized tools.

The process of 3DP involves the utilization of AM techniques to create a diverse range of components. On March 9, 1983, Charles W. Hull successfully produced the first 3DP component through the process of stereolithography. This groundbreaking achievement involved the utilization of an original printing machine to fabricate a teacup (Jimenez et al., 2019). The 3DP technology operates on the fundamental principle of incremental material deposition, wherein successive layers are added to fabricate the desired component. A comprehensive comprehension of the manufacturing principles associated with each method is an essential prerequisite in the process of identifying the underlying cause of every tribological issue. Furthermore, it is imperative to acknowledge that the selection of materials employed during the manufacturing process significantly influences the TPs exhibited by the end product.

Therefore, it is anticipated that this chapter will offer a comprehensive analysis of the research conducted over the past 10 years, focusing on the various 3D-printed AM NFRCs. Additionally, it will examine the performance of these composites in terms of surface roughness, friction, wear, contact conditions, test parameters, and their respective applications. In accordance with the established writing process, this discourse aims to adequately address the limitations and shortcomings present in research endeavors pertaining to the field of tribology on 3D-printed AM FRCs. The intention is to provide a foundation for future investigations in the field of tribology pertaining to 3D-printed AM fibers and their composites.

6.2 NATURAL FIBERS

NFs are received from organic sources and can be transformed into filaments, threads, or ropes, as well as utilized in various forms such as mats, woven fabrics, knitted textiles, or bindings. NFs can be categorized into three distinct groups, such as mineral, plant, and animal fibers (Tesinova et al., 2011). The automotive industry favors the use of plant-based NFs as reinforcements for structural applications owing to their superior strength (Holbery & Houston, 2006). Figure 6.2 illustrates the classification of NFs. PCs, which are fortified with fibers, exhibit a superior strength-to-weight ratio when compared to metals. Fiber-reinforced PCs have been widely utilized in the aerospace industry due to these crucial properties. Bamboo fiber, with a production rate of 30,000 metric tons, and sugarcane bagasse, with a production rate of 75,000 metric tons, are two of the most abundant NFs globally. In addition to ramie (100), abaca (70), flax (830), hemp (214), coir (100), sisal (375), jute (2,300), and kenaf (970) are among the various fibers that are readily accessible on a global scale (Rajak et al., 2019; Zagho et al., 2018; Milosevic et al., 2020).

The NFRC consists of a polymer matrix that has been infused with a diverse range of NFs, including sisal, flax, jute, coir, hemp, kenaf, and betelnut. In general, polymers can be classified into two distinct categories: thermosetting and thermoplastics. Both of these polymers are utilized in the fabrication process of the composite (Singh et al., 2021). Various polymers exhibit distinct properties, which subsequently determine their diverse range of applications. In a thermoset's polymers, only a few kinds of polymers with diverse applications are included. Polyester is utilized in the field of composite materials (CMs), specifically in bearing applications. On the other hand, epoxide finds its application in boat hulls and CMs. Polyurethane materials

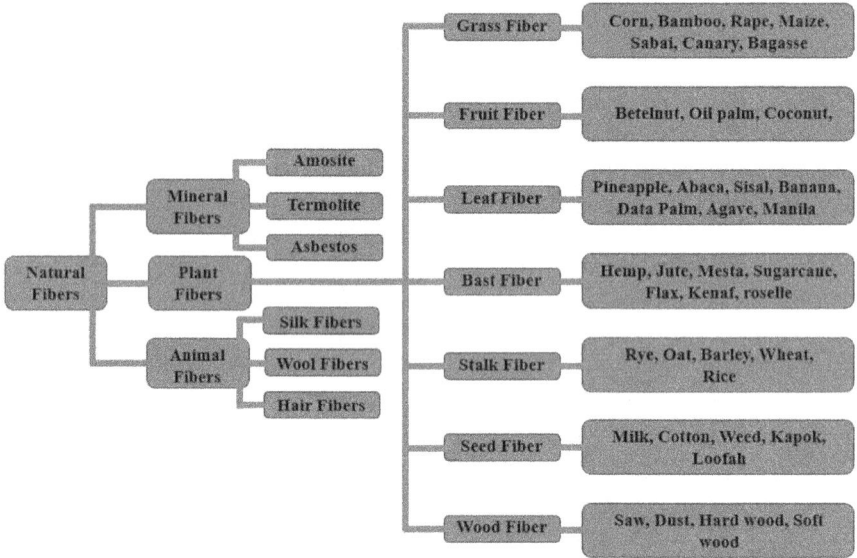

FIGURE 6.2 Classification of NFs (Milosevic et al., 2020; Sathish et al., 2021).

are utilized in the composition of refrigerator parts. Polyimides have also been utilized in the construction of circuit boards and the manufacturing of airplane bodies
(Parikh and Gohil, 2017; Nirmal et al., 2015; Chaudhary et al., 2018; Chand et al.,
2008). Similarly, in the category of thermoplastic polymers, there is a significant
multitude of available polymers. Certain thermoplastic polymers, such as acrylics,
find utility in various applications, including lenses for automotive lights and name
tags. Polyethylene is commonly utilized in various applications such as wire and
cable insulation, car interiors, containers, ballpoint tubing, and fans. Polyvinyl chloride is commonly utilized for the purpose of piping and floor tiles. Polystyrene is
commonly employed in various applications, such as cups, vending machines, and
packaging. Another variant of polymer that is accessible is referred to as an elastomer. Within the classification of elastomers, there exists a range of polymers that are
employed for diverse applications. Natural rubber is commonly utilized in various
industries for the production of hoses, belts, mats, gloves, and other related products.
Silicone rubber is utilized in various applications, including adhesive bonding and
the manufacturing of O-rings. Polybutadiene is commonly employed in the manufacturing of automobile tires. Epichlorohydrin is employed in the process of water
purification for the purpose of handling explosives. Polyether block amide (PEBA)
is utilized in various applications such as footwear, insulators, and components of
damping systems (Nirmal et al., 2015; Chaudhary et al., 2018; Chand et al., 2008;
Keerthiveettil et al., 2022).

The study examines the hydrophilic properties of NFs (Shehu et al., 2014) and
investigates the impact of fiber stacking on the properties of the CM (Medalia
et al., 1980). Typically, a higher level of fiber loading is necessary in order to attain

enhanced performance from NFRC (Holbery & Houston, 2006). The chemical composition of NFs significantly influences the characteristics of the composite, which are calculated by the relative amounts of cellulose, hemicellulose, pectin, lignin, wax, ash, and moisture. The chemical composition of commonly found NFs is listed in Table 6.1. On the basis of this table, one can infer that cellulose, being a primary constituent of NFs, has a direct influence on the flexural modulus and Young's modulus of the material.

The effectiveness of NFRC is influenced by various factors, including its physical properties, interaction between the fiber and the matrix microfibrillar angle, structural characteristics, presence of defects, cell dimension, and chemical properties (Ramesh et al., 2014; Al-Oqla & Sapuan, 2014; Dai & Fan, 2014; Weyenberg et al., 2003). Several studies have indicated that the mechanical performance of NFRC is dependent on the boundary conditions established by the matrix–fiber interface, specifically stress transfer. This stress transfer mechanism involves the transportation of stress from the matrix to the fiber, as reported by various researchers (Jimenez et al., 2019; Holbery & Houston, 2006; Zagho et al., 2018). The mechanical properties (MPs) of NFRC are influenced by various characteristic components of NFs, including moisture absorption, orientation, volume fraction, physical appearance, and impurities (Jayamani et al., 2014; Ren et al., 2012; Boopathi et al., 2012). These factors are crucial in determining the MPs of NFRCs. The MPs of NFs are documented in Table 6.2.

TABLE 6.1
Typical Chemical Constituents of Prominent NFs (Sathish et al., 2021; Nirmal et al., 2015)

Natural Fiber	Cellulose (%)	Hemicellulose (%)	Pectin (%)	Lignin (%)	Wax (%)	Ash (%)	Moisture (%)
Jute	67	16	0.2	9	0.5	0.1	10
Flax	64.1	16.7	1.8	2	1.5	13.1	10
Sisal	65.8	11.5	0.8	9.9	0.3	4.2	10
Pineapple	80.5	17.5	4	8.3	3.3	0.9–4	–
Cotton	82.7	5.7	5.7	28.2	0.6	–	10
Ramie	68.6	13.1	1.9	0.6	0.3	4.2	10
Bamboo	34.5	20.5	0.37	26	–	2.3	11.7
Oil palm	42.7–65	17.1–33.5	–	13.2–25	0.6	1.3–6	–
Abaca	56–63	15–17	0.3	7–9	0.1	3.2	–
Sugarcane	28.3–55	20–36.3	–	21.2–24	0.9	1–4	–
Coir	19.9–36	11.9–15.9	4.7–7	32.7–53	–	–	0.4
Banana	48–60	10.2–15.9	2.1–4.1	14.4–21	3–5	2.1	2–3
Betelnut	35–64.8	29–33.1	9.2–15	13–26	0.5–0.7	1.1–2	–
Hemp	55–80.2	12–22.4	0.9–3	2.6–13	0.2	0.5–1	6.5
Kenaf	37–49	18–24	8.9	15–21	0.5	2.4–5	–
Bagasse	37	21	10	22	–	–	–
Nettle	86	10	–	4	4	–	–

TABLE 6.2
MPs of Some NFs (Sathish et al., 2021; Nirmal et al., 2015)

Natural Fiber	Density (g/cm³)	Elongation (%)	Tensile Strength (MPa)	Young' Modulus (GPa)	Specific Gravity
Jute	1.3–1.5	1.4–2.1	393–773	13–26.5	1.3–1.5
Flax	1.3–1.54	1.1–3.3	340–1,600	25–81	1.5
Sisal	1.3–1.6	3–7	468–640	9.4–22	1.3
Pineapple	0.8–1.6	14.5	400–627	1.44	1.4–1.6
Cotton	1.5–1.6	7–8	287–800	5.5–12.6	1.5
Ramie	1.4–1.5	1.2–3.8	400–938	61–128	1.5–1.6
Bamboo	1.2–1.5	1.9–3.2	500–575	27–40	0.4–0.8
Oil palm	0.7–1.6	4–18	50–400	0.6–9	1.1–1.2
Abaca	1.5	1.2–1.5	430–815	31.1–33.6	–
Sugarcane	1.1–1.6	6.3–7.9	170–350	5.1–6.2	1.4–1.5
Coir	1.2–1.6	14–30	170–230	3–7	1.2–1.4
Banana	0.5–1.5	2.4–3.5	711–789	4–32.7	1.1–1.2
Betelnut	0.2–0.4	22–24	120–166	1.3–2.6	1.3–1.4
Hemp	1.1–1.6	0.8–3	285–1,735	14.4–44	1.5
Kenaf	0.6–1.5	1.6–4.3	223–1,191	11–60	1.1–1.4
Bagasse	0.8–1	6.3–7.9	250–300	17–20	1.4–1.5
Nettle	–	1.7	650	38	–

6.3 TRIBOLOGY

Tribology is the field of study that encompasses the examination of wear, friction, and lubrication phenomena occurring between interacting surfaces. Currently, the primary concern in design applications is the degradation caused by wear and tear when exposed to relative motion. Furthermore, it is important to note that factors such as wear, friction, and heat cannot be entirely eliminated, although efforts can be made to minimize their impact. During the early 1960s, heavy industries experienced a significant increase in the failure of machine parts due to excessive wear and tear on the equipment. This phenomenon led to substantial financial losses for these industries (Pan and Zhong, 2015; Mahesha et al., 2017). The wear and performance of PCs are influenced by operational parameters such as sliding distance (SD), sliding velocity (SV), and applied load (AL). In some cases, an addition of NFs to the polymer can increase the wear rate (WR) of the NFRCs. There are several types of tribometers available that are used for the purpose of conducting wear tests. These include the pin-on-disk, pin-on-drum, block-on-ring, block-on-disk, linear tribo machine, and dry sand rubber wheel (Nirmal et al., 2015; Kim et al., 2002; Low et al., 2010; Mergler et al., 2004).

6.3.1 PIN ON DRUM

Figure 6.3a demonstrates the representation of a pin-on-drum tribological equipment. It is designed on the basis of the ASTM A514 standard. The movement of the specimen is achieved in a linear fashion with the assistance of a power screw,

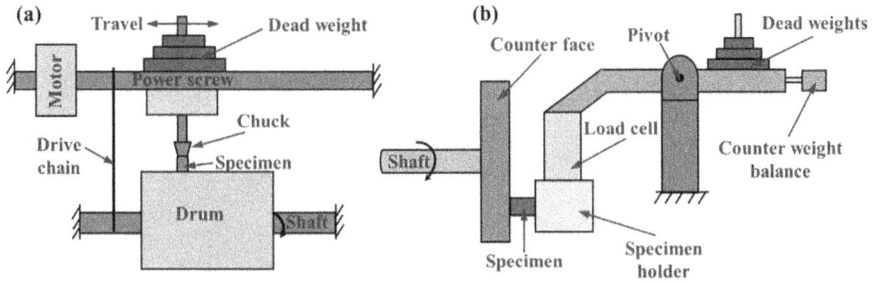

FIGURE 6.3 A block diagram to show a layout of (a) pin on drum and (b) pin on disk (Nirmal et al., 2015).

whereas the drum rotates at a predetermined speed with the use of a drive chain. The tests undertaken may exhibit abrasive properties when the drum is coated with abrasive paper, while in the absence of such a coating, the test demonstrates adhesive characteristics. The utilization of testing involves a range of methodologies, including the implementation of rotational rollers or analogous approaches to assess the performance of products. This method is used to evaluate the effects of high-stress, two-body abrasive wear. The experimental setup involves the translation of a cylindrical pin specimen across an abrasive paper, resulting in the removal of material from the specimen and the simultaneous crushing of the stationary abrasive grains. The wear phenomenon is hypothesized to replicate the wear experienced during the crushing and grinding of ore, processes in which the abrasive particles are subjected to crushing forces. This type of wear is commonly referred to as high-stress abrasive wear (Nirmal et al., 2015; Kim et al., 2002).

6.3.2 PIN ON DISK

The depiction of a pin-on-disk tribometer is illustrated in Figure 6.3b. The framework has been implemented in accordance with the specifications outlined in the ASTM G99 standard. The dimensions suggested for the specimen are $10 \times 10 \times 20 \text{mm}^3$. During the sliding process, the specimen experiences both perpendicular and horizontal forces exerted on the counterface. It is important to note that the contact area between the specimen and the counterface remains constant during the period of sliding. The operational parameters of the system display are similar to those observed in the block on ring arrangement. The utilization of testing encompasses a diverse array of materials, with a specific emphasis on the assessment of sliding wear (Nirmal et al., 2015; Low et al., 2010).

6.3.3 BLOCK ON RING

Figure 6.4a depicts the schematic representation of a block-on-ring tribo machine. The building of the mentioned structure adheres to the guidelines outlined in ASTM G77 and G137-95, which provide a recommended specimen dimension of

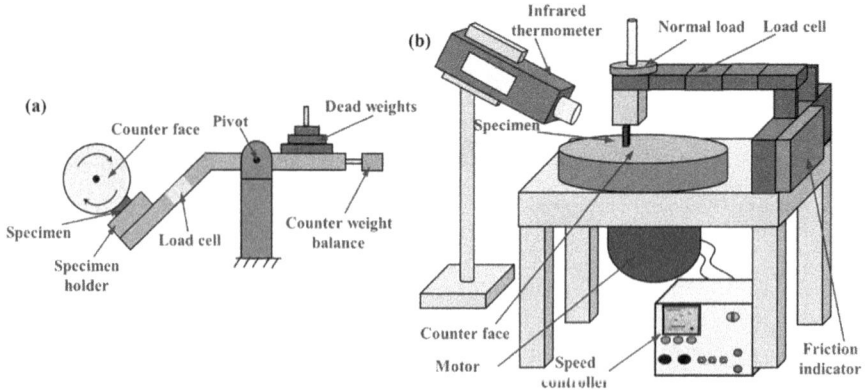

FIGURE 6.4 A block diagram to show a layout of (a) block on ring and (b) block on disk (Nirmal et al., 2015).

$10 \times 20 \times 50 \, \text{mm}^3$. The specimen is in parallel contact with the side of the rotating counterface, and the contact area of the specimen shows fluctuations based on the sliding time. A load cell is employed for the purpose of quantifying the frictional forces exerted by a counterface. The counterface can be constructed using different types of materials, such as stainless steel, cast iron, and aluminum. The utilization of testing encompasses a multitude of facets, encompassing but not limited to the assessment of sliding or rolling wear properties in rubber tires, bearings, camshafts, and pulleys (Nirmal et al., 2015; Mergler et al., 2004).

6.3.4 BLOCK ON DISK

Figure 6.4b depicts the representation of a block-on-disk tribological equipment. The specimen's design adheres to the ASTM G99 standard, specifically following the dimensions of $10 \times 10 \times 20 \, \text{mm}^3$. The test sample is subjected to a vertically oriented counterface, with the contact area maintained at a constant value. To facilitate the monitoring of temperatures at various interfaces during the testing procedure, it is possible to incorporate a portable infrared thermometer into the block-on-disk machine. The adhesive and abrasive characteristics of a test specimen were assessed in a dry sliding environment. The motor is connected to a speed controller unit, which is utilized to regulate the speed of an interface. The utilization of testing encompasses a broad spectrum of applications that extend beyond the confines of evaluating sliding wear in diverse materials, with a specific emphasis on the examination of the constant contact area (Nirmal et al., 2015; Mergler et al., 2004).

6.3.5 LINEAR TRIBO MACHINE

Figure 6.5a emphasizes the representation of a linear tribo machine. The counterface, made of stainless steel, experiences linear motion through the use of a power screw. Simultaneously, the specimens move within a container that contains

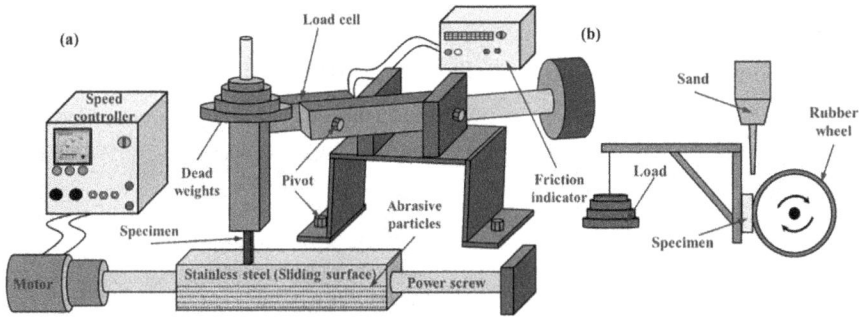

FIGURE 6.5 A block diagram to show a layout of (a) linear tribo machine and (b) dry sand rubber wheel (Nirmal et al., 2015).

abrasive particles for conducting an abrasive test. The experimental configuration entails the coupling of a frictional indicator with a load cell for the purpose of quantifying frictional forces. Furthermore, the implementation of a speed controller is utilized in order to regulate the set SV of the counter surface. The application of testing encompasses a range of aspects, including the evaluation of linear sliding mechanisms such as windows, panels, door rails, hinges, and drawers (Nirmal et al., 2015; Low et al., 2010).

6.3.6 DRY SAND RUBBER WHEEL

The representation of a dry sand rubber wheel is depicted in Figure 6.5b. The construction of this product is based on the guidelines outlined in the ASTM G65 standard. The suggested dimensions for the specimen are approximately $70 \times 20 \times 7\,mm^3$. The rubber wheel makes contact with the test specimen when a load is applied. In order to perform the abrasive wear test on the specimen, a controlled flow rate of sand is fed into the rubbing interfaces during the testing procedure. When sand is not present, the test is limited to a solely adhesive mode. The utilization of testing encompasses multiple facets, including the assessment of roller performance, tire performance, bush performance, and bearing performance (Nirmal et al., 2015; Kim et al., 2002).

6.4 DESCRIPTION OF AM 3DP TECHNIQUE

Presently, polymers are accessible in several physical forms, such as resins, reactive monomers, filaments, and powders. These forms are utilized by diverse AM technologies to fabricate PCs. The most extensively studied AM technologies for the production of PCs include direct write (DW), fused filament fabrication (FFF), binder jetting (BJ), selective laser sintering (SLS), and stereolithography (SLA) procedures. Several AM techniques have been employed based on factors such as complexity, part shape, feedstock materials, processing limits such as resolution, and the ability

TABLE 6.3

An Overview of AM Technique (Balla et al., 2019)

Process	Matrix	Reinforcement	Feed Form
FFF	Amorphous thermoplastics: Acrylonitrile butadiene styrene (ABS), polycarbonates (PC), polylactic acid (PLA), polyethyleneimine, polyether ether ketone (PEEK), thermoplastic: polyurethane	Particles: Cu, Fe, W, graphene, hydroxyapatite (HA), carbon nano tube (CNT), Al_2O_3, TiO_2, $CaTiO_3$, $BaTiO_3$, tricalcium phosphate (TCP), polymer blends (PC + ABS) Fibers: Continuous carbon fiber (CCF), glass fiber, carbon fiber	Filament
DW	PLA, epoxy, polycaprolactone, hydrogel, elastomer	Short carbon fiber, bioactive glass, graphene, Fe_3O_4, CNT, HA, SiC whisker, TCP, silica	Slurry
SLS	Semicrystalline form of polymers such as polypropylene, PEEK, polyamide (PA)	Glass, carbon, Al, Al_2O_3, TiO_2, silica, $CaSiO_3$	Powder
BJ	Any polymer in powder form	SiC, gypsum	Powder
SLA	Epoxy, acrylates acrylics	Al_2O_3, CNT, graphene oxide, TiO_2, $BaTiO_3$, HA, TCP	Photocurable liquid

to deposit multiple materials. A more comprehensive analysis of these processes is elaborated on in the subsequent sections. A comprehensive study of the AM technique is listed in Table 6.3.

6.4.1 Fused Filament Fabrication

This technology, also known as fused deposition modeling (FDM), is widely recognized as a predominant method for manufacturing PCs. During this procedure, a feedstock material with a filament structure undergoes a melting process. Subsequently, it is placed on a construction sheet in a predefined pattern after being extruded through a nozzle. This technique aims to fabricate a 3D object that corresponds to the design represented in a computer-aided design (CAD) model. Amorphous thermoplastics are frequently employed in FFF owing to their wide temperature range and elevated viscosity. These properties provide convenient processing using extrusion nozzles with a diameter ranging from 0.2 to 0.5 mm. The technique employs two distinct materials: one for the primary component and another for the supporting structures. These supporting structures can be eliminated either by hand removal or by melting or dissolving. Figure 6.6a illustrates the procedure, which begins by introducing the feedstock filament into the extrusion head. Subsequently, heat is supplied to the filament and it is extruded via a nozzle in a partially solidified condition using a solid filament and roller mechanism. Finally, the extruded material is deposited onto a constructed sheet. A software application has been developed for supporting tool paths and structures that are crucial for minimizing customization of internal architecture, supporting build time, and improving the MPs

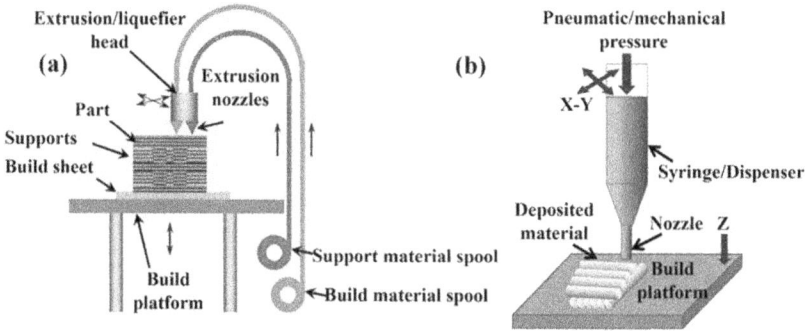

FIGURE 6.6 Block diagrams of (a) fused filament fabrication and (b) direct write (Balla et al., 2019).

of components. A solidification of the extruded material occurs through the process of bonding with both the surrounding and previously deposited materials within a temperature-controlled build chamber. This controlled environment serves the purpose of maintaining the warmth of the existing deposits, thereby facilitating optimal bonding. After the completion of each layer, the build platform undergoes downward movement by a thickness ranging from 0.1 to 0.5 mm, followed by the subsequent deposition of the following layer. Simultaneously, a distinct extruder is employed to construct the support structures for each layer (Balla et al., 2019).

6.4.2 DIRECT WRITE

The process is alternatively referred to as direct ink writing, 3D planning, and robocasting. The procedure bears resemblance to the FFF technique, wherein a high-viscosity solution is administered via a computer-controlled precision syringe in accordance with the CAD model, as depicted in Figure 6.6b. The apertures of the nozzles can vary from 1 μm to greater than 1 mm, enabling the attainment of both high resolution and rapid fabrication rates. The deposited component undergoes a hardening process using post-fabrication methods such as heating or exposure to UV radiation, utilizing reactive feedstock. The feedstock materials have the potential to exist in various forms, such as hydrogel, paste, slurry, and solution; hence, this method is highly adaptable. Nevertheless, the construction of overhang structures is a significant challenge due to the inherent difficulty in maintaining the structural integrity of the deposited material, which tends to be excessively soft and prone to collapse. Hence, the inclusion of support structures may be necessary in instances where components exhibit complex shapes (Xu et al., 2007).

6.4.3 STEREOLITHOGRAPHY

This technique requires the polymerization of a monomer, which is subjected to electromagnetic radiation. Polymerization occurs sequentially, progressing from one point to another, followed by a line-by-line process, and ultimately culminating in a

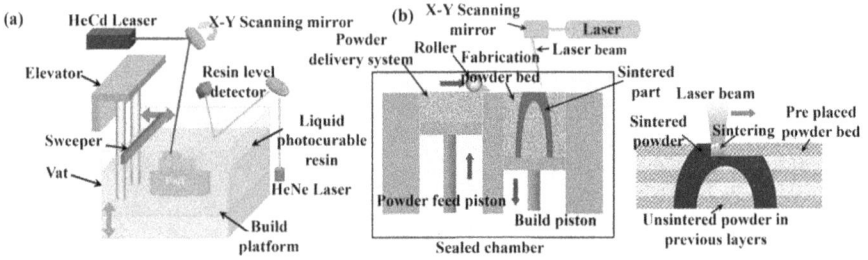

FIGURE 6.7 Block diagrams of (a) stereolithography and (b) selective laser sintering (Balla et al. 2019).

layer-by-layer formation, all of which transpire at room temperature. The first step involves lowering the construction platform to a level that corresponds to the thickness of each layer to be cured. Subsequently, a laser beam is guided onto a surface of the liquid to initiate the curing process, as illustrated in Figure 6.7a. The beam is rastered in order to finish a single layer or cross-section, following the CAD model. Subsequently, the constructed region experiences a decrease in elevation due to the inclusion of each layer, and this iterative procedure is subsequently repeated. In order to achieve effective bonding between layers and scans, it is necessary to carefully regulate the cure depth, ranging from 25 to 500 μm, as well as the width. This can be accomplished by employing a suitable scan speed and beam size. In contrast to conventional AM approaches, in the context of SLA, the construction of the part follows a bottom-up approach, as depicted in Figure 6.7a. Once the build process is finished, the components undergo a curing procedure, which involves the application of heat or photo-curing techniques. This curing step serves to enhance the MPs of the parts (Manapat et al., 2017).

6.4.4   Selective Laser Sintering

In this technique, a fine coating of powder is evenly distributed onto a constructed region, often within a chamber with a controlled environment, utilizing the spreading device. The fusion of the powder layer is achieved by employing a laser beam with high power. This laser beam is then scanned over the surface of the bed using an X-Y scanner, following the cross-section of the CAD model as depicted in Figure 6.6b. The thermal energy produced by the interaction between a powder and a laser is sufficient to induce the melting of the powder, resulting in the formation of a solid cross-sectional structure. The unaltered loose powder has the potential to provide structural reinforcement for overhanging structures. The aforementioned procedure is iterated for each individual cross-section subsequent to the application of a new layer of powder to the constructed region. This is achieved by adjusting the constructed region and feed box incrementally by a thickness equivalent to one layer or slice, approximately 100 μm. After the construction of all layers, the components are subjected to a cooling process within a controlled environment, during which any excess powder is eliminated (Athreya et al., 2011).

6.4.5 BINDER JETTING

This procedure shows notable similarities to SLS and is alternatively referred to as 3DP. The procedure involves the utilization of an inkjet printer to selectively print liquid binder onto a powder bed, following the cross-section of a CAD model. Following this, the construction platform quickly moves in response to a heat source in order to remove any moisture present and aid in the drying process of the binding agent, as illustrated in Figure 6.8. Proper drying is of utmost importance in order to prevent the binder from permeating further into preceding layers, as this might have adverse effects on the overall quality and attributes of the component. The printing process involves the sequential layering of each subsequent layer, with loose powder being expelled upon completion. The component involved in this particular procedure is referred to as the "green component," which requires cautious handling in order to prevent any instances of damage or fracture. These green parts are subjected to a heating process within an oven in order to induce solidification of the binder, with the specific temperature and duration of the heating process depending upon the type of binder employed (Utela et al., 2008).

6.5 WEAR PERFORMANCE OF 3D AM COMPOSITES

The following section presents a comprehensive overview of the TPs of 3D AM FRCs.

6.5.1 BIOGENIC CARBON/PLA COMPOSITE

Ertane et al. (2018) first fabricated the biocarbon-reinforced PLA filament composites by employing FDM. In this study, the incorporation of biocarbon powders into PLA was investigated at three distinct volumetric ratios: 5%, 15%, and 30%. The objective was to develop PLA filaments with enhanced mechanical properties while maintaining a consistent diameter of 1.75 mm, as demonstrated in Figure 6.9. Biocarbon refers

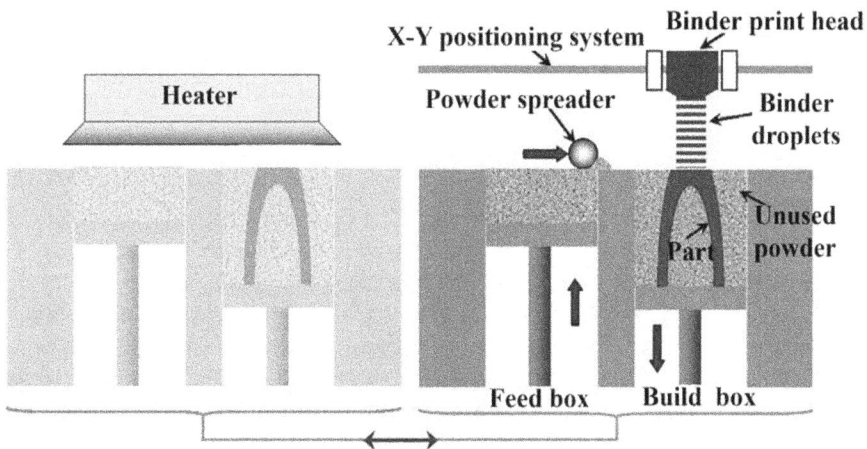

FIGURE 6.8 Block diagrams of binder jetting (Balla et al. 2019).

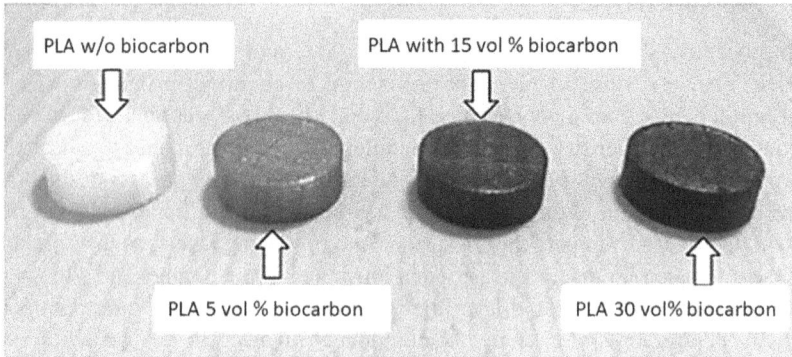

FIGURE 6.9 3DP biopolymers for tribological tests (Ertane et al., 2018).

to the carbon derived from vegetation, including trees, plants, and soils, which serves as a natural mechanism for the absorption and sequestration of carbon dioxide from the Earth's atmosphere. PLA is a widely utilized bioplastic that is primarily generated from the starch found in various agricultural crops, including corn, sugarcane, sugar beets, and wheat. In this study, they produced biocarbon through the pyrolysis of wheat stems. The production of the filaments involved the polymer extrusion process. After the manufacturing of the filaments, the CAD designs were created using SolidWorks software. Subsequently, the design data was converted into a standard template library, which was then processed into a 3D printable format using the Simplify 3D software. It was observed that the PLA sample, which was fabricated using a carbon volume fraction of 30%, has empirically demonstrated minimal wear volume. Various wear mechanisms, such as fatigue wear and abrasion, were observed. The behavior of both mechanisms is subject to the influence of the material properties of the biocomposite, primarily characterized by alterations in stiffness and embrittlement resulting from the reinforcement. Moreover, the coefficient of friction (COF) is relatively higher at approximately 0.5, whereas no significant change is observed in friction values when biocarbon is reinforcing. The stability of friction is observed for a composite reinforced with 30% biocarbon in PLA as compared to neat PLA.

6.5.2 Flex Yarn/PLA Composite

Kanakannavar et al. (2021) investigate the TPs by reinforcing the flax yarn (NF 3D-braided woven fabric) (NEBF) into the PLA at different weight percentages (wt.%) of fiber content, such as 18, 26, and 35. The analysis of the COF and WR of the NFRCs is conducted using a pin-on-disk tribometer in a dry contact sliding condition. This analysis is performed under various operating conditions, including different SV and AL, while maintaining a fixed SD of 3,000 m. The research indicates that the incorporation of NFBF into PLA results in a significant improvement in both the COF and WR. The incorporation of NFBF in PLA improved frictional force and diminished the height loss of the NFRC. The WR and SWR of both pure PLA and NFBF/PLA composites exhibit an upward trend as the AL increases. The incorporation of

NFBF reinforcement has been found to significantly decrease the SWR of PLA. And a reduction of almost 95% is achieved when using a 35 wt.% reinforcement. They state that common failure modes in composites include fiber fracture, fiber peeling, debonding, and the formation of microcracks.

6.5.3   GREWIA/NETTLE/SISAL/PLA COMPOSITE

Bajpai et al. (2013) fabricated laminated biocomposites by incorporating *Grewia optiva*, nettle, and sisal fiber mats into a PLA and conducted tribological investigations on these composites, with an emphasis on dry contact conditions. The fabrication of all specimens was conducted using a film stacking approach using hot compression. It has been found that the incorporation of NFs did not impact the average COF of PLA. Nevertheless, the COF of the produced composites exhibited a decrease in its maximum value when compared to the maximum COF of the neat PLA, as depicted in Figure 6.10a. However, the addition of NFs to PLA led to a notable enhancement in WR. This improvement was shown by a considerable reduction in the SWR of the NFRCs compared to that of pure PLA, as shown in Figure 6.10b.

6.5.4   CORN COB/PLA COMPOSITE

Fouly et al. (2022) examine the TPs of a corn cob (CC)/PLA composite for their potential application in the artificial transplant of the bones and joints. For the sake of this, they fabricated a new polymer green composite by reinforcing the natural filler extracted from CC in a PLA matrix at different wt.%, such as 5%, 10%, 15%, and 20%, by employing the FDM 3DP technique, as illustrated in Figure 6.11. The experimental parameters consisted of AL ranging from 5 to 20 N and SD spanning from 100 to 500 m. They reported that the incorporation of the CC into PLA by up to 10% enhanced the WRE as well as minimized the COF of the PLA-CC composite.

6.5.5   DATE PARTICLE/PLA COMPOSITE

Albahkali et al. (2023) fabricated the green composite by reinforcing the date particle into the PLA at different wt.%, such as 2, 4, 6, 8, and 10, and kept all the samples for

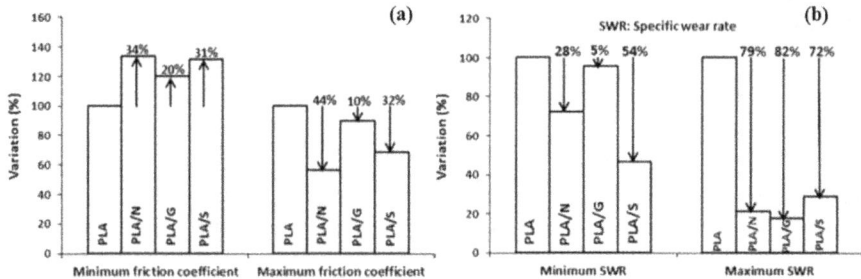

FIGURE 6.10   Variation of minimum and maximum: (a) COF and (b) SWR of different types of reinforced NFs (Bajpai et al., 2013).

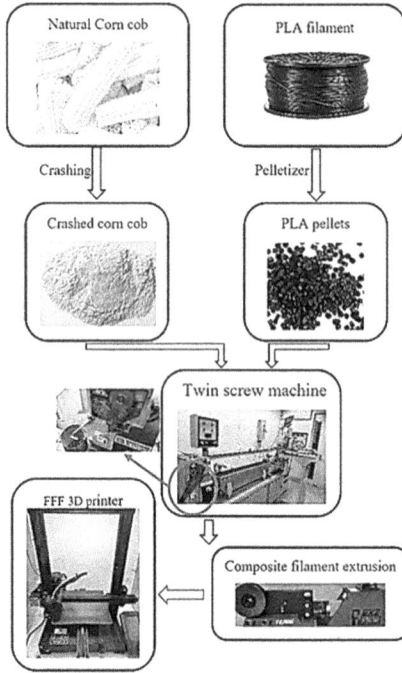

FIGURE 6.11 Process of 3DP AM technique for the fabrication of PLA-CC composite (Fouly et al., 2022).

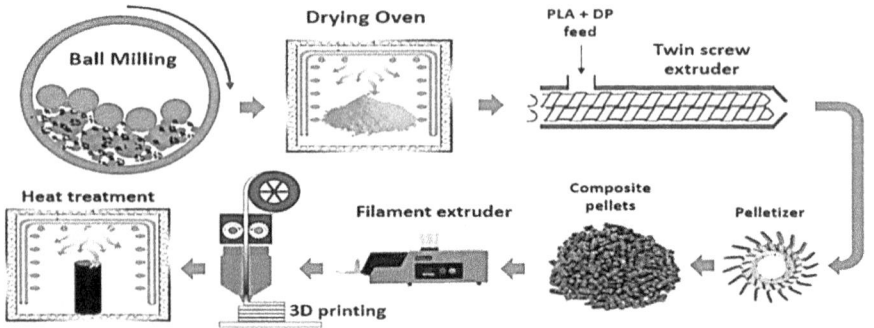

FIGURE 6.12 Preparation of date/PLA composite by employing 3DP AM technique (Albahkali et al., 2023).

the heat treatment for different time spans, such as 2.5, 5, 10, and 20 hours. Figure 6.12 demonstrates the fabrication process of the PC with the help of the 3DP technique for the biomedical application.

The samples underwent annealing for different durations in order to mitigate residual stresses produced by the 3DP process. Subsequently, tribological tests were conducted on the samples, wherein normal ALs and SDs were systematically varied.

This study emphasizes that the heat treatment of the green composites enhances tribological performance, making it a potential candidate for biomedical applications.

6.6 CONCLUSION

The present study aimed to examine the wear and friction characteristics of 3D AM NFRC, specifically focusing on various types of fibers, including biogenic carbon, CCF, flex yarn, grewia, nettle, sisal, CC, silica, and bronze. This chapter speculates on the potential impact of fiber volume fraction, fiber orientation, fiber treatment, different variable constraints such as AL, SV, and SD, and the effect of heat treatment on the tribological performance of NFRCs. As explained in this review, the use of AM for PCs has exhibited its capability in fabricating complex, functioning components with precise dimensions suitable for immediate application. Moreover, one notable advantage of employing AM in the production of NFRCs is its capacity to create functionally graded NFRCs that exhibit customized functionality and performance tailored to specific locations. Apart from this, we can create and produce structures with altered fiber alignment by modifying the deposition paths inside each layer. Nevertheless, AM of these NFRCs presents notable difficulties in various aspects. These challenges include the preparation of composite filaments suitable for FFF, the inherent tendency of NFs to agglomerate, the presence of a significant amount of moisture, and the formation of voids, as well as the complications encountered during the 3DP process of NFRCs, such as non-uniform curing, nozzle clogging, and fiber degradation or breakage. Two critical challenges in AM NFRCs are inadequate bonding between layers and insufficient bonding between fibers and the matrix. This is mostly due to the lower pressures involved in AM compared to conventional processing methods. Hence, it is imperative to allocate greater attention towards resolving these concerns by optimizing the process parameters of the AM technique, improving the feedstock quality, and enhancing hardware. The chapter yields the following conclusions:

- The distribution of fibers and the contact between the matrix and fibers within the filament were identified as the key characteristics that influenced the overall tribological performance of the 3DP AM NFRCs.
- The presence of voids in the 3DP NFRCs was also noted, which can reduce the TPs. Therefore, post-processing techniques, such as CM, have been used to significantly decrease the void content and enhance the WRE and load carrying capacity of NFRCs.
- The enhancement of WRE properties in NFRCs can be achieved through the alteration of factors such as fiber concentration, fiber orientation, and fiber content.
- It has been widely acknowledged that NFRCs possess favorable wear properties, rendering them highly suitable for utilization in various industries such as biomedical, marine, automotive, and chemical.

ACKNOWLEDGMENTS

The authors express gratitude to Indian Institute of Technology Indore, Indore, Madhya Pradesh, India and Sant Longowal Institute of Engineering and Technology, Longowal, Punjab, India for technical support.

REFERENCES

F.M. Al-Oqla, S.M. Sapuan, Natural fiber reinforced polymer composites in industrial applications: Feasibility of date palm fibers for sustainable automotive industry. *J. Clean. Prod.* 66 (2014) 347–354.

T. Albahkali, H.S. Abdo, O. Salah, A. Fouly, Adaptive neuro-fuzzy-based models for predicting the tribological properties of 3D-printed PLA green composites used for biomedical applications. *Polymers* 15(14) (2023) 3053.

S.R. Athreya, K. Kalaitzidou, S. Das, Mechanical and microstructural properties of Nylon-12/carbon black composites: Selective laser sintering versus melt compounding and injection molding. *Compos. Sci. Technol.* 71(4) (2011) 506–510.

P.K. Bajpai, I. Singh, J. Madaan, Tribological behavior of natural fiber reinforced PLA composites. *Wear* 297(1–2) (2013) 829–840.

V.K. Balla, K.H. Kate, J. Satyavolu, P. Singh, J.G.D. Tadimeti, Additive manufacturing of natural fiber reinforced polymer composites: Processing and prospects. *Compos. B. Eng.* 174 (2019) 106956.

U.S. Bongarde, V.D. Shinde, Review on natural fiber reinforcement polymer composites. *Int. J. Eng. Innov. Technol.* 3(2) (2014) 431–436.

L. Boopathi, P.S. Sampath, K. Mylsamy, Investigation of physical, chemical and mechanical properties of raw and alkali treated Borassus fruit fiber. *Compos. Part B: Eng.* 43(8) (2012) 3044–3052.

N. Chand, U.K. Dwivedi, Sliding wear and friction characteristics of sisal fibre reinforced polyester composites: Effect of silane coupling agent and applied load. *Polym. Compos.* 29(3) (2008) 280–284.

V. Chaudhary, P.K. Bajpai, S. Maheshwari, An investigation on wear and dynamic mechanical behavior of jute/hemp/flax reinforced composites and its hybrids for tribological applications. *Fibers Polym.* 19 (2018) 403–415.

D. Dai, M. Fan, Wood fibres as reinforcements in natural fibre composites: Structure, properties, processing and applications. In: *Natural Fibre Composites*, Edited by Alma Hodzic, Robert Shanks, Woodhead Publ. (2014) 3–65.

E.G. Ertane, A. Dorner-Reisel, O. Baran, T. Welzel, V. Matner, S. Svoboda, Processing and wear behaviour of 3D printed PLA reinforced with biogenic carbon. *Adv. Tribol.* 2018 (2018) 1–11.

O. Faruk, A.K. Bledzki, H.P. Fink, M. Sain, Biocomposites reinforced with natural fibers: 2000-2010. *Prog. Polym. Sci.* 37(11) (2012) 1552–1596.

A. Fouly, A.K. Assaifan, I.A. Alnaser, O.A. Hussein, H.S. Abdo, Evaluating the mechanical and tribological properties of 3D printed polylactic-acid (PLA) green-composite for artificial implant: Hip joint case study. *Polymers* 14(23) (2022) 5299.

J. Holbery, D. Houston, Natural-fiber-reinforced polymer composites in automotive applications. *JOM.* 58 (2006) 80–86.

Q. Huimin, G. Zhang, L. Chang, F. Zhao, T. Wang, Q. Wang, Ultralow friction and wear of polymer composites under extreme unlubricated sliding conditions. *Adv. Mater. Interfaces* 4(13) (2017) 1601171.

E. Jayamani, S. Hamdan, M.R. Rahman, M.K. Bakri, Investigation of fiber surface treatment on mechanical, acoustical and thermal properties of betelnut fiber polyester composites. *Procedia. Eng.* 97 (2014) 545–554.

M. Jimenez, L. Romero, I.A. Dominguez, M.D.M. Espinosa, M. Dominguez, Additive manufacturing technologies: An overview about 3D printing methods and future prospects. *Complexity* 2019 (2019) 1–30.

S. Kanakannavar, J. Pitchaimani, M.R. Ramesh, Tribological behaviour of natural fibre 3D braided woven fabric reinforced PLA composites. *Proc. Inst. Mech. Eng. J: J. Eng. Tribol.* 235(7) (2021) 1353–1364.

R.S. Keerthiveettil, K. Vijayananth, G.P. Muthukutti, P. Spatenka, A. Arivendan, S.P. Ganesan, The effect of various composite and operating parameters in wear properties of epoxy-based natural fiber composites. *J. Mater. Cycles Waste Manag.* 24(2) (2022) 667–679.

Y.S. Kim, J. Yang, S. Wang, A.K. Banthia, J.E. McGrath, Surface and wear behavior of bis-(4-hydroxyphenyl) cyclohexane (bis-Z) polycarbonate/polycarbonate-polydimethylsiloxane block copolymer alloys. *Polymer* 43(25) (2002) 7207–7217.

T. Liu, X. Tian, Y. Zhang, Y. Cao, D. Li, High-pressure interfacial impregnation by micro-screw in-situ extrusion for 3D printed continuous carbon fiber reinforced nylon composites. *Compos. A: Appl.* 130 (2020) 105770.

K.O. Low, J.L. Lim, K.J. Wong, An experimental study on the scratch characteristics of bamboo fibre-reinforced epoxy composite. *Adv. Compos.* 19(4) (2010) 096369351001900403.

C.R. Mahesha, N.M. Shivarudraiah, R. Suprabha, Three body abrasive wear studies on nanoclay/nanoTiO2 filled basalt-epoxy composites. *Mater. Today: Proc.* 4(2) (2017) 3979–3986.

J.Z. Manapat, Q. Chen, P. Ye, R.C. Advincula, 3D printing of polymer nanocomposites via stereolithography. *Macromol. Mater. Eng.* 302(9) (2017) 1600553.

A.I. Medalia, On structure property relations of rubber. *Proc. Int. Conf.* 13 (1980).

Y.J. Mergler, R.P. Schaake, A.J. Huis, Material transfer of POM in sliding contact. *Wear* 256(3–4) (2004) 294–301.

M. Milosevic, P. Valasek, A. Ruggiero, Tribology of natural fibers composite materials: An overview. *Lubricants* 8(4) (2020) 42.

U. Nirmal, J. Hashim, M.M. Ahmad, A review on tribological performance of natural fibre polymeric composites. *Tribol. Int.* 83 (2015) 77–104.

U. Nirmal, J. Hashim, S.T. Lau, Testing methods in tribology of polymeric composites. *Int. J. Mech. Mater. Eng.* 6(3) (2011) 367–73.

T. Nishino, K. Hirao, M. Kotera, K. Nakamae, H. Inagaki, Kenaf reinforced biodegradable composite. *Compos Sci Technol.* 63(9) (2003) 1281–1286.

Y. Pan, Z. Zhong, A micromechanical model for the mechanical degradation of natural fiber reinforced composites induced by moisture absorption. *Mech. Mater.* 85 (2015) 7–15.

H.H. Parikh, P.P. Gohil, Experimental investigation and prediction of wear behavior of cotton fiber polyester composites. *Friction* 5 (2017) 183–193.

D.K. Rajak, D.D. Pagar, P.L. Menezes, E. Linul, Fiber-reinforced polymer composites: Manufacturing, properties, and applications. *Polym. J.* 11(10) (2019) 1667.

M. Ramesh, T.S. Atreya, U.S. Aswin, H. Eashwar, C. Deepa, Processing and mechanical property evaluation of banana fiber reinforced polymer composites. *Procedia. Eng.* 97 (2014) 563–572.

B. Ren, T. Mizue, K. Goda, J. Noda, Effects of fluctuation of fibre orientation on tensile properties of flax sliver-reinforced green composites. *Compos. Struct.* 94(12) (2012) 3457–3464.

S. Sathish, N. Karthi, L. Prabhu, S. Gokulkumar, D. Balaji, N. Vigneshkumar, T.S. Ajeem Farhan, A.A. Kumar, V.P. Dinesh, A review of natural fiber composites: Extraction methods, chemical treatments and applications. *Mater. Today: Proc.* 45 (2021) 8017–8023.

U. Shehu, H.I. Audu, M.A. Nwamara, A.F. Ade-Ajayi, U.M. Shittu, M.T. Isa, Natural fiber as reinforcement for polymers: A review. *Pac. J. Sci. Technol.* 2(1) (2014) 238–253.

F.F. Shuhimi, M.F.B. Abdollah, M.A. Kalam, H.H. Masjuki, A.E. Mustafa, S.E. Mat Kamal, H. Amiruddin, Effect of operating parameters and chemical treatment on the tribological performance of natural fiber composites: A review. *Part. Sci. Technol.* 35(5) (2017) 512–524.

P.K. Singh, R.P. Modanwal, D. Kumar, Fabrication and mechanical characterization of glass fiber/Al 2 O 3 hybrid-epoxy composite. *Sadhana* 46(1) (2021) 1–10.

P. Tesinova. *Advances in Composite Materials: Analysis of Natural and Man-made Materials.* 2011, 1–437. IntechOpen.

B. Utela, D. Storti, R. Anderson, M. Ganter, A review of process development steps for new material systems in three dimensional printing (3DP). *J. Manuf. Process.* 10(2) (2008) 96–104.

I.D. Weyenberg, J. Ivens, A.D. Coster, B. Kino, E. Baetens, I. Verpoest, Influence of processing and chemical treatment of flax fibres on their composites. *Compos. Sci. Technol.* 63 (2003) 1241–1246.

M. Xu, J.A. Lewis, Phase behavior and rheological properties of polyamine-rich complexes for direct-write assembly. *Langmuir* 23(25) (2007) 12752–12759.

M.M. Zagho, E.A. Hussein, A.A. Elzatahry, Recent overviews in functional polymer composites for biomedical applications. *Polym. J.* 10(7) (2018) 739.

7 Study on the Effect of Carbon-Fiber-Reinforced Composites on Tribological Properties

Shalini Mohanty and Adrian Murphy
Queen's University

Rashi Tyagi
Chandigarh University

7.1 INTRODUCTION

The upsurging need to upgrade the tribo-mechanical properties of machine components experiencing relative motion calls for careful consideration of high wear resistance and low friction coefficients (Mohanty & Gokuldoss Prashanth, 2023). The configuration of these requirements relies on specific wear rate that controls the friction properties, thereby the overall system performance. Subsequently, understanding the tribological characteristics of such parts in extreme environmental conditions is of utmost importance for expanding their lifespan and decreasing their maintenance needs (Kichloo et al., 2022; Kumar & Singh, 2022). Thus, designing and manufacturing parts is crucial for industries, especially those made from composites. The tribological study of such composite products is equally essential when used at interfacial parts. Within this context, a profound understanding of the composites' tribological characteristics contributes towards improving the operational longevity of parts, mitigating the maintenance, and bolstering the reliability of machine components exposed to interfacial friction (Kumar et al., 2020; Wong & Tung, 2016).

Composite materials are considerably the most promising contenders for aerospace applications due to their exceptional strength-to-weight ratios and remarkable fatigue-resistant properties (Banakar et al., 2012; Kundachira Subramani & Siddaramaiah, 2015). Polymer-based composite systems offer a multitude of functionalities by blending functional fillers with easily processable polymers, thus spreading out their application window (Ranjan et al., 2023a; Tyagi et al., 2023; Prashanth et al., 2017). Basically, composites for structural applications fall under the regime of high-performance systems, predominantly constituting synthetic materials offering high strength-to-weight ratios. However, they would require controlled manufacturing environmental conditions to achieve optimal performance (Ranjan et al., 2023b; Tyagi and Tripathi, 2023). Fiber-reinforced polymers emerged as highly advanced materials in military and electronic applications (Oladele et al., 2020). Their application has primarily gained attention in smaller components like reinforcement bars, girders, staying cables, bridge decks, and handrails ("Structural Properties of Composites in Fire," n.d.; Subramanian & Senthilvelan, 2011; Zhang et al., 1994). More recently, the scope of reinforced plastics has extended to encompass building, construction, transportation, and electronics. Fiber-reinforced polymers (FRPs) in the transportation sector find applications in automotive, aviation, shipping, and related fields. FRPs are instrumental in creating cryostats, dry transformers, high-voltage switches, and more in the energy and electronics sector. Notably, carbon-fiber-reinforced polymers (CFRPs) have also found usage in advanced technical functions, including rocket nozzles (McClelland et al., 2022). The market size growth of carbon fiber reinforcements is estimated to grow by 10.9% from 2017 to 2025 (Rajak et al., 2021). The prediction also reveals that there would be an uprising demand for composites made from carbon fibers for different industries. The main applications of such composites are shown in Figure 7.1.

Carbon-fiber-reinforced composites are a novel category of high-strength and high-performance materials, characterized by a blend of graphitic and non-crystalline regions. Carbon fibers stand out among the spectrum of reinforcing fibers available due to their exceptional specific modulus and strength. Particularly, these fibers outcast other reinforcements owing to their versatile properties, such as high tensile strength and moisture resistance at extreme temperatures (Ramulu & Kramlich, 2004). On the contrary, glass and organic polymer fibers tend to break easily at stress concentration

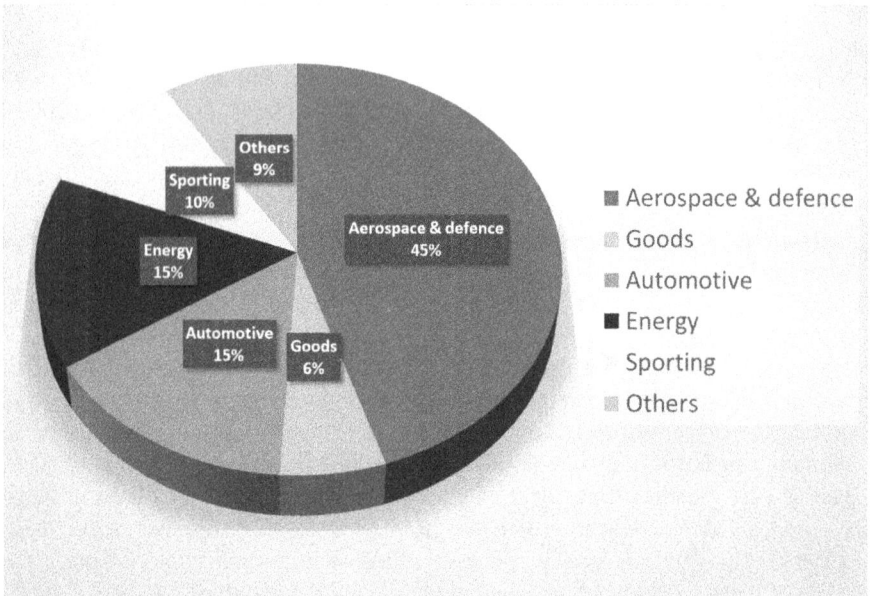

FIGURE 7.1   Industries acquiring carbon fibers for composite applications.

regions. Moreover, carbon-fiber-reinforced composites exhibit remarkable thermo-electrical conductivity with low coefficient of thermal expansion (Das, 2011; El-Hofy & El-Hofy, 2019). These striking characteristics of carbon fibers make them an optimal candidate for aerospace, micro-electronics, and automobile industries. Carbon fibers can attain a maximum strength of 7 GPa in terms of mechanical attributes. The axial compressive strength of carbon fiber alone ranges from 10% to 60% of their tensile strength, while the transverse compressive strength constitutes 12%–20% of the axial compressive strength. Such fibers can be utilized in two forms: pre-finished part or finished product. The finished products comprise twisted tubes or pultruded sections, while the pre-products encompass twisted or non-twisted yarns, short fibers, continuous filaments, and tows as per the application.

Although carbon fibers have several merits alone, the carbon-fiber-reinforced composites prepared from different methods might be subjected to tribological concerns wherein friction comes into the picture. There are situations wherein the composites might encounter environmental impact and the material failure occurs when subjected to wear. Specifically, polymer matrix composites, being quite popular in several industries, face difficulty under conditions in which they may be subjected to friction (Sudarshan Rao et al., 2015). For instance, the CFRP composites used in pipelines carrying petroleum slurries, helicopter and impeller blades, water turbines, aircraft operating in deserts, etc. are subjected to constant wear and tear. Thus, it has been necessary to address the issues concerning the tribological aspect of CFRPs.

The continuing innovation in 3D printing, also known to be additive manufacturing (AM), has propelled the fabrication of fiber-reinforced polymer (FRP) composites into the spotlight. This surge in interest can largely be attributed to the technology's capacity to convert intricate geometries into their final forms without the requirement for molds or specialized tools. Notably, there exist three primary methods for printing continuously reinforced FRP composites i.e., the in situ fusion of thermoplastic resins prior to extrusion, the in situ fusion of liquid thermoplastics with fibers at the tip of the nozzle, and the extrusion of pre-impregnated fibers (Man et al., 2021). Of many available fabrication processes, AM has gained attention among the research fraternity to manufacture carbon-fiber-reinforced composites. Some of the widely used methods to assess such composites are fused deposition modeling (FDM), selective laser sintering, stereolithography, selective laser melting, etc. The use of such composites in tribo-applications such as in gears, bearings, and rollers forced the researchers to assess the tribological studies (Kichloo et al., 2022; Man et al., 2021).

In the current framework of the chapter, a brief discussion on the tribological properties of the CFRP composites is conducted. Envisioning the importance of CFRP composites, the research fraternity has carried out much work to assess their frictional characteristics. Some of which is discussed in the succeeding sections.

7.2   TRIBOLOGICAL ANALYSIS OF CARBON-FIBER-REINFORCED COMPOSITES

To date, many research studies have been conducted on additively manufactured CFRP composites with regard to their mechanical properties (strength, fracture toughness, stiffness, etc.), as well as touching the tribological aspect. CFRP composites account for an emerging class of materials within the realm of tribological materials. There are failures in several engineering material systems due to extensive wear and friction. These composites' friction and wear characteristics are liable to chosen reinforcement and resin materials, manufacturing processes, operational conditions, fiber volume fraction, length, orientation, and surface treatment mechanisms (Bijwe et al., 2001). It is thus essential to acknowledge that no material alone is perfect or ideal, irrespective of the wear modes. The tribological assessment of composites can only be envisioned accurately through small-scale laboratory evaluations under diverse operating conditions (Parikh & Gohil, 2015). The research fraternity has systematically investigated the solid particle erosion tendencies in polymer composites constituting fibers and particulate materials (Boggarapu et al., 2020; Patnaik et al., 2010; Tarodiya & Levy, 2021). Their studies delved into the complications of solid particle erosion, aiming at the processes, modes, and discrete characteristics within the polymer matrix composites. Various proposed models have also been discussed to elucidate erosion rates and their applicability.

The past decade of research work constitutes investigations on the effect of process parameters on the mechanical characteristics, analysis on the failure mode mechanisms, microstructural effect, development of novel fiber-matrix combinations and innovative printing techniques (Chacón et al., 2019; Díaz-Rodríguez et al., 2021; Naranjo-Lozada et al., 2019; Penumakala et al., 2020; Sharma et al., 2011; Suresha & Kumar, 2009; van de Werken et al., 2020; Wickramasinghe et al., 2020). However, there remains notable dearth of research related to wear and frictional properties of these materials. Within the tribology realm, pioneering work has revolved around particle and short FRP composites. For instance, Dawoud et al. (2015) explored the tribological characteristics of the AM polymers bolstered with supplementary reinforcements, such as the incorporation of graphite flakes into an Acrylonitrile Butadiene Styrene (ABS) matrix. Some interesting tribological results were observed with the inclusion of graphite flakes (solid lubricant) into the ABS matrix. An increase in the frictional coefficient and wear rate was observed for the composites which might be due to inadequate interfacial bond amidst the fillers and the polymeric matrix, a commonly encountered challenge in the printed composite materials. Consequently, it was recommended that further exploration be directed towards fabrication parameters and filler selection. Indeed, several studies have underscored the substantial impact of processing parameters—such as raster angle, layer thickness, and printing orientation—on the mechanical properties and wear behavior of printed components (Palma et al., 2019). Lin et al. (2019) employed 3D printing to create a coating of short carbon-fiber-reinforced polyether ether ketone (PEEK) on authentic PEEK substrates. Their study revealed that the friction performance of the coated material could be significantly enhanced through the incorporation of nano-silica particles.

Jeon et al. (2020) achieved wear resistance in 3D-printed thermoplastic elastomers by incorporating surface-modified carbon black and graphene fillers that were electrochemically exfoliated. Notably, even a minimal concentration of 1 wt.% nanofillers yielded substantial enhancements in mechanical properties and wear resistance, thanks to graphene's favorable dispersibility and lubricating properties.

Furthermore, it's worth highlighting the innovative recycling approach introduced by Liu et al. (2021), who utilized AM to convert FRP waste into 3D composite components possessing desirable tribo-mechanical properties. This research showcased the potential of 3D printing as a method for recycling polymeric waste, hence contributing to improved environmental sustainability. Collectively, these examples underscore the promising prospects of AM technologies in crafting complex geometric tribo-parts and customizing their tribological performance through the incorporation of diverse fillers. Such capabilities are increasingly vital for meeting the demanding design requirements of modern industries. For high performance of additively manufactured carbon-fiber-reinforced composites, it is necessary to assess the factors that affect friction and wear. A stable coefficient of friction and low wear rate attest to the tribological performance indicators for CFRP composites. Some of the case studies have been presented in the succeeding section.

7.3 CASE STUDIES

Several studies interpreted the influence of adding carbon fiber as reinforcement in additively manufactured composites on the tribological studies. Some of the studies are illustrated in this section.

In a study conducted by Kichloo et al. (2022), Polyethylene Terephthalate Glycol (PETG)-based composites were fabricated by incorporating carbon fibers using FDM. The investigation shredded light on to how the addition of carbon fibers and processing parameters (such as layer thickness, infill pattern, and infill percentage) affected the tensile and flexural strength and the tribology of the polymeric composites. The findings indicated that an inclusion of 20 wt.% fibers in PETG led to significant improvements in the tensile strength of the composite. The maximum enhancement, amounting to 114%, was observed with triangular infill pattern, while the minimum increase of 43.7% was recorded with full honeycomb infill pattern. Similarly, the bending strength of the CFPETG composite was notably enhanced, with a maximum increase of 25% (for the full honeycomb infill pattern). Moreover, the results from tribological testing demonstrated a substantial reduction in the coefficient of friction with carbon fiber addition. Compared to PETG, the COF decreased by approximately 47.3% at minimal speeds (100 rpm) and 44.79% at above-average speeds (500 rpm). Fractographic and worn surface analysis revealed discrete fracture modes and wear-out systems for different composite samples, highlighting the role of carbon fiber in enhancing the surface properties of the acquired composites. Regarding unreinforced PETG, the analysis reveals delamination, indicating plastic deformation in the sliding direction. Some intense grooves were also present on the worn surface, along with visible cracks. Notably, the worn surfaces exhibit a smoother appearance, underscoring the contribution of carbon fibers in reducing friction.

In a similar study conducted by Luo et al. (2022), fabrication of short carbon-fiber-reinforced nylon (SCFRN) composites was done using FDM. The research delves into these printed materials' friction and wear characteristics, examining dry sliding and lubricated (water) conditions. The findings reveal that the introduction of short fibers significantly reduces the friction coefficient and enhances wear-resistant properties compared to pure nylon in all tested scenarios. The printed SCFRN exhibits a consistently low friction coefficient within the water regime, owing to water's cooling and lubricating properties. However, at lower loads, the specific wear rate of printed specimens can be higher under these conditions, particularly when compared to dry sliding conditions. The study also explores the impact of a square textured surface incorporated during printing to enhance the materials' tribological performance. It is observed that this textured surface improves the wear resistance of printed SCFRN in dry sliding conditions, attributed to its ability to collect or clean debris from the surface. However, this cleaning effect is less pronounced in lubricated conditions, as liquid facilitates effective surface cleaning. Conversely, surface textures soar the surface area subjected to water lubricant, leading to surface tempering due to higher water absorption rates. Consequently, samples with surface textures exhibit high wear rates under water lubrication (Luo et al., 2022). Figure 7.2 shows the effect of carbon fiber on tribological property of prepared carbon-reinforced composite which confirms that the friction coefficient changes slightly with increasing fiber content from 1 wt.% to 5 wt.%. Above 5 wt.% fiber content, the friction coefficient goes up with further increasing content of carbon fibers to 7 wt.% (Midan & Luo, 2015). Overall, this research provides valuable insights into the design of wear-resistant polymeric materials using 3D printing technologies, considering various sliding conditions.

Man et al. (2021) explored the Fused Filament Fabrication (FFF) 3D printing technique to fabricate polyamide 6 (PA6) composites reinforced with continuous carbon fibers (CCF). They focused on the sliding friction and wear behavior of the printed CCF/PA6 composites, considering various factors such as fiber alignments (normal, parallel, and antiparallel) and deposited layer orders (parallel and perpendicular) relative to the sliding route. They found that the orientation of the fibers enacted a critical role in deciding the wear resistance of the 3D-printed composites. When the fibers were oriented normal to the sliding direction, they observed severe fiber breakage, resulting in a higher wear rate compared to composites with fibers oriented in parallel or anti-parallel directions. To gain a deeper insight into the

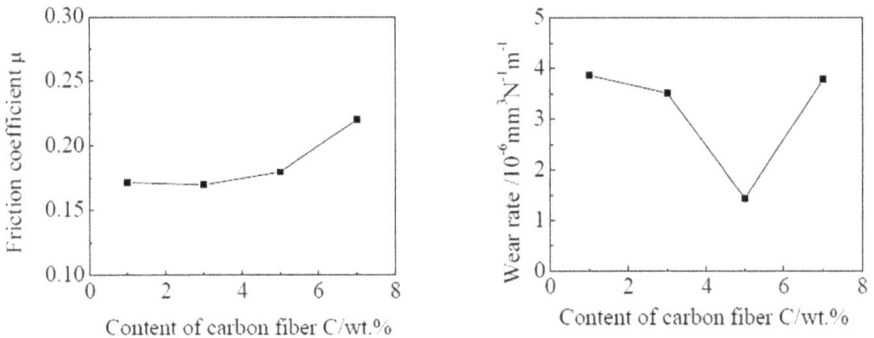

FIGURE 7.2   Plot of effect of carbon fiber weight % on (a) friction coefficient and (b) wear rate (Midan & Luo, 2015).

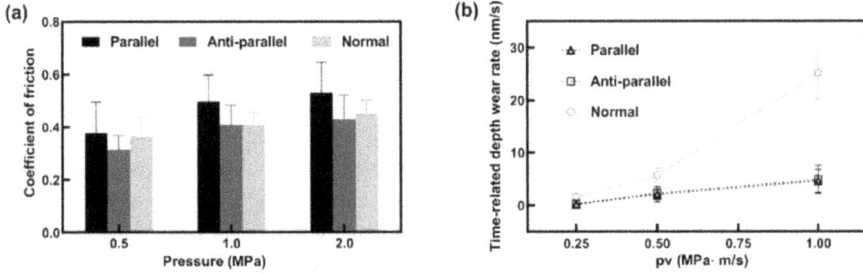

FIGURE 7.3 Tribological performance illustrated through (a) COF versus nominal pressure and (b) time-associated depth wear rate (Man et al., 2021).

impact of defects on the worn-out mechanism, they conducted additional tribo-tests on specimens that had undergone compression molding (CM) post-treatment. These tests confirmed that internal defects could significantly affect wear performance depending on the fiber orientations. The plots in Figure 7.3 indicate that as the normal loads increase, both friction and wear also tend to increase. This trend is consistent with observations made in conventionally manufactured polymer composites (Chen et al., 2017; Duan et al., 2021). Generally, critical sliding conditions, such as higher load pressure and/or sliding momentum, exacerbate the wear-out process. This is because the elevated stress and/or contact temperature accelerate thermal-mechanical failure. Furthermore, their investigation highlighted the importance of fiber distribution and the fiber/matrix interface quality within the filament as crucial parameters influencing the overall tribological performance of printed structures. These findings contribute to realizing the design and production of wear-resistant polymer composites using AM technologies.

7.4 APPLICATIONS AND FUTURE IN 3D-PRINTED CARBON FIBER COMPOSITES

The manufacturing process of carbon-fiber-reinforced AM composites and their mechanical and tribological properties holds strong appeal for various industries such as aerospace, automotive, and energy. When it comes to tools and fixtures, those constructed from 3D-printed composites can surpass machined parts in both performing and cost-efficiency. The inclusion of carbon fibers has the potential to enhance the mechanical and thermal characteristics of printed components, opening up the possibility of utilizing them as molds for manufacturing polymer-based parts.

Numerous researchers have explored the use of AM in aerospace applications. Initially, the focus was on single-material systems or metal alloys such as Al, Ti, or thermoplastics to create basic, non-structural components like brackets, enclosures, and fixtures. Examples include the production of titanium waveguide brackets for NASA's Juno spacecraft, aluminum brackets for communication satellites, and thermal blanket preservation brackets on Boeing satellites (Lim et al., 2016; van de Werken et al., 2020). As AM processes improved in quality and reliability, these techniques began to be employed for structural and high-performance components in jet and rocket engines, and spacecrafts, from preliminary assembly to maintenance and restoration (Hsissou et al., 2021; Penumakala et al., 2020). Today, AM is revolutionizing the production of substantial fragments of high-performance systems like aircraft, auto-aerial vehicles, and launching aircraft. For instance, NASA has realized an 80% reduction in part computation by employing AM for rocket engines, and Aerojet Rocketdyne has unveiled wholly 3D-printed rocket engines, condensing loads of traditionally mass-produced parts just through AM. While most of these developments have revolved around metal AM processes, featuring aluminum and titanium alloys, thermoplastics have mainly been limited to non-structural aerospace applications due to material limitations. However, the use of CFRPs in AM has the potential to significantly enhance structural performance and the applicability of polymeric AM components. AM CFRPs offer advantages similar to traditional manufacturing processes, including reduced mass and improved strength-to-weight and stiffness-to-weight ratios. Additionally, they provide two key benefits over AM metals: minor cost, facilitating innovative concepts like drone swarms and disposable UAVs, and lower processing temperatures, enabling multi-material AM for multi-functional components. Multifunctional AM holds the potential to transform aerospace systems by integrating electronic sensors or antennas into structures, offering mass saving, cost reductions, and performance enhancements.

Lastly, the use of AM in CFRPs has the potential to revolutionize space exploration. NASA and Made in Space have proven the competence to print thermoplastic jobs in a micro-gravity environment aboard the International Space Station (Goh et al., 2017; Saleh Alghamdi et al., 2021; van de Werken et al., 2020). While this has enormous potential for manned space search, it represents the beginning of the possibilities for on-orbit manufacturing. The drawback of the current approach is the use of single-material, thermoplastic parts with mechanical properties similar to terrestrial AM thermoplastics, limiting their applications. However, AM CFRPs can potentially enable the on-orbit manufacturing of large space assemblies, including antennas, solar arrays, reflectors, and space stations, for many applications (Espalin et al., 2014). NASA is hounding on-orbit manufacture theories through its in-space assembly technological efforts.

7.5 CONCLUSIONS

Carbon-fiber-reinforced composites are sketchily utilized in automobile, aerospace, military, weaponry, and chemical industries due to their caressing properties of being lightweight, corrosion-resistant, and high-strength. The chapter provides a comprehensive overview of research conducted on 3D printing of carbon-fiber-reinforced composites and their effect on tribological properties. It delves into various parameters that influence the tribological properties through case studies. The paper offers a thorough exploration of the mechanical properties of 3D-printed parts using different composites. The insights presented in this study pave the approach for developing a design support system tailored to components produced through AM. As industries increasingly adopt this technology for the production of complex small-batch parts, the knowledge generated by this chapter will aid in formalizing methods and tools to optimize the design of composite parts under various tribological constraints.

ACKNOWLEDGMENTS

The authors are thankful to Queen's University, Belfast, United Kingdom and Chandigarh University for technical support.

REFERENCES

Banakar, P., Shivanand, H. K., & Niranjan, H. B. (2012). Mechanical properties of angle ply laminated composites-A review. *International Journal of Pure and Applied Sciences and Technology*, *9*(2). www.ijopaasat.in.

Bijwe, J., Indumathi, J., John Rajesh, J., & Fahim, M. (2001). Friction and wear behavior of polyetherimide composites in various wear modes. *Wear*, *249*(8), 715–726. https://doi.org/10.1016/S0043-1648(01)00696-2.

Boggarapu, V., Gujjala, R., & Ojha, S. (2020). A critical review on erosion wear characteristics of polymer matrix composites. *Materials Research Express*, *7*(2), 022002. https://doi.org/10.1088/2053-1591/ab6e7b.

Chacón, J. M., Caminero, M. A., Núñez, P. J., García-Plaza, E., García-Moreno, I., & Reverte, J. M. (2019). Additive manufacturing of continuous fibre reinforced thermoplastic composites using fused deposition modelling: Effect of process parameters on mechanical properties. *Composites Science and Technology*, *181*, 107688. https://doi.org/10.1016/j.compscitech.2019.107688.

Chen, B., Li, X., Li, X., Jia, Y., Yang, J., Yang, G., & Li, C. (2017). Friction and wear properties of polyimide-based composites with a multiscale carbon fiber-carbon nanotube hybrid. *Tribology Letters*, *65*(3), 111. https://doi.org/10.1007/s11249-017-0891-z.

Das, S. (2011). Life cycle assessment of carbon fiber-reinforced polymer composites. *The International Journal of Life Cycle Assessment*, *16*(3), 268–282. https://doi.org/10.1007/s11367-011-0264-z.

Dawoud, M., Taha, I., & Ebeid, S. J. (2015). Effect of processing parameters and graphite content on the tribological behaviour of 3D printed acrylonitrile butadiene styrene. *Materialwissenschaft Und Werkstofftechnik*, *46*(12), 1185–1195. https://doi.org/10.1002/mawe.201500450.

Díaz-Rodríguez, J. G., Pertúz-Comas, A. D., & González-Estrada, O. A. (2021). Mechanical properties for long fibre reinforced fused deposition manufactured composites. *Composites Part B: Engineering*, *211*, 108657. https://doi.org/10.1016/j.compositesb.2021.108657.

Duan, J., Zhang, M., Chen, P., Li, Z., Pang, L., Xiao, P., & Li, Y. (2021). Tribological behavior and applications of carbon fiber reinforced ceramic composites as high-performance frictional materials. *Ceramics International*, *47*(14), 19271–19281. https://doi.org/10.1016/j.ceramint.2021.02.187.

El-Hofy, M. H., & El-Hofy, H. (2019). Laser beam machining of carbon fiber reinforced composites: A review. *The International Journal of Advanced Manufacturing Technology*, *101*(9–12), 2965–2975. https://doi.org/10.1007/s00170-018-2978-6.

Espalin, D., Muse, D. W., MacDonald, E., & Wicker, R. B. (2014). 3D Printing multifunctionality: Structures with electronics. *The International Journal of Advanced Manufacturing Technology*, *72*(5–8), 963–978. https://doi.org/10.1007/s00170-014-5717-7.

Goh, G. D., Agarwala, S., Goh, G. L., Dikshit, V., Sing, S. L., & Yeong, W. Y. (2017). Additive manufacturing in unmanned aerial vehicles (UAVs): Challenges and potential. *Aerospace Science and Technology*, *63*, 140–151. https://doi.org/10.1016/j.ast.2016.12.019.

Hsissou, R., Seghiri, R., Benzekri, Z., Hilali, M., Rafik, M., & Elharfi, A. (2021). Polymer composite materials: A comprehensive review. *Composite Structures*, *262*, 113640. https://doi.org/10.1016/j.compstruct.2021.113640.

Jeon, H., Kim, Y., Yu, W.-R., & Lee, J. U. (2020). Exfoliated graphene/thermoplastic elastomer nanocomposites with improved wear properties for 3D printing. *Composites Part B: Engineering*, *189*, 107912. https://doi.org/10.1016/j.compositesb.2020.107912.

Kichloo, A. F., Raina, A., Haq, M. I. U., & Wani, M. S. (2022). Impact of carbon fiber reinforcement on mechanical and tribological behavior of 3D-printed polyethylene terephthalate glycol polymer composites-An experimental investigation. *Journal of Materials Engineering and Performance*, *31*(2), 1021–1038. https://doi.org/10.1007/s11665-021-06262-6.

Kumar, S., & Singh, K. K. (2022). Tribological characteristics of glass/carbon fibre-reinforced thermosetting polymer composites: A critical review. *Journal of the Brazilian Society of Mechanical Sciences and Engineering*, *44*(11). https://doi.org/10.1007/s40430-022-03817-z.

Kumar, S., Singh, K. K., & Ramkumar, J. (2020). Comparative study of the influence of graphene nanoplatelets filler on the mechanical and tribological behavior of glass fabric-reinforced epoxy composites. *Polymer Composites*, *41*(12), 5403–5417. https://doi.org/10.1002/pc.25804.

Kundachira Subramani, N., & Siddaramaiah. (2015). Opto-electrical characteristics of poly (vinyl alcohol)/cesium zincate nanodielectrics. *The Journal of Physical Chemistry C*, *119*(35), 20244–20255. https://doi.org/10.1021/acs.jpcc.5b03652.

Lim, C. W. J., Le, K. Q., Lu, Q., & Wong, C. H. (2016). An overview of 3-D printing in manufacturing, aerospace, and automotive industries. *IEEE Potentials*, *35*(4), 18–22. https://doi.org/10.1109/MPOT.2016.2540098.

Lin, L., Ecke, N., Huang, M., Pei, X.-Q., & Schlarb, A. K. (2019). Impact of nanosilica on the friction and wear of a PEEK/CF composite coating manufactured by fused deposition modeling (FDM). *Composites Part B: Engineering*, *177*, 107428. https://doi.org/10.1016/j.compositesb.2019.107428.

Liu, W., Huang, H., Zhu, L., & Liu, Z. (2021). Integrating carbon fiber reclamation and additive manufacturing for recycling CFRP waste. *Composites Part B: Engineering*, *215*, 108808. https://doi.org/10.1016/j.compositesb.2021.108808.

Luo, M., Huang, S., Man, Z., Cairney, J. M., & Chang, L. (2022). Tribological behaviour of fused deposition modelling printed short carbon fibre reinforced nylon composites with surface textures under dry and water lubricated conditions. *Friction*, *10*(12), 2045–2058. https://doi.org/10.1007/s40544-021-0574-5.

Man, Z., Wang, H., He, Q., Kim, D. E., & Chang, L. (2021). Friction and wear behaviour of additively manufactured continuous carbon fibre reinforced PA6 composites. *Composites Part B: Engineering*, *226*, 109332. https://doi.org/10.1016/j.compositesb.2021.109332.

McClelland, J., Murphy, A., Jin, Y., & Goel, S. (2022). Correlating tool wear to intact carbon fibre contacts during drilling of continuous fibre reinforced polymers (CFRP). *Materials Today: Proceedings*, *64*, 1418–1432. https://doi.org/10.1016/j.matpr.2022.03.727.

Midan, L. I., & Luo, R. (2015). Effect of carbon fiber content on mechanical and tribological properties of carbon/phenolic resin composites. In *5th International Conference on Information Engineering for Mechanics and Materials*. pp. 539–542.

Mohanty, S., & Gokuldoss Prashanth, K. (2023). Metallic coatings through additive manufacturing: A review. *Materials*, *16*(6), 2325. https://doi.org/10.3390/ma16062325.

Mouritz, A. P., & Gibson, A. G. (2006). Structural properties of composites in fire. (n.d.). In Fire Properties of Polymer Composite Materials (pp. 163–213). Springer, Netherlands. https://doi.org/10.1007/978-1-4020-5356-6_6

Naranjo-Lozada, J., Ahuett-Garza, H., Orta-Castañón, P., Verbeeten, W. M. H., & Sáiz-González, D. (2019). Tensile properties and failure behavior of chopped and continuous carbon fiber composites produced by additive manufacturing. *Additive Manufacturing*, *26*, 227–241. https://doi.org/10.1016/j.addma.2018.12.020.

Oladele, I. O., Omotosho, T. F., & Adediran, A. A. (2020). Polymer-based composites: An indispensable material for present and future applications. *International Journal of Polymer Science, 2020*, 1–12. https://doi.org/10.1155/2020/8834518.

Palma, T., Munther, M., Damasus, P., Salari, S., Beheshti, A., & Davami, K. (2019). Multiscale mechanical and tribological characterizations of additively manufactured polyamide 12 parts with different print orientations. *Journal of Manufacturing Processes, 40*, 76–83. https://doi.org/10.1016/j.jmapro.2019.03.004.

Parikh, H. H., & Gohil, P. P. (2015). Tribology of fiber reinforced polymer matrix composites - A review. *Journal of Reinforced Plastics and Composites, 34*(16), 1340–1346. https://doi.org/10.1177/0731684415591199.

Patnaik, A., Satapathy, A., Chand, N., Barkoula, N. M., & Biswas, S. (2010). Solid particle erosion wear characteristics of fiber and particulate filled polymer composites: A review. *Wear, 268*(1–2), 249–263. https://doi.org/10.1016/j.wear.2009.07.021.

Penumakala, P. K., Santo, J., & Thomas, A. (2020). A critical review on the fused deposition modeling of thermoplastic polymer composites. *Composites Part B: Engineering, 201*, 108336. https://doi.org/10.1016/j.compositesb.2020.108336.

Prashanth, S., Subbaya, K. M., Nithin, K., & Sachhidananda, S. (2017). Fiber reinforced composites - A review. *Journal of Material Science & Engineering, 06*(03), 1000341. https://doi.org/10.4172/2169-0022.1000341.

Rajak, D. K., Wagh, P. H., & Linul, E. (2021). Manufacturing technologies of carbon/glass fiber-reinforced polymer composites and their properties: A review. *Polymers, 13*(21), 3721. https://doi.org/10.3390/polym13213721.

Ramulu, M., & Kramlich, J. C. (2004). *Machining of Fiber Reinforced Composites: Review of Environmental and Health Effects Hydrothermal Destruction of Hazardous Wastes View project*. https://www.researchgate.net/publication/303155598.

Ranjan, N., Tyagi, R., Kumar, R., & Babbar, A. (2023a). 3D printing applications of thermo-responsive functional materials: A review. *Advances in Materials and Processing Technologies. 2023*, 1–17. https://www.tandfonline.com/doi/abs/10.1080/2374068X.2023.2205669.

Ranjan, N., Tyagi, R., Kumar, R., & Kumar, V. (2023b). On fabrication of acrylonitrile butadiene styrene-zirconium oxide composite feedstock for 3D printing-based rapid tooling applications. *Journal of Thermoplastic Composite Materials*, 08927057231186310. https://doi.org/10.1177/089270572311863.

Saleh Alghamdi, S., John, S., Roy Choudhury, N., & Dutta, N. K. (2021). Additive manufacturing of polymer materials: Progress, promise and challenges. *Polymers, 13*(5), 753. https://doi.org/10.3390/polym13050753.

Sharma, M., Bijwe, J., & Singh, K. (2011). Studies for wear property correlation for carbon fabric-reinforced PES composites. *Tribology Letters, 43*(3), 267–273. https://doi.org/10.1007/s11249-011-9805-7.

Subramanian, C., & Senthilvelan, S. (2011). Joint performance of the glass fiber reinforced polypropylene leaf spring. *Composite Structures, 93*(2), 759–766. https://doi.org/10.1016/j.compstruct.2010.07.015.

Sudarshan Rao, K., Varadarajan, Y. S., & Rajendra, N. (2015). Erosive wear behaviour of carbon fiber-reinforced epoxy composite. *Materials Today: Proceedings, 2*(4–5), 2975–2983. https://doi.org/10.1016/j.matpr.2015.07.280.

Suresha, B., & Kumar, K. N. S. (2009). Investigations on mechanical and two-body abrasive wear behaviour of glass/carbon fabric reinforced vinyl ester composites. *Materials & Design, 30*(6), 2056–2060. https://doi.org/10.1016/j.matdes.2008.08.038.

Tarodiya, R., & Levy, A. (2021). Surface erosion due to particle-surface interactions - A review. *Powder Technology, 387*, 527–559. https://doi.org/10.1016/j.powtec.2021.04.055.

Tyagi, R., Singh, G., Kumar, R., Kumar, V., & Singh, S. (2023). 3D-Printed sandwiched acrylonitrile butadiene styrene/carbon fiber composites: Investigating mechanical, morphological, and fractural properties. *Journal of Materials Engineering and Performance, 0(0)*, 1–14. https://doi.org/10.1007/s11665-023-08292-8.

Tyagi, R., & Tripathi, A. (2023). Coating/cladding based post-processing in additive manufacturing. In *Handbook of Post-Processing in Additive Manufacturing: Requirements, Theories, and Methods*, Edited by: Gurminder Singh, Ranvijay Kumar, Kamalpreet Sandhu, Eujin Pei, Sunpreet Singh, pp. 127, CRC Press.

van de Werken, N., Tekinalp, H., Khanbolouki, P., Ozcan, S., Williams, A., & Tehrani, M. (2020). Additively manufactured carbon fiber-reinforced composites: State of the art and perspective. *Additive Manufacturing, 31*, 100962. https://doi.org/10.1016/j.addma.2019.100962.

Wickramasinghe, S., Do, T., & Tran, P. (2020). FDM-based 3D printing of polymer and associated composite: A review on mechanical properties, defects and treatments. *Polymers, 12*(7), 1529. https://doi.org/10.3390/polym12071529.

Wong, V. W., & Tung, S. C. (2016). Overview of automotive engine friction and reduction trends-Effects of surface, material, and lubricant-additive technologies. *Friction, 4*(1), 1–28. https://doi.org/10.1007/s40544-016-0107-9.

Zhang, Q., Liang, Y., & Warner, S. B. (1994). Partial carbonization of aramid fibers. *Journal of Polymer Science Part B: Polymer Physics, 32*(13), 2207–2220. https://doi.org/10.1002/polb.1994.090321308.

8 Impact of 3D Printing Process Parameters on Tribological Behaviour of Polymers

Sehra Farooq and Nishant Ranjan
Chandigarh University

8.1 INTRODUCTION

Charles Hull invented 3D printing processes in the year 1980 (Shahrubudin et al., 2019). As of now, 3D printing is primarily used to manufacture artificial heart pumps, jewellery, 3D-printed human corneas, food, aviation-related products, and many rocket engine parts. 3D printing technology evolved explicitly from computer-aided design (CAD) drawings through layer-wise fabrication of the 3D model (Low et al., 2017). 3D printing is a truly revolutionary expertise that has surfaced as a dynamic technology stage (Ranjan et al., 2023; Tyagi et al., 2023). It creates new possibilities and brings hope to many people. Many businesses that want to improve their manufacturing efficiency are showing some sincere appreciation for 3D printing (Saheb & Kumar, 2020). Traditional thermoplastic polymers, ceramic materials, metals, and other carbon-based materials can now be printed utilizing the method of 3D printing (Rahman et al., 2021). 3D printing is poised to transform industries and the fabrication processes. The applications of 3D printing technology will upsurge production while dwindling costs. Concurrently, consumer demand will have a prodigious effect on production. Consumers round about the final good or service and can call for an entity manufactured according to their disclaimers (Thomas & Gilbert, 2015). In the interim, 3DP facilities will be in higher proximity to customers, enabling added adaptability and quick production line, in addition to greater efficiency. Besides that, the utilization of 3D printing technology reduces the demand for global carriage (Singh, 2020). Finally, the use of 3D printing may alter the company's logistics. The company's logistics can manage the entire process, providing more extensive and end-to-end services. 3D printing is now extensively utilized all over the globe (Wong & Hernandez, 2012). This technology is rapidly being used for product specialization and the creation of free and open-source creations in the agriculture, medical services, manufacturing, and aviation industries. To summarize, the use of 3D printing has risen to prominence as a versatile and powerful tool in the advanced manufacturing industry (Kundu & Mandal, 2022). This technology is widely utilized in numerous nations, particularly in the

DOI: 10.1201/9781003400523-8

manufacturing base. As a result, this paper provides an overview of various types of 3D printing technologies, their applications, materials used in 3DP and, finally, the tribological effect of these materials used in 3DP.

8.2 TYPES OF 3D PRINTING

8.2.1 Fused Deposition Modelling Printing

Fused deposition modelling (FDM) is what most people think of as 3D printing as it ranks as the most common and, in several ways, the simplest of the options (Bryll et al., 2018). FDM employs a wide range of plastic materials with varying melting temperatures that liquefy and resolidify, the most prevalent among which are polylactic acid (PLA) polymer and acrylonitrile butadiene styrene (ABS) (Finnes & Letcher, 2015). The most common configuration for an FDM Printing machine is known as a Cartesian print engine since the printed objects are created using basic Coordinates (X, Y, Z). The printhead of this printer is a metallic tube and contains a thermistor inside for temperature control. The heat from the printhead melts the material substrate and then a pressure developed inside takes more material in, exacerbating certain amounts of melted material to eject through a tiny, holed nozzle usually 0.2–0.5 mm in size (Popescu et al., 2018). An FDM printer print commences with one plastic layer that is applied very finely to the printing platform, followed by the nozzle progressing across the printing surface and stashing the material in the form of the item being created. This serves as the object's base layer; the second coating will be stashed immediately on top of it. Both layers will fuse due to their inherent properties. Again, when the second layer is finished, the third layer will be added, and so on, gradually building the product (Bhavar et al., 2017). Figure 8.1 shows the schematic diagram of the fused deposition printer.

8.2.2 Stereolithography

Stereolithography (SLA) allows for the creation of extremely precise and delicate objects, thus making it perfect for the building of complicated designs and prototypes (Lee et al., 2015). In SLA printer, a liquid material results in a notably smooth surface, making subsequent processing unnecessary. This minimizes labour and expense while also contributing to increased material consumption efficiency (Kumar et al., 2012). The accuracy of the object is just limited by the device's quality. The finer the laser beam diameter, the significantly larger the accuracy of the printed objects. SLA is a member of the vat photopolymerization clan of AM methods, also known as resin 3D printing technology (Weng et al., 2016). All of these machines work according to the same principle: they use a source of light laser or light source to cure thermoplastic material into hardened plastic. The key parts, such as the source of light, build framework, and resin tank, are the primary physical distinction. Light-reactive materials known as "resin" are used in SLA 3D printers. When SLA resins are subjected to specific light wavelengths, it polymerizes monomers into solid rigid or versatile geometries (Sano et al., 2018). Figure 8.2 gives a brief representation of a stereolithographic printer.

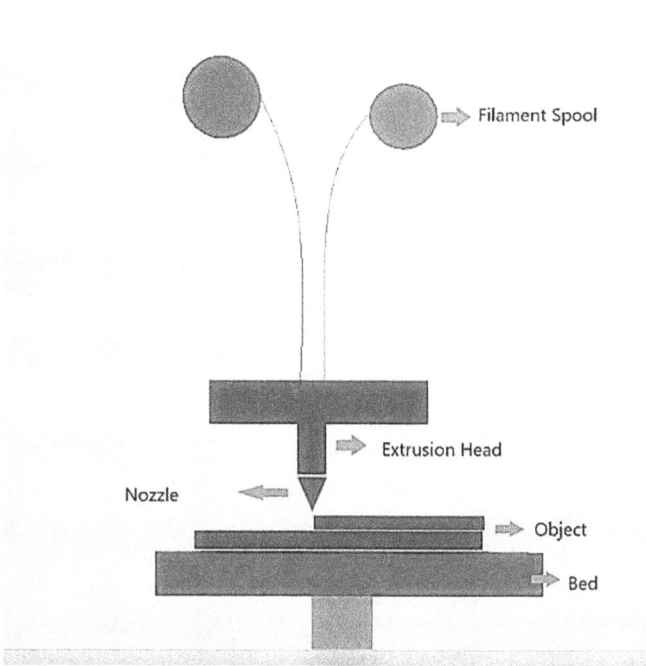

FIGURE 8.1 Schematic working diagram of fused deposition printer.

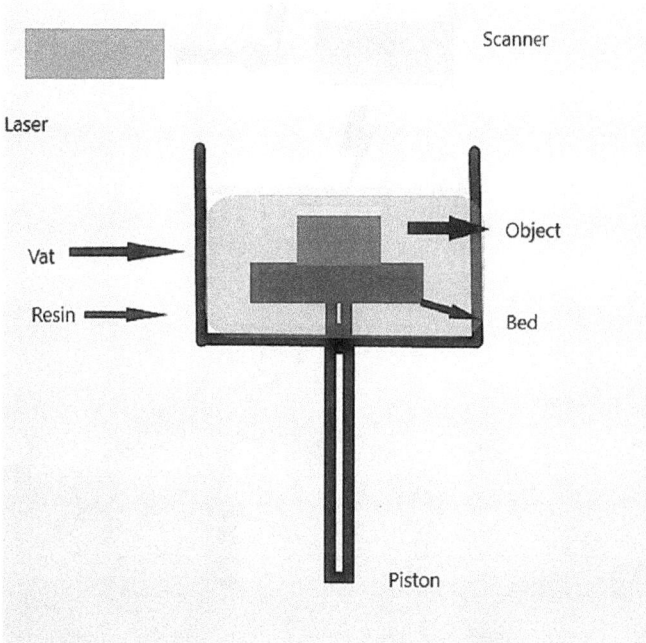

FIGURE 8.2 Schematic working diagram of stereolithographic printer.

8.2.3 SELECTIVE LASER SINTERING

Selective Laser Sintering (SLS) is yet another technique used by today's 3D printers. The high temperature from a powered laser beam joins particles of material or polymer to create a solid throughout this process (Simchi, 2006). According to the data source, the beam is tracked down across a powdered bed. The powder within the powdered bed is densely packed. The beam moves in two different directions: X and Y. The laser then strikes the outer layer of the powder material, sintering the particles together to create a solid (Olakanmi et al., 2015). Once a layer is finished, the bed descends in the Z direction, and the levelling roller smoothens the powder across the bed (Zeng et al., 2012). Figure 8.3 shows the schematic working of a SLS printer.

8.2.4 3D INKJET PRINTER

The Layer-wise deposition of 3D inkjet printing is parallel to that of other 3D printing processes. Printing directions are obtained from a 3D modelling tool of your product, just like in common FDM processes (Noguera et al., 2005). This well-established method of AM, on the other hand, is unique in that it can be used with an extensive variety of liquids or suspensions, producing insulative or conductive structures with extremely high printing precision (Guo et al., 2018). Inkjet printing, unlike methods that involve the merging of plastic or metal materials, does not require any post-curing process—the final printing product is completely operational as soon as it exits the printing system (Guo et al., 2018). Figure 8.4 gives the representation of an inkjet printer.

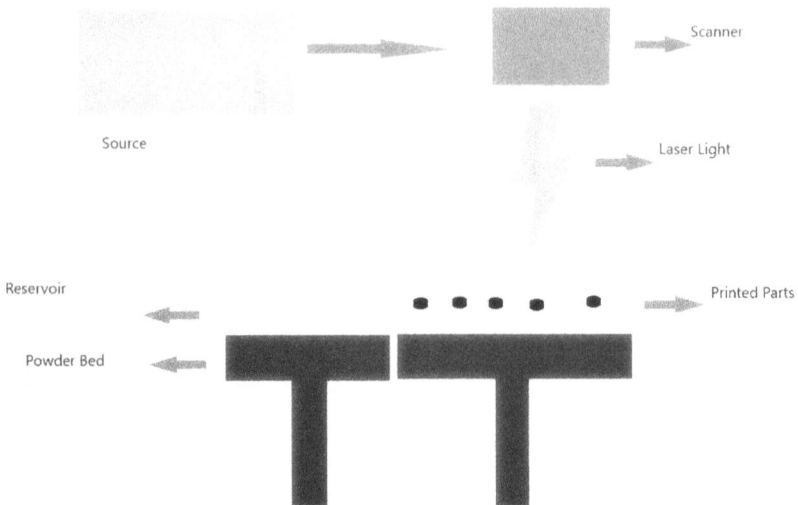

FIGURE 8.3 Schematic working diagram of selective laser sintering printer.

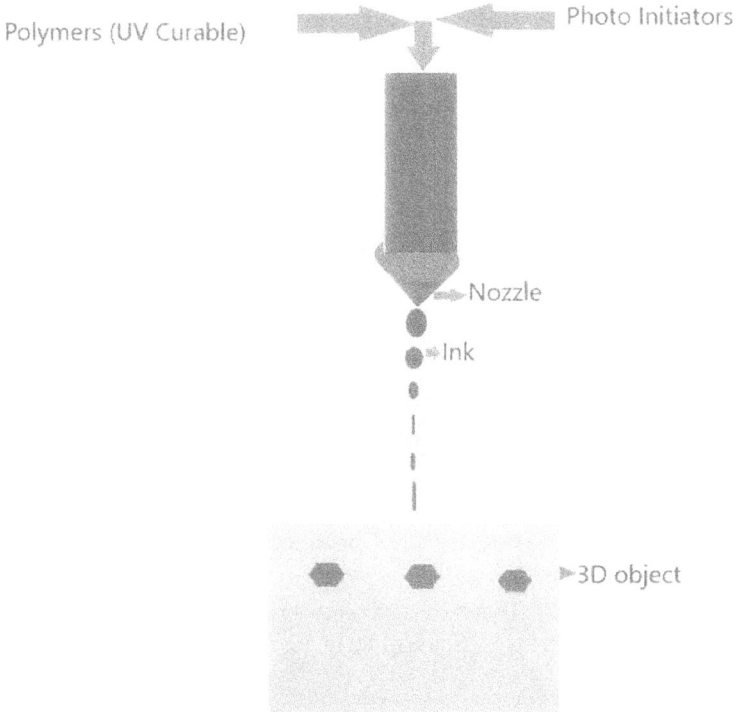

Polymers (UV Curable)   Photo Initiators

Nozzle

Ink

3D object

FIGURE 8.4   Schematic working diagram of 3D inkjet printer.

8.2.5 BINDER JETTING PRINTER

Binder jetting is another AM process technique that involves depositing a binding agent selectively upon a thin layer of powdered particles, be it foundry sand, ceramic materials, metal, or composite materials, to create personalized and extensive kind of parts and tooling (Gonzalez et al., 2016). The procedure is repeated successively layer by layer, equivalent to printing on paper, by utilizing a guide from a CAD file, until the product is complete (Bai & Williams, 2018). Binder jetting continues to stand out among 3D printing techniques due to its high rate of speed as well as material versatility as a method with the potential to convert regular high-volume production and deliver the design, cost, and sustainability benefits of 3D printers to the masses (Gaytan et al., 2015). Figure 8.5 gives the working of a Binder jetting printer.

8.3   3DP PROCESS PARAMETERS

The process parameters have a significant impact on the specificity, efficiency, and attributes of the manufactured part. As a result, fundamental studies into various process parameters ought to be part of any effort at creating functionally reliable

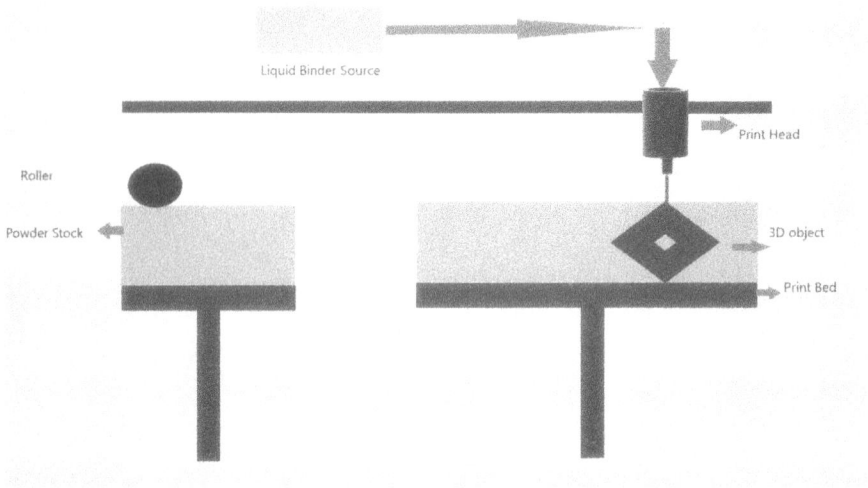

FIGURE 8.5   Schematic working diagram of binder jetting printer.

parts (Zhang et al., 2007). There are several determining parameters kept in mind while working with 3D printers. Some of them are mentioned below:

- Printing Speed

    Print speed, as the term suggests, is the speed at which the motor of the printer moves. It consists of the coordinate (X, Y) control motors, as well as the extruder motor meters (Christiyan et al., 2016).

- Temperature

    This parameter includes the temperature of the bed and the melting temperature of the material. Heated beds are necessary for 3D printing. It must always be fixed to a specific temperature. For instance, the ideal temperature range is 55°C–70°C, while the melting temperature of the material is the temperature at the exit of the extruder (Christiyan et al., 2016).

- Size of Nozzle and Size of Filament

    In a 3D printer, the nozzle is the component that protrudes the filament to generate the part. The diameter of the nozzle can vary from 10 to 100 mm. The filament in 3DP is a thermoplastic raw material used in the production of 3D parts. There are various filaments with different properties that require different printing temperatures. The filaments are available in two standard diameters: 1.75 and 2.85 mm (Popescu et al., 2018).

- Layer Thickness and Number of Layers

    With each successive addition of material, the 3D printing process takes place. The layer height is the vertical level of the z-axis which is a significant technical characteristic of a 3D printer. The number of layers determines the successive addition of layers (Popescu et al., 2018).

- Infill Density and Geometry

  The amount of material taking up the article's interior. The amount of material is typically adjusted by adjusting programmes from the vacuous part (0%) to the filled solid part (100%). In 3D printing, pattern refers to infill. Printing duration, strength, speed, and mechanical properties are all affected by patterns. Infill patterns are classified into four types: triangular, rectangular, hexagonal or honeycomb, and wiggle (Ramaswamy & Blunt, 2004).

- Tribological Properties

  Tribology is an essential technology of industries. It is used in healthcare engineering aspects, magnetic tracking head-media connectors, LSI production systems, and space systems (Wang et al., 2013). Wear and friction fluctuations degrade device or tool performance in these fields. Tribological characteristics, such as friction and wear, are not inherent material characteristics but rather depend on the entire tribological system. As a result, the prototype of any type of rheological material must be determined by the demands of the specific application (Wakabayashi et al., 2007). Interfacial damage in tribological materials can be easily triggered by frequent rolling or sliding contact under protracted tribological processes in the form of surface microcracks or surface wear. Polymers and polycarbonate materials have grown into an acceptable alternative for metallic materials; as a result, they must be capable of providing decent tribological behaviour, which implies several features favourable to the application's reliable operation (Field et al., 2004). This work will be focused towards highlighting the effects of 3D printing on the tribological properties of polymers and metals.

8.4   POLYMER ADDITIVE MANUFACTURING

Polymer AM is a form of 3DP technology that uses polymer as printing material. It is a method of building 3D objects by laying one layer placed above another one at a time to create the entire part. One layer is fabricated over the other, which can be either entirely or partially melted (Alghamdi et al., 2021). Previously, AM focused on models that were used to visualize the final part as a prototype. Recently, AM has been widely used to fabricate end-use products in the fields of aeroplanes, medical and dental implants, automotive, and the nutrition and fashion industries (Ramezani Dana et al., 2019). Figure 8.6 gives a brief overview of a polymer AM.

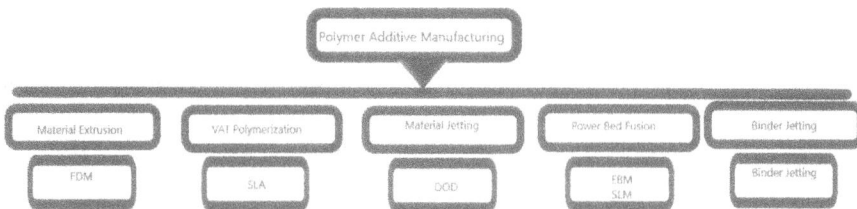

FIGURE 8.6   Types of polymer additive manufacturing.

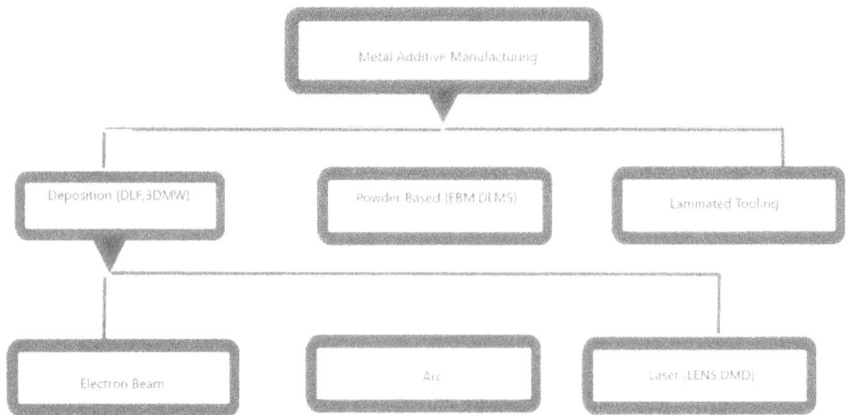

FIGURE 8.7   Types of metal additive manufacturing.

8.5   METAL ADDITIVE MANUFACTURING

Metal AM, also known as Metal 3DP, is the process of creating a 3D solid of almost any contour or style from a computer-aided framework (Wu et al., 2018). 3D metal objects are frequently created by stacking fine metal powders with a bonding material between layers (Liu et al., 2022). As the design and template have been created, they are uploaded to the device to initiate printing; however, the exact printing varies depending on the machine. The ability to generate 3D objects of any shape differentiates AM from all other manufacturing methods (Herzog et al., 2016). Figure 8.7 summarizes the process of metal AM.

8.6   COMPOSITE ADDITIVE MANUFACTURING

Composite AM is a process in which layers of a fibre-reinforced composite material like Carbon Fibre are tumbled beneath a 3D printer which extrudes a liquid solution over the sheet in the form of layers (Turk et al., 2016). The usage of fibre-reinforced polymer composites opens up possibilities for lightweight, switching from metal to polymers and compressing assembly, much like AM (Yakout & Elbestawi, 2017). Similar to AM, composites production requires layering components; however, in the instance of composites, this typically entails designing a unique tool. Composites have established themselves in many applications nowadays. They offer tried-and-true materials and processes for producing a wide range of useful components, especially significant aeroplane parts. However, the use of composites continues to grow, and today 3D printing is speeding up that development (Holmes, 2019). Figure 8.8 gives a brief overview of composite AM.

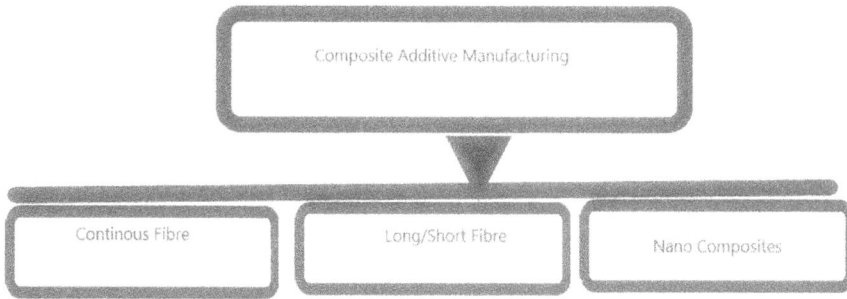

FIGURE 8.8 Types of composite additive manufacturing.

8.7 MATERIALS USED IN 3D PRINTING

The materials used in 3DP are as diversified as that of the products that are created. As an upshot, 3DP is pliable enough to enable manufacturers to tailor a product's structure, contour, and strength. Some of the novel materials used in 3DP are as follows:

- Polymers

  Polymers serve as the most commonly used material for 3DP today. The appearance of polymer is simple to comprehend given its suppleness, adaptability, softness, and wide range of colour options. Polymer, as a fairly affordable option, is relatively gentle in terms of cost for both consumers and manufacturers (Shahrubudin et al., 2019). Some of the popular types of polymers materials used in 3D technology are ABS, PLA, Polyethylene terephthalate (PET), Polycarbonate (PC), High-impact polystyrene (HIPS), Polyether ether ketone (PEEK), and Thermoplastic polyester (TPC) (Alghamdi et al., 2021). Polymers have played an important role in medical goods, devoted to device performance, inert materials, and a mechanical support in several orthopaedic applications (Ramezani Dana et al., 2019).
- Metals

  A metal 3D printer's primary aim remains to liquefy metallic 1c substance into the shape of wire or sometimes powder with the help of an electron or laser beam as a source of energy. The melted material is subsequently utilized to stack metal or alloy layers on top of each other following a CAD model to create a compact product (Rahman et al., 2021). In comparison to traditional manufacturing techniques, 3DP of metals enables the manufacturing of complicated shapes (Zeng et al., 2012). 3DP of metallic materials is primarily employed in small-scale production in the engineering and manufacturing, aviation, medical, and defence industries. Some commonly used metals or alloys for 3DP are Stainless steel, Aluminium Alloy, Nickel Alloy, and cobalt alloy (Johnson et al., 2019). 3DP technology can create nickel-based alloy aircraft parts. The objects product (Moritz & Maleksaeedi, 2018) made from nickel-based alloys is suitable for use in hazardous environments. This is because of its excellent resistance to corrosion and capacity to withstand temperatures of up to 2192°F (Behera et al., 2019).

- Ceramics

    Ceramics are hard, brittle, resistant to heat, and noncorrosive materials. 3DP techniques can now produce 3D-printed objects without cracks or substantial porosities by trying to adjust process variables; as a matter of fact, the manufactured product has superior mechanical characteristics (Galante et al., 2019). Ceramic materials enact a foremost role in dentistry and aeronautics. A few examples of ceramics used in 3DP are Zirconia, Alumina Powder, and Bioactive glasses. The 3DP technique can create complicated alumina products of elevated density (Moritz & Maleksaeedi, 2018). Bio-active glass/ glass-ceramics created employing an SLA machine increased products' strength properties. The improved mechanical properties allow for the use of bio-glass in therapeutic formations such as scaffolding and bone configurations. The scarcity of 3D-printable ceramics, as well as dimensional imprecision and deficient surface polish, are major problems (Deckers et al., 2014).

- Composites

    Composites stand as substances that are made up of multiple materials with distinguishable physical and chemical attributes. When compared to individual material properties, composite materials have superior physical, mechanical, and chemical properties (Yakout & Elbestawi, 2017). Composites are growing in popularity in 3D printing over conventional methods because of their accessibility, improved functionalities, and outlay product customization with extreme accuracy (Kundu & Mandal, 2022). Composites are now used in a variety of sectors, including production, aerospace, health sciences, institutional, and automotive. Glass-fibre-reinforced composite materials and carbon fibre polymer composites are examples of composite materials (Bryll et al., 2018). Carbon-fibre-reinforced polymer structures are widely used in the aviation industry due to their higher strength properties, rigidity, and corrosion resistance efficiency. Glass-reinforced composite materials, on the other side, are widely used in 3D printing for a variety of purposes (Holmes, 2019).

- Biomaterials

    Biomaterials are the most frequently used substances in the medical sector to create customized services. According to the present study, the use of biomaterials in 3DP is increasing as a result of advances in the field, such as enhanced printing accuracy, the ability to print complex shapes, reduced wastage, and relatively inexpensive products (Bose et al., 2018a). Advances in 3DP have allowed for significant advances in medical services including medical devices and implant materials. A variety of biomaterials are suitable for employing 3DP technology (Bose et al., 2018b). Chitosan, fibrin, PEEK, and collagen are all examples of biomedical polymer materials. These materials are completely biocompatible and meet all medical requirements. Fibrin, PEEK, and biomedical ceramics such as TCP and HA are examples of such materials. These substances are entirely biocompatible and meet all medical requirements (Zadpoor & Malda, 2017).

8.8 TRIBOLOGICAL PROPERTIES OF POLYMERS AND COMPOSITES

Polymers and composite materials (PCM) are extensively used as friction units in equipment and machinery due to their considerable specific strength, outstanding capacity for absorbing vibration and noise, high durability against abrasive wear, impact, and contamination, and simple fabrication (Bagsik & Schöppner, 2011). Polymeric materials with self-lubricating characteristics are gaining popularity because of their outstanding tribological performance in preventing contamination caused by the use of lubricating grease or oils in industrial communities (Norani et al., 2021).

Polyimides, fluoroplastics, polyoxymethylene, and poly aryl ether ketones are examples of self-lubricating polymers that are widely used in industry (Aher et al., 2020). However, their properties may not always be adequate for satisfying the increasing technological prerequisites for tribological materials. The use of different additive particles and fibres, as well as nanoparticles, is the main technique in improving the tribological and mechanical characteristics of polymers. Polymers and PCM are more convenient to use in friction units than metals and ceramics because they can continue operating in sliding constraints without requiring dry friction or lubricants. It is especially important when lubricants are unable to be employed for fear of contamination of the unit or when the unit cannot be serviced. Some of the primary advantages and disadvantages of polymers considering their tribological properties are mentioned in Table 8.1.

TABLE 8.1
Advantages and Disadvantages of Polymers Based on Tribological Properties

Properties	Advantages	Disadvantages
Coefficient of friction	Low	–
Thermal conductivity	–	Low
Stiffness	–	Low
Corrosion resistance	High	–
Surface energy	Low	–
Seizure resistance	High	–
Strength	–	Low

8.9 PARAMETERS AFFECTING THE TRIBOLOGICAL PROPERTIES OF POLYMERS

The factors that influence the state of the polymer's molecular structure determine its tribological properties, some of which are mentioned below:

8.9.1 STRUCTURE OF THE POLYMER

- Branching
- Cross-linking
- Direction of the molecules
- Crystallinity
- Chemical Affinity

8.9.2 VISCOELASTICITY

It is the state of a material that has undergone deformation and exhibits both characteristics of Liquid (viscosity) and Solid (elasticity).

Elastic materials deform in direct proportion to the pressure (stress) placed on them. Strain builds immediately when the tension is applied since this linear relation is time-independent. Under a certain amount of strain, as well as an elasticity limit, metals and ceramics exhibit elastic behaviour (Georgescu et al., 2014).

Viscous substances (liquids) withstand rapid deformation when under load. A viscous substance flows, altering its physical characteristics for a while after changing the load, compared to an elastic material, which recovers to its previous shape instantly once the force is removed (time-dependent process) (Parameswaranpillai et al., 2023).

8.9.3 TRANSFER FILM

Polymer particles stick to the metal's countersurface and create a transfer film when the polymer substance rubs against a tougher material (like metal). The interaction between the polymeric-metal adhesive work and cohesive work governs the development of the transfer film (Ray et al., 2023).

Transferred polymer reduces the mating surface's roughness by filling in its micro defects, which lowers its coefficient of wear and friction. As the transfer film is created, the polymer component rubs against the metal counterpart's polymer film rather than the metal itself. The value of the coefficient of resistance with the polymer transfer sheet is also reduced due to the reduced surface energy of polymers (Sarath et al., 2023).

8.9.4 POLYMER WEAR

- **Adhesive Wear:** This is a consequence of binding between the opposite imperfections on the rubbing faces of the counter bodies, resulting in micro-junctions. Because of the significant stress placed on the interacting asperities, adhesion connections are created when they distort and adhere to one another. The micro-joints burst as a result of the action of the friction between the counter bodies. Micro-asperities in the bonded polymer break off and spread across the counterface (Jain, 2023).

- **Abrasive Wear:** This occurs when the abrasive process of the tougher substance wears down the softer material. Cutting is the primary abrasive wear mode, and during this process, a chip develops in front of the cutting grit. The material is taken away from the worn track's surface. Another form of abrasion operation is ploughing, which, unlike cutting, does not cause wear or material loss. The material is moved to the wear groove's sidewalls during ploughing. Nothing is done to take it off the surface. Two-body wear is the term used to describe wear if there are just two contacting parts engaged in the friction process. It is referred to as three-stage wear if the wear is brought on by a hard material (grit) caught between the contacting surfaces (Ray et al., 2023).
- **Chemical Wear:** This is a rubbing-induced chemical reaction-accelerated wear on the surface of the polymer (Sarath et al., 2023).
- **Fatigue Wear:** During friction, cycle loads result in a material's fatigue wear. Surface-to-subsurface propagation of fatigue cracks occurs in materials. The material parts could separate and delaminate as a result of the fissures connecting. Breaking the chain of molecules in the surface layer of a polymer may also happen from cycling loads given to the material, which lowers its degree of crystallinity (Norani et al., 2021).

8.10 SIGNIFICANCE OF TRIBOLOGICAL PROPERTIES IN ADDITIVE MANUFACTURING

Tribological qualities share a significant impact on part functioning and can determine in case a manufactured part will perform adequately or not in critical profused industries like the aerospace, transport, and automotive industries. Their range extends from the field of tribology to AM. This is particularly true for energy-intensive components, where wear and friction are important to determine how much energy is being consumed in a system (Ralls et al., 2021). Since wear and friction are system traits, one method of improving tribological effectiveness can be to change the system's inherent characteristics, such as the average grain size (Sarath et al., 2023).

Some researchers have shown a correlation between grain size wear and friction since friction and wear are related directly to the mechanical characteristics of a system. Different build directions were investigated which included 0°, 45°, and 90°. The examination revealed that of the three build directions, the 0° build had the most aligned grains, having the lowest mean grain size (Ray et al., 2023). Scientists mostly blame this phenomenon on how the laser's heat energy is dispersed. There was evidence of a reduction in convective radiation across the melt zone as the laser's radial distance (or scan angle) rose, which in turn had a detrimental effect on grain size (Shah et al., 2022). According to Hall-Petch, both tribological and mechanical strength significantly rose as grain size dropped. Given that this condition is well-known and well-established, it is clear that a system's friction and wear characteristics do share an inverse association, which can be very advantageous in a large number of applications (Guenther et al., 2021).

Significant research has been done on alloy development as the usage of metal AM becomes mainstream. The most researched metal AM alloys among the numerous powdered raw materials include steels, in addition to Al-, Ti-, and Ni-based alloys. Typically, Titanium-6 aluminium-4 alloys are used in AM research (Rahman et al., 2021). Following processing, the crystalline structure incorporates an impact on the tensile characteristics of these AM components. For example, SLM Manufactured Ti-6Al-4V could be developed possessing a higher tensile strength compared to parts made conventionally with the use of an optimized microstructure (Ralls et al., 2021).

Due to large part/high strength against weight ratio, durability against corrosion, and simplicity in casting as well as shaping, aluminium alloys are the most frequently employed alloys in manufacturing environments (Türk et al., 2016). As a result, AM is becoming more and more interested in Al alloys. In metal AM, Aluminium {10–12}{Silicon-Magnesium} alloy remains the most widely utilized Aluminium-based alloy. Considering the possibility of extensive use in the medical sector, iron alloys and stainless steel alloys are also utilized. Popular iron-alloy feedstock 316L is an austenitic material of stainless steel that is renowned for its great ductility, corrosion resistance, and strength (Low et al., 2017).

Polymer-based materials demonstrate outstanding self-lubricating characteristics, giving them an edge in tribe-set applications like lip fasteners, bearings, and artificial joints, despite having relatively diminished physical or structural characteristics compared to most metallic and ceramic composites (Alghamdi et al., 2021). In tribological applications, a strongly crosslinked PBM is essential for avoiding loss of material and breakage upon surface contact. However, it has been noted that the influence of printing circumstances on the tribological properties and wearing mechanism of FDM parts presents a new set of difficulties when characterizing both the wear and friction characteristics of these materials (Bryll et al., 2018). It is widely acknowledged that not a single polymer operates best in all tribological situations. Understanding the tribological properties of objects made by AM under various wear regimes is therefore essential (Deleanu et al., 2021).

8.11 EFFECT OF POST-PROCESSING ON TRIBOLOGICAL PROPERTIES

Despite being a promising technique, metal AM has its drawbacks. The surface quality of metal AM components is the most frequently raised issue. The additively made components are vulnerable to surface imperfections because of the layer-after-layer process (Saheb & Kumar, 2020). This issue can be resolved through post-processing.

The layer imperfection can be removed using many surface treatment techniques (Babu et al., 2015). It should be mentioned that the most important disadvantage of AM is the requirement for post-processing (Liu et al., 2022). The time and energy needed may make metal AM inefficient until a widely applicable and accessible approach to post-processing is developed (Herzog et al., 2016).

The final surface finish concerning AM products can also be substantially connected with tribological performance. Researchers provide evidence in support of the effects of surface texturing approaches for tribological optimization. It is particularly true in lubricated environments, in which textured creases might reduce other body wear by trapping any worn debris inside the sacks (Žarko et al., 2017). Given that the textured system experiences some wear over time, the surface characteristics of a part have a significant impact on how long it will last, improve its overall quality, and have long-term sustainable advantages. The exterior surface of the part is crucial for assessing the tribological integrity of manufactured parts about part quality (Wolfs et al., 2019). In general, several surface traits are employed to characterize this. Manufacturers can determine various things related to the created part by analyzing its surface. A part's surface characteristics can reveal the broad tolerance of complicated components and whether they can be accurately joined together (Johnson et al., 2019). Some writers emphasize it because of how crucial it is from an industrial perspective. They clarify this idea by providing examples as follows: Manufacturers of aerospace-related components, for example, blades for turbines, won't be not fly these blades unless they had complete trust in their metrological qualities (Thomas & Gilbert, 2015).

To reduce material waste when clearances are higher, post-processing techniques like near-net-shape manufacturing might be required (Singh & Gill, 2019). In particular fit-for-purpose operations, the optimization of having a precisely fitting component can have significant effects (Saheb & Kumar, 2020). By allowing lubricants and other additives to enter, possessing a proper and precise clearance would make it possible for a reduction in friction in the case that the bearing's heat from the friction causes early part failure. Second, a better surface finish will reduce the need for further post-processing steps for smoothing the surface. This enhances the effectiveness of energy and time used in AM processes. The effectiveness with which the AM process has fused the particles will be determined by a high-quality surface finish (Johnson et al., 2019). The likelihood of porosity and contaminants on the surface will be minimized if the metallic or polymeric particles are successfully melted. The mechanical as well as tribological characteristics of the system will be improved if the quality of the surface is improved since there might be lower chances for any sort of cracks to occur when any outside force is applied (Ralls et al., 2021).

Removal of material is the majorly used subset of the post-processing. The goal of all surface finishing methods is to remove a small layer of material from an uneven surface (Olakanmi et al., 2015). These procedures can be laser-based, mechanical in nature, or chemical. Although chemical treatment methods can be used on both the internal and external surfaces of an AM part, they are advantageous alternatives for post-processing. Etching, polishing, and brightening are common chemical processing techniques. Depending on the required level of substance removal, an AM part is typically submerged in chemical baths of varying temperatures and lengths (Kundu & Mandal, 2022).

Micromachining with lasers is a typical surface preparation (Zeng et al., 2012). The adaptability of the elimination of material centred on the frequency optimization and time of the pulse is the main benefit of laser-based treatments (Liu et al., 2022). Laser polishing, which involves passing a laser over a pre-deposited layer to slowly melt its external layer, is the most viable laser-based technique (Deckers et al., 2014). By using a laser to burn any leftover material after each layer has been placed before adding the next, the technique can also be used during the manufacturing process (Ralls et al., 2021).

8.12   CONCLUSION AND FUTURE SCOPE

The demand for technological advancements in AM has surged as a result of the 4th Industrial Revolution (also called Industry 4.0), which can produce complex parts and prints of excellent quality at just a fraction of the cost. The literature investigating the impact of various process printing parameters, composite materials, structural geometries, post-processing, and enhancement on the tribological characteristics of metal, polymer, and composite is the focus of this paper. Due to the larger contact area and asperities, cracks under particular sliding conditions, 3D-printed polymers have reduced friction and wear. It is found that scratching, asperity recess, ploughing, deformation, and cracking are the mechanisms of wear and deterioration. To accurately assess whether these parts are capable of meeting the precise mechanical specifications for which they are produced, it is crucial to collect and comprehend information regarding the relationship between parameters related to processing and composite substances and the physical and tribological properties of 3D-printed polymer samples and parts. The mechanical characteristics of metal AM parts are very different from those of metal parts that are manufactured conventionally. AM components exhibit decreased ductility; however, they are tougher and more durable compared to cast/wrought ones. According to tribological laboratory testing, some SLM parts have COFs that are more suitable.

Technologies to improve tribological qualities are growing together with the growth in interest in AM. Techniques for post-processing that are based on chemicals, mechanics, and lasers look to be an exciting potential for tribological improvement. Chemical coatings should get extra consideration because metal AM is the best method for creating complicated components. For components with complicated geometry, the method of complete submersion in chemical treatments offers a greater surface quality. Similarly, there is a lot of room for advancement in laser polishing. Surface roughness can be significantly reduced by running a laser across each layer before adding the next. These options present potentials that can be immediately incorporated into AM industrial lines, preventing the need for expensive post-processing.

ACKNOWLEDGEMENTS

The authors gratefully acknowledge the University Centre for Research and Development, Chandigarh University for their assistance and support.

REFERENCES

Aher, V. S., Shirsat, U. M., Wakchaure, V. D., & Venkatesh, M. A. (2020). *An experimental investigation on tribological performance of UHMWPE composite under textured dry sliding conditions. Jurnal Tribologi*, *24*(December 2019), 110–125.

Alghamdi, S. S., John, S., Choudhury, N. R., & Dutta, N. K. (2021). Additive manufacturing of polymer materials: Progress, promise and challenges. *Polymers*, *13*(5), 1–39. https://doi.org/10.3390/polym13050753.

Babu, S. S., Love, L., Dehoff, R., Peter, W., Watkins, T. R., & Pannala, S. (2015). Additive manufacturing of materials: Opportunities and challenges. *MRS Bulletin*, *40*(12), 1154–1161. https://doi.org/10.1557/mrs.2015.234.

Bagsik, A., & Schöppner, V. (2011). Mechanical properties of fused deposition modeling parts manufactured with Ultem*9085. *Annual Technical Conference - ANTEC, Conference Proceedings*, *2*(January 2011), 1294–1298.

Bai, Y., & Williams, C. B. (2018). Binder jetting additive manufacturing with a particle-free metal ink as a binder precursor. *Materials and Design*, *147*, 146–156. https://doi.org/10.1016/j.matdes.2018.03.027.

Behera, M. P., Dougherty, T., & Singamneni, S. (2019). Conventional and additive manufacturing with metal matrix composites: A perspective. *Procedia Manufacturing*, *30*, 159–166. https://doi.org/10.1016/j.promfg.2019.02.023.

Bhavar, V., Kattire, P., Patil, V., Khot, S., Gujar, K., & Singh, R. (2017). A review on powder bed fusion technology of metal additive manufacturing. In: *Additive Manufacturing Handbook: Product Development for the Defense Industry*, Edited by Adedeji, B., Vhance, V., & Liu, D., pp. 251–261. CRC Press (Taylor & Francis Group). https://doi.org/10.1201/9781315119106.

Bose, S., Ke, D., Sahasrabudhe, H., & Bandyopadhyay, A. (2018a). Additive manufacturing of biomaterials. *Progress in Materials Science*, *93*, 45–111. https://doi.org/10.1016/j.pmatsci.2017.08.003.

Bose, S., Robertson, S. F., & Bandyopadhyay, A. (2018b). Surface modification of biomaterials and biomedical devices using additive manufacturing. *Acta Biomaterialia*, *66*, 6–22. https://doi.org/10.1016/j.actbio.2017.11.003.

Bryll, K., Piesowicz, E., Szymański, P., Slaczka, W., & Pijanowski, M. (2018). Polymer composite manufacturing by FDM 3D printing technology. *MATEC Web of Conferences*, *237*, 0–6. https://doi.org/10.1051/matecconf/201823702006.

Christiyan, K. G. J., Chandrasekhar, U., & Venkateswarlu, K. (2016). A study on the influence of process parameters on the mechanical properties of 3D printed ABS composite. *IOP Conference Series: Materials Science and Engineering*, *114*(1). https://doi.org/10.1088/1757-899X/114/1/012109.

Deckers, J., Vleugels, J., & Kruth, J.-P. (2014). Additive manufacturing of ceramics: A review author names and affiliations Jan Deckers*. *Additive Manufacturing of Ceramics: A Review Author*, *5*(4), 1–51.

Deleanu, L., Botan, M., & Georgescu, C. (2021). Tribological behavior of polymers and polymer composites. In: *Tribology in Materials and Manufacturing - Wear, Friction and Lubrication*, December, Edited by Patnaik, A., Singh, T., & Kukshal, V. Intech Open. https://doi.org/10.5772/intechopen.94264.

Field, S. K., Jarratt, M., & Teer, D. G. (2004). *Tribological properties of graphite-like and diamond-like carbon coatings. Tribology International*, *37*(11–12 SPEC.ISS.), 949–956. https://doi.org/10.1016/j.triboint.2004.07.012.

Finnes, T., & Letcher, T. (2015). High definition 3D printing-comparing SLA and FDM printing technologies. *The Journal of Undergraduate Research*, *13*, 3. https://openprairie.sdstate.edu/jurhttps://openprairie.sdstate.edu/jur/vol13/iss1/3HIGHDEFINITION3DPRINTING.

Galante, R., Figueiredo-Pina, C. G., & Serro, A. P. (2019). Additive manufacturing of ceramics for dental applications: A review. *Dental Materials*, *35*(6), 825–846. https://doi.org/10.1016/j.dental.2019.02.026.

Gaytan, S. M., Cadena, M. A., Karim, H., Delfin, D., Lin, Y., Espalin, D., MacDonald, E., & Wicker, R. B. (2015). Fabrication of barium titanate by binder jetting additive manufacturing technology. *Ceramics International*, *41*(5), 6610–6619. https://doi.org/10.1016/j.ceramint.2015.01.108.

Georgescu, C., Botan, M., & Deleanu, L. (2014). Influence of adding materials in PBT on tribological behaviour. *Materiale Plastice*, *51*(4), 351–354.

Gonzalez, J. A., Mireles, J., Lin, Y., & Wicker, R. B. (2016). Characterization of ceramic components fabricated using binder jetting additive manufacturing technology. *Ceramics International*, *42*(9), 10559–10564. https://doi.org/10.1016/j.ceramint.2016.03.079.

Guenther, E., Kahlert, M., Vollmer, M., Niendorf, T., & Greiner, C. (2021). Tribological performance of additively manufactured aisi h13 steel in different surface conditions. *Materials*, *14*(4), 1–10. https://doi.org/10.3390/ma14040928.

Guo, Y., Peters, J., Oomen, T., & Mishra, S. (2018). Control-oriented models for ink-jet 3D printing. *Mechatronics*, *56*(April), 211–219. https://doi.org/10.1016/j.mechatronics.2018.04.002.

Herzog, D., Seyda, V., Wycisk, E., & Emmelmann, C. (2016). Additive manufacturing of metals. *Acta Materialia*, *117*, 371–392. https://doi.org/10.1016/j.actamat.2016.07.019.

Holmes, M. (2019). Additive manufacturing continues composites market growth. *Reinforced Plastics*, *63*(6), 296–301. https://doi.org/10.1016/j.repl.2018.12.070.

Jain, A. (2023). Tribology of carbon nanotubes/polymer nanocomposites. In: *Tribology of Polymers, Polymer Composites, and Polymer Nanocomposites*, Edited by George, S. C., Haponiuk, J. T., Thomas, S., Reghunath, R., & Sarath, P. S. Elsevier Inc. https://doi.org/10.1016/b978-0-323-90748-4.00003-0.

Johnson, L., Mahmoudi, M., Zhang, B., Seede, R., Huang, X., Maier, J. T., Maier, H. J., Karaman, I., Elwany, A., & Arróyave, R. (2019). Assessing printability maps in additive manufacturing of metal alloys. *Acta Materialia*, *176*, 199–210. https://doi.org/10.1016/j.actamat.2019.07.005.

Kumar, S., Hofmann, M., Steinmann, B., Foster, E. J., & Weder, C. (2012). Reinforcement of stereolithographic resins for rapid prototyping with cellulose nanocrystals. *ACS Applied Materials and Interfaces*, *4*(10), 5399–5407. https://doi.org/10.1021/am301321v.

Kundu, M., & Mandal, A. (2022). Additive manufacturing process (3D printing): "A critical review of techniques, applications & future scope." *International Journal of Engineering and Technical Research*, *11*(10), 171–193.

Lee, M. P., Cooper, G. J. T., Hinkley, T., Gibson, G. M., Padgett, M. J., & Cronin, L. (2015). Development of a 3D printer using scanning projection stereolithography. *Scientific Reports*, *5*. https://doi.org/10.1038/srep09875.

Liu, Z., Zhao, D., Wang, P., Yan, M., Yang, C., Chen, Z., Lu, J., & Lu, Z. (2022). Additive manufacturing of metals: Microstructure evolution and multistage control. *Journal of Materials Science and Technology*, *100*, 224–236. https://doi.org/10.1016/j.jmst.2021.06.011.

Low, Z. X., Chua, Y. T., Ray, B. M., Mattia, D., Metcalfe, I. S., & Patterson, D. A. (2017). Perspective on 3D printing of separation membranes and comparison to related unconventional fabrication techniques. *Journal of Membrane Science*, *523*(October 2016), 596–613. https://doi.org/10.1016/j.memsci.2016.10.006.

Moritz, T., & Maleksaeedi, S. (2018). Ceramic components. In: *Additive Manufacturing: Materials, Processes, Quantifications and Applications*, Edited by Zhang, j., and Jung, Y.-G., PP. 105–161. Elsevier Inc. https://doi.org/10.1016/B978-0-12-812155-9.00004-9.

Noguera, R., Lejeune, M., & Chartier, T. (2005). 3D fine scale ceramic components formed by ink-jet prototyping process. *Journal of the European Ceramic Society*, *25*(12 SPEC. ISS.), 2055–2059. https://doi.org/10.1016/j.jeurceramsoc.2005.03.223.

Norani, M. N. M., Abdullah, M. I. H. C., Abdollah, M. F. B., Amiruddin, H., Ramli, F. R., & Tamaldin, N. (2021). Mechanical and tribological properties of fff 3d-printed polymers: A brief review. *Jurnal Tribologi*, *29*(March), 11–30.

Olakanmi, E. O., Cochrane, R. F., & Dalgarno, K. W. (2015). A review on selective laser sintering/melting (SLS/SLM) of aluminium alloy powders: Processing, microstructure, and properties. *Progress in Materials Science*, *74*, 401–477. https://doi.org/10.1016/j.pmatsci.2015.03.002.

Parameswaranpillai, J., Jacob, J., Vijayan, S., Midhun Dominic, C. D., Muthukumar, C., Thiagamani, S. M. K., Krishnasamy, S., Salim, N. V., & Hameed, N. (2023). Tribological behavior of natural fiber-reinforced polymeric composites. In: *Tribology of Polymers,*

Polymer Composites, and Polymer Nanocomposites. Edited by George, S. C., Haponiuk, J. T., Thomas, S., Reghunath, R., & Sarath, P. S., pp. 153–171. Elsevier Inc. https://doi.org/10.1016/b978-0-323-90748-4.00014-5.

Popescu, D., Zapciu, A., Amza, C., Baciu, F., & Marinescu, R. (2018). FDM process parameters influence over the mechanical properties of polymer specimens: A review. *Polymer Testing*, *69*(April), 157–166. https://doi.org/10.1016/j.polymertesting.2018.05.020.

Rahman, M. S., Miah, Y., & Hasan, S. (2021). Additive manufacturing using metal 3D printers. *Iarjset*, *8*(2). https://doi.org/10.17148/iarjset.2021.8217.

Ralls, A. M., Kumar, P., & Menezes, P. L. (2021). Tribological properties of additive manufactured materials for energy applications: A review. *Processes*, *9*(1), 1–33. https://doi.org/10.3390/pr9010031.

Ramasawmy, H., & Blunt, L. (2004). Effect of EDM process parameters on 3D surface topography. *Journal of Materials Processing Technology*, *148*(2), 155–164. https://doi.org/10.1016/S0924-0136(03)00652-6.

Ramezani Dana, H., Barbe, F., Delbreilh, L., Azzouna, M. Ben, Guillet, A., & Breteau, T. (2019). Polymer additive manufacturing of ABS structure: Influence of printing direction on mechanical properties. *Journal of Manufacturing Processes*, *44*(May), 288–298. https://doi.org/10.1016/j.jmapro.2019.06.015.

Ranjan, N., Tyagi, R., Kumar, R., & Kumar, V. (2023). On fabrication of acrylonitrile butadiene styrene-zirconium oxide composite feedstock for 3D printing-based rapid tooling applications. *Journal of Thermoplastic Composite Materials*. In Press. https://doi.org/10.1177/08927057231186310.

Ray, S. K., Banerjee, A., Bhangi, B. K., Pyne, D., & Dutta, B. (2023). Tribological analysis-general test standards. In: *Tribology of Polymers, Polymer Composites, and Polymer Nanocomposites*, Edited by George, S. C., Haponiuk, J. T., Thomas, S., Reghunath, R., & Sarath, P. S., pp. 17–50. Elsevier Inc. https://doi.org/10.1016/b978-0-323-90748-4.00001-7.

Saheb, S. H., & Kumar, J. V. (2020). A comprehensive review on additive manufacturing applications. *AIP Conference Proceedings*, *2281*(April 2019). https://doi.org/10.1063/5.0026202.

Sano, Y., Matsuzaki, R., Ueda, M., Todoroki, A., & Hirano, Y. (2018). 3D printing of discontinuous and continuous fibre composites using stereolithography. *Additive Manufacturing*, *24*(October), 521–527. https://doi.org/10.1016/j.addma.2018.10.033.

Sarath, P. S., Reghunath, R., Haponiuk, J. T., Thomas, S., & George, S. C. (2023). Introduction: A journey to the tribological behavior of polymeric materials. In: *Tribology of Polymers, Polymer Composites, and Polymer Nanocomposites*, Edited by George, S. C., Haponiuk, J. T., Thomas, S., Reghunath, R., & Sarath, P. S., pp. 1–16. Elsevier Inc. https://doi.org/10.1016/b978-0-323-90748-4.00010-8.

Shah, R., Pai, N., Rosenkranz, A., Shirvani, K., & Marian, M. (2022). Tribological behavior of additively manufactured metal components. *Journal of Manufacturing and Materials Processing*, *6*(6), 138, 1–15. https://doi.org/10.3390/jmmp6060138.

Shahrubudin, N., Lee, T. C., & Ramlan, R. (2019). An overview on 3D printing technology: Technological, materials, and applications. *Procedia Manufacturing*, *35*, 1286–1296. https://doi.org/10.1016/j.promfg.2019.06.089.

Simchi, A. (2006). Direct laser sintering of metal powders: Mechanism, kinetics and microstructural features. *Materials Science and Engineering A*, *428*(1–2), 148–158. https://doi.org/10.1016/j.msea.2006.04.117.

Singh, H. (2020). A comprehensive study on 3D prnting. *Journal of Emerging Technologies and Innovative Research (JETIR)*, *7*(6), 151–160.

Singh, L., & Gill, H. (2019). Performance parameters in fdm 3d printed parts. *International Journal of Innovative Technology and Exploring Engineering*, *9*(1), 367–370. https://doi.org/10.35940/ijitee.A4130.119119.

Thomas, D. S., & Gilbert, S. W. (2015). Costs and cost effectiveness of additive manufacturing: A literature review and discussion. In: *Additive Manufacturing: Costs, Cost Effectiveness and Industry Economics*, Edited by Thomas, D. S., & Gilbert, S. W., pp. 1–96. NIST Special Publication.

Türk, D. A., Kussmaul, R., Zogg, M., Klahn, C., Spierings, A., Ermanni, P., & Meboldt, M. (2016). Additive manufacturing with composites for integrated aircraft structures. *International SAMPE Technical Conference*, 2016, Janua.

Tyagi, R., Singh, G., Kumar, R., Kumar, V., & Singh, S. (2023). 3D-printed sandwiched acrylonitrile butadiene styrene/carbon fiber composites: Investigating mechanical, morphological, and fractural properties. *Journal of Materials Engineering and Performance*, 1–4. In press. https://doi.org/10.1007/s11665-023-08292-8

Wakabayashi, T., Suda, S., Inasaki, I., Terasaka, K., Musha, Y., & Toda, Y. (2007). Tribological action and cutting performance of MQL media in machining of aluminum. *CIRP Annals - Manufacturing Technology*, *56*(1), 97–100. https://doi.org/10.1016/j.cirp.2007.05.025.

Wang, Y., Wang, L., Li, J., Chen, J., & Xue, Q. (2013). Tribological properties of graphite-like carbon coatings coupling with different metals in ambient air and water. *Tribology International*, *60*, 147–155. https://doi.org/10.1016/j.triboint.2012.11.014.

Weng, Z., Zhou, Y., Lin, W., Senthil, T., & Wu, L. (2016). Structure-property relationship of nano enhanced stereolithography resin for desktop SLA 3D printer. *Composites Part A: Applied Science and Manufacturing*, *88*, 234–242. https://doi.org/10.1016/j.compositesa.2016.05.035.

Wolfs, R. J. M., Bos, F. P., & Salet, T. A. M. (2019). Hardened properties of 3D printed concrete: The influence of process parameters on interlayer adhesion. *Cement and Concrete Research*, *119*(February), 132–140. https://doi.org/10.1016/j.cemconres.2019.02.017.

Wong, K. V., & Hernandez, A. (2012). A review of additive manufacturing. *ISRN Mechanical Engineering*, *2012*, 1–10. https://doi.org/10.5402/2012/208760.

Wu, B., Pan, Z., Ding, D., Cuiuri, D., Li, H., Xu, J., & Norrish, J. (2018). A review of the wire arc additive manufacturing of metals: Properties, defects and quality improvement. *Journal of Manufacturing Processes*, *35*(August), 127–139. https://doi.org/10.1016/j.jmapro.2018.08.001.

Yakout, M., & Elbestawi, M. A. (2017). Additive manufacturing of composite materials: An overview. *6th International Conference on Virtual Machining Process Technology (VMPT), Montréal, May*, pp. 1–8. https://www.researchgate.net/profile/Mostafa-Yakout/publication/316688880_Additive_Manufacturing_of_Composite_Materials_An_Overview/links/590c9667458515978182e951/Additive-Manufacturing-of-Composite-Materials-An-Overview.pdf.

Zadpoor, A. A., & Malda, J. (2017). Additive manufacturing of biomaterials, tissues, and organs. *Annals of Biomedical Engineering*, *45*(1), 1–11. https://doi.org/10.1007/s10439-016-1719-y.

Žarko, J., Vladić, G., Pál, M., & Dedijer, S. (2017). Influence of printing speed on production of embossing tools using FDM 3d printing technology. *Journal of Graphic Engineering and Design*, *8*(1), 19–27. https://doi.org/10.24867/jged-2017-1-019.

Zeng, K., Pal, D., & Stucker, B. (2012). A review of thermal analysis methods in laser sintering and selective laser melting. *23rd Annual International Solid Freeform Fabrication Symposium - An Additive Manufacturing Conference, SFF 2012*, pp. 796–814.

Zhang, H. W., Zhang, Z., & Chen, J. T. (2007). 3D modeling of material flow in friction stir welding under different process parameters. *Journal of Materials Processing Technology*, *183*(1), 62–70. https://doi.org/10.1016/j.jmatprotec.2006.09.027.

9 Effect of the Tribological Properties on Structural Applications of 3D-Printed Thermoplastic Composites

Vinay Kumar
Chandigarh University

9.1 INTRODUCTION

In the past two decades, a significant change has been observed in the need of conventional industrial materials due to increase in the role of thermoplastics for various consumer applications (Mahieux and Reifsnider, 2002). This may be due to the availability of durable and cost-effective polymer-based products such as channels, pipes, tubes, and electrical goods for functional applications (Wu et al., 2021). Some review studies reported by the researchers have outlined the novel advantages of thermoplastic composite welding in aerospace engineering, naval transportation, and automobile applications (Costa et al., 2012). The increasing number of applications of thermoplastic polymers has also raised the environmental issue that may be answered very effectively by adopting the suitable methods of polymer recycling and reusing. This may be implemented by developing innovative processes and recycled materials with value-added properties for novel industrial applications (Grigore, 2017). The studies reported on recycling of one of the highly used commercial plastics like polyethylene, and acrylonitrile butadiene styrene (ABS) thermoplastic for the fabrication of smart thermoplastic composites (for 4D printing applications) have outlined the innovative industrial applications of thermoplastic composite materials for modern manufacturing processes like additive manufacturing (AM) as a sustainable solution (Kumar et al., 2021; Kumar et al., 2022b). Based on Scopus database, it was observed that a total of 3,705 research publications have been reported on thermoplastic polymers and their composites for various structural and non-structural engineering applications. Figure 9.1 shows the increase in number of articles reported on thermoplastic polymers and their composites for various structural and non-structural engineering applications in the past years since 1972.

The increase in number of published documents per year on thermoplastics and their utilities (as shown in Figure 9.1) indicated that thermoplastic and thermosetting

DOI: 10.1201/9781003400523-9

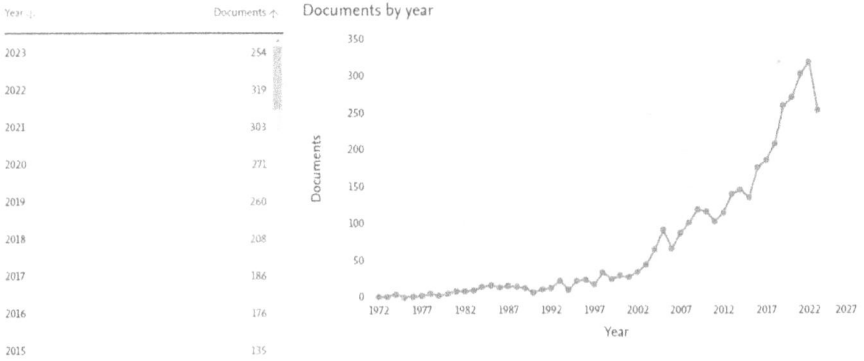

Year ↓	Documents ↑
2023	254
2022	319
2021	303
2020	271
2019	260
2018	208
2017	186
2016	176
2015	135

FIGURE 9.1 Year-wise distribution of number of documents published (since 1972) on thermoplastic polymers and their composites for various engineering applications.

polymers hold a good potential in the manufacturing sector for high-end industrial requirements. Some recent reports have evidently outlined the usefulness of recycled polypropylene-based thermoplastic for cable insulation applications (Huang et al., 2017). Also, some studies have been reported on thermoplastic-based nanocomposites for pressure sensing applications (Coiai et al., 2015). Some researchers have reported the advantages of 3D printing and thermoplastic composites for the fabrication of partially absorbable smart implants (Husain et al., 2023). Polylactic acid (PLA)-based smart composite matrix have also been reported in recent past for biomedical-based scaffolding applications (Singh et al., 2022). Very fine coatings of thermoplastic composites have also been investigated by some researchers for microfluidic applications (Studer et al., 2002). Similar to such surface and coating utilities of polymeric materials and their composites, the research articles published by various publication sources since 1978 are shown in Figure 9.2.

The significant increment in the publication of various research articles (such as reviews, conference articles, and encyclopedia) related to thermoplastics in a variety of sources (as shown in Figure 9.2) highlights the turbulence in this research area for the development of innovative products for consumer applications. The introduction of polymeric materials at large scale in 3D printing based manufacturing practices (in past two decades) has been observed as a boost factor that pushed the role of composites for very large industrial, mechanical, production, and structural engineering applications (Hofmann, 2014). The recent studies reported on advancements in 3D printing of polymer composites matrix highlighted the existence of variety of materials that may be used for 3D printing based rapid prototyping for metal casting, welding, and manufacturing finished/final products (Valino et al., 2019; Singh et al., 2017). Based on such review of literature, some investigators have reported the rapid tooling applications of thermoplastic composite materials. The product so prepared using rapid tooling may be customized for different implant fabrication purposes (Junk and Tränkle, 2011; Melgoza et al., 2014). Due to wide engineering and biomedical applications, many countries have reported a significant number of studies on polymeric materials. Figure 9.3 shows the documents published under the

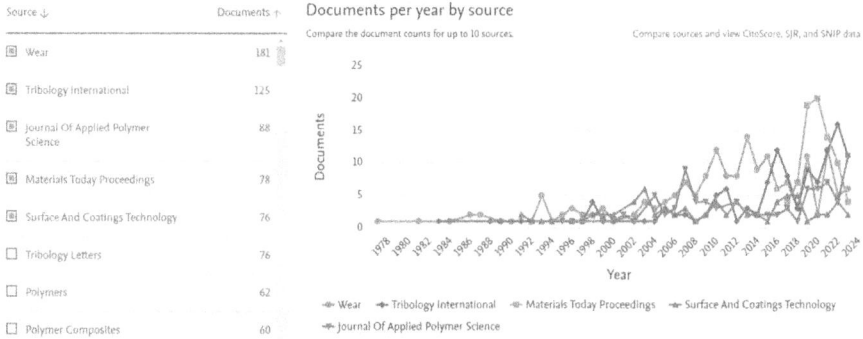

Source ↓	Documents ↑
☑ Wear	181
☑ Tribology International	125
☑ Journal Of Applied Polymer Science	88
☑ Materials Today Proceedings	78
☑ Surface And Coatings Technology	76
☐ Tribology Letters	76
☐ Polymers	62
☐ Polymer Composites	60

FIGURE 9.2 Research publication reported in various publication sources on thermoplastics and their composites since 1978 (for wear, tribological, and other applications).

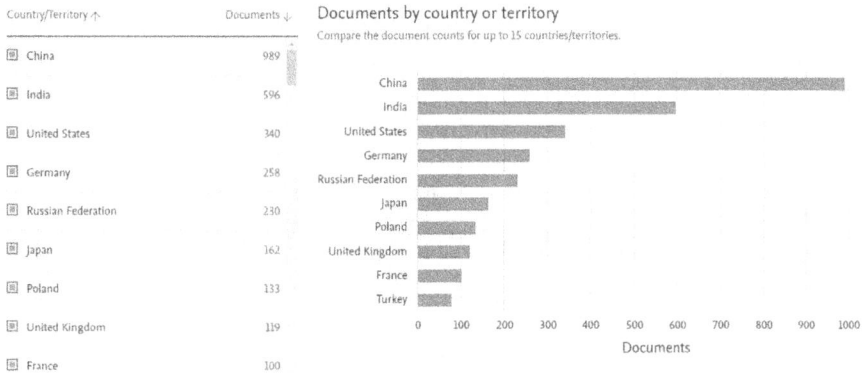

Country/Territory ↑	Documents ↓
☑ China	989
☑ India	596
☑ United States	340
☑ Germany	258
☑ Russian Federation	230
☑ Japan	162
☑ Poland	133
☑ United Kingdom	119
☑ France	100

FIGURE 9.3 Number of research documents published by various countries on thermoplastic and their composites.

affiliations of different countries around the globe in which high number of studies are reported on thermoplastic materials and their composites for conventional and advance engineering applications.

In line with international scenario of development for smart polymer matrix based biomedical implants and scaffolds, some investigations have highlighted the usefulness of PVDF polymer matrix based composite for non-enzymatic glucose sensing applications. The Cu-doped ZnO-reinforced PVDF composite has been reportedly outlined acceptable properties to prepare biomedical sensors at large scale by 3D printing processes like fused deposition modeling (Kumar et al., 2022a). Some researchers have reported the fabrication of PLA nanofibers-reinforced smart scaffolds for fracture repair of human patients, whereas some studies have outlined the fabrication of self-healing thermoplastic composites for non-structural crack repair applications in heritage buildings (Kumar and Singh 2023; Kumar et al., 2022c). Thermoplastic composite matrix of ABS and PLA has been also reported as a novel tool for digital

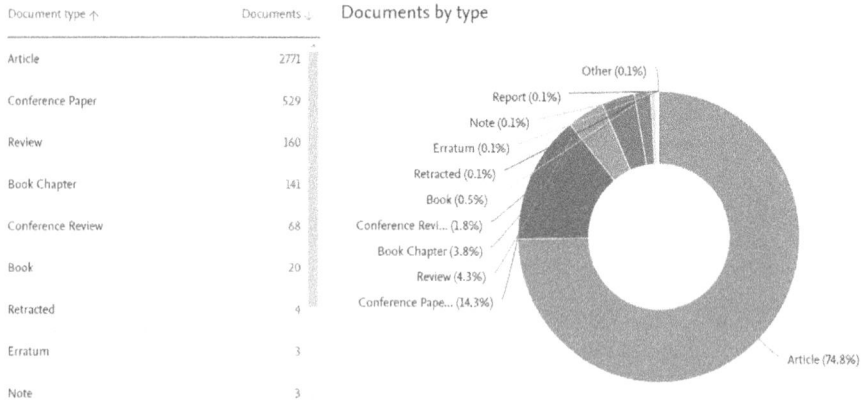

Document type ↑	Documents ↓
Article	2771
Conference Paper	529
Review	160
Book Chapter	141
Conference Review	68
Book	20
Retracted	4
Erratum	3
Note	3

Documents by type

Other (0.1%)
Report (0.1%)
Note (0.1%)
Erratum (0.1%)
Retracted (0.1%)
Book (0.5%)
Conference Revi... (1.8%)
Book Chapter (3.8%)
Review (4.3%)
Conference Pape... (14.3%)
Article (74.8%)

FIGURE 9.4 Distribution of reported literature on thermoplastic composites for structural and non-structural applications in terms of document types.

twinning applications of heritage structures (Kumar et al., 2023b). The reports published on such innovative composite materials indicated that a variety of documents are published on advantages and limitations of thermoplastics in the form of research articles, books, reports, book chapters, etc. Figure 9.4 shows the % distribution of type of documents published in various forms mentioned on thermoplastic composites for different structural and non-structural engineering applications. It may be stated that nearly 75% of the document types published on composites are covered by the research article category followed by conference papers with 14.3% share in overall documents published so far.

The review studies reported on mechanical and tribological characteristics of 3D-printed polymeric materials such as ABS, PLA, PVDF, and nylon have highlighted the role of good wear-resistant 3D-printed components as useful friction materials for rapid prototyping and rapid tooling processes (Norani et al., 2021; Gbadeyan et al., 2021). Some investigations performed on tribological properties specific biomedical materials on the basis of systematic review of dental studies have outlined the machining capabilities of such 3D printable materials in surgical practices (Rudnik et al., 2022; Prause et al., 2022). Besides the reported literature on dental studies, many other subject areas also have been explored by the researchers to utilize the tribological properties of thermoplastic composite materials for consumer needs. Figure 9.5 shows the pie chart for subject area wise distribution of documents reported on thermoplastic and their composites. It was observed that a maximum of studies on thermoplastic composites are reported on materials science subject area followed by engineering-, physics-, and chemistry-based research areas.

The experimental studies performed on fabrication of recycled ABS-based composite for 4D printing outlined that nano graphene powder reinforced ABS composite prepared by tertiary (3°) recycling or chemical-assisted mechanical blending process has good mechanical and bonding properties for 3D/4D printing. Further investigations of wear properties of ABS-graphene composites indicated that uniform blending of reinforcements contributed in better wear properties of the ABS composite

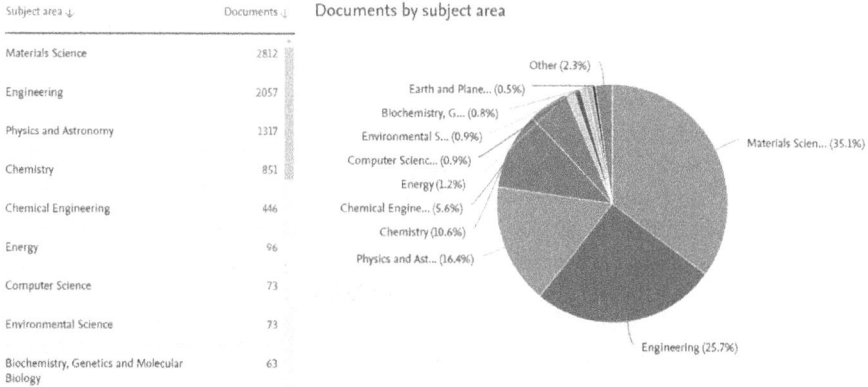

Subject area ↓	Documents ↓
Materials Science	2812
Engineering	2057
Physics and Astronomy	1317
Chemistry	851
Chemical Engineering	446
Energy	96
Computer Science	73
Environmental Science	73
Biochemistry, Genetics and Molecular Biology	63

Documents by subject area

Other (2.3%)
Earth and Plane... (0.5%)
Biochemistry, G... (0.8%)
Environmental S... (0.9%)
Computer Scienc... (0.9%)
Energy (1.2%)
Chemical Engine... (5.6%)
Chemistry (10.6%)
Physics and Ast... (16.4%)
Materials Scien... (35.1%)
Engineering (25.7%)

FIGURE 9.5 Pie chart distribution of research documents published by different research areas on thermoplastic composites for structural applications.

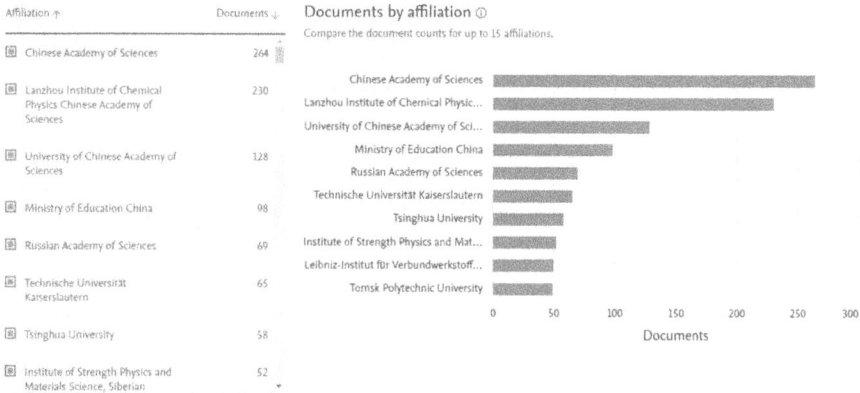

Affiliation ↑	Documents ↓
Chinese Academy of Sciences	264
Lanzhou Institute of Chemical Physics Chinese Academy of Sciences	230
University of Chinese Academy of Sciences	128
Ministry of Education China	98
Russian Academy of Sciences	69
Technische Universität Kaiserslautern	65
Tsinghua University	58
Institute of Strength Physics and Materials Science, Siberian	52

Documents by affiliation ⓘ
Compare the document counts for up to 15 affiliations.

FIGURE 9.6 Number of documents reported by various affiliations on thermoplastic composites for structural applications.

(Kumar et al., 2022d,e). In a separate study, the researchers have outlined that graphene particles contributed very effectively to improve the wear properties of the elastomer-based polymeric composite (Jeon et al., 2020). Such investigations highlighted that various funding agencies affiliated to different research and academic organizations are keen to explore the new possibilities in composites for tribological properties based structural and non-structural applications. Figure 9.6 shows the documents published on tribological properties of composites by various researchers under different affiliations for structural and non-structural engineering applications.

The tribological analysis based on mechanical properties and fretting wear resistance of 3D metal printed functional prototypes has been reported by some researchers, which highlighted the structural applications of commercial metals and alloys for 3D printing based batch production activities (Gao and Zhou, 2018). Wear properties of 3D-printed low-density polyethylene also have been reported for non-structural applications

(Olesik et al., 2019). The wear property investigations performed on Cobalt (Co)-based alloys further paved the way for investigating the tribological properties of thermoplastic composite matrix for structural applications (Enneti and Prough, 2019).

In the past five years, significant studies have been reported by the researchers on utilization of 3D printing technology for structural applications in automobile engineering, friction stir welding, and aerospace engineering (Velu et al., 2019; You et al., 2023). The results of these review and experimental studies indicated a significant potential in thermoplastic composite matrix based materials for innovative structural/non-structural applications, whereas wear properties of the same are least reported by utilizing the 3D printing processes (Ashraf et al., 2018). Some researchers have reported the development of 3D printable smart thermoplastic composites with customizable and programmable properties for maintenance of heritage structures (Kumar et al., 2022f). The solution so developed has also been explored for ascertaining the online health monitoring capabilities for long-term conservation practices along with weathering control properties because weathering defect is highly faced by the heritage structures due to various environmental issues (Kumar et al., 2022g,h). Such studies outline investigation needed to be performed on tribological properties of the smart thermoplastic composites for effective preservation practices of structures that may not only increase the lifespan of heritage buildings but also increase the structural impact of the solutions developed.

9.2   RESEARCH GAP AND PROBLEM FORMULATION

The literature review reveals that very significant studies have been reported on novel industrial applications of thermoplastic polymers and their composites for engineering and biomedical applications along with the integration of modern manufacturing practices like 3D printing (Tay et al., 2017). Also, good number of studies has been performed on investigating the structural and non-structural engineering applications of 3D printable thermoplastic composites like in the area of heritage structures conservation, smart civil structures. In contrary, tribological properties of conventional engineering materials like metals and their alloys along with variety of coatings have been extensively reported by many researchers for rapid prototyping and rapid tooling (Bedarf et al., 2021). The role of smart thermoplastic composites in various structural applications (like in the smart civil structures) has been increased significantly in past one decade. Many researchers have also outlined the advantages of thermoplastic composite matrix based smart solutions for structural as well as non-structural applications of heritage structures. But hitherto very less has been reported on investigating the tribological properties of such smart thermoplastic composite matrices for ascertaining the acceptable wear-resistant properties in the composite solution for desired structural application. The research gap analysis based on 3,705 research publications was performed using VOS viewer software package. The analysis of the reported literature highlighted a total of 8,404 research keywords explored by the researchers for different characteristic studies, research areas, applications, and product/ process developments. By considering the keywords such as thermoplastic composites, tribological properties, structural applications, smart solutions, customizable materials, and 3D printing, the most relevant keywords were processed further to obtain the web of keywords. This web of keywords highlights various highly relevant studies performed on 3D-printed thermoplastic composite matrix for tribological properties based structural applications. Table 9.1 shows the top 50 highly relevant keywords out of 152 threshold keywords studied by more than 60% researchers around the world.

TABLE 9.1
List of Keywords Studied on 3D-Printed Thermoplastic Composite Matrix for Tribological Properties Based Structural Applications

Id	Keywords	Occurrences	Relevance Score
1	Abrasive wear	26	0.7455
2	ABS	11	0.6328
3	Adhesive wear	20	1.1103
4	Aerospace	10	1.8601
5	Agent	20	0.8524
6	Alumina	11	0.8953
7	Article	34	0.3445
8	Ball wear	20	0.6735
9	Bearing	18	1.2567
10	Block	15	1.1323
11	Break load	12	1.3638
12	Carbon nanotube	29	1.3856
13	Case studies	19	0.5957
14	Carbon nanotubes	11	1.5511
15	Coefficient of friction	13	0.7923
16	Comparison	19	0.7672
17	Compatibilizers	10	1.5793
18	Composite material	35	0.4986
19	Contact length	25	1.2672
20	Degrees Celsius	36	0.5259
21	Development of composites	48	0.48
22	Differential scanning calorimetry	10	1.3614
23	Dispersion rate	31	0.521
24	Distance covered	32	0.5177
25	Diameter of track	10	1.3169
26	Dynamic mechanical analysis	11	1.2638
27	Elastic modulus	16	0.495
28	Electron microscope	15	0.6686
29	Elongation	18	1.1985
30	Environment	21	0.6235
31	Ether ketone	18	0.6635
32	Experiment	32	0.4998
33	Experimental investigation	10	1.2775
34	Fabrication	17	0.3613

(Continued)

TABLE 9.1 (*Continued*)
List of Keywords Studied on 3D-Printed Thermoplastic Composite Matrix for Tribological Properties Based Structural Applications

Id	Keywords	Occurrences	Relevance Score
35	Field	31	0.695
36	Flexural strength	23	0.8502
37	Form	16	0.4365
38	Formation	39	0.4976
39	Fourier	11	1.2002
40	Glass	14	0.7823
41	Glass fiber	19	0.6015
42	Glass transition temperature	11	1.0756
43	Graphene	13	2.3201
44	Graphite	22	0.503
45	HDPE/LDPE composites	13	3.1466
46	High-density polyethylene	16	2.7714
47	High temperature	14	0.8559
48	Hybrid composite	17	0.6598
49	Impact	17	0.4192
50	Impact strength	23	1.0266

Based on the mentioned keywords (like shown in Table 9.1), Figure 9.7 shows the web of keywords which indicated that good studies have been reported on development of composite materials and their characterization for better tribological properties and 3D printing applications. Further analyses of keywords composite materials, wear mechanisms, and wear tests as independent nodal points shown in Figures 9.8, 9.9, and −9.10 respectively outlined that, although a good number of investigations have been performed on thermoplastic composites for variety of engineering applications, very less is performed on investigations of tribological properties for 3D printing integrated structural applications of thermoplastic composites. The partially hidden texts in Figures 9. 8−and 9.10 reflect the gap in the previous studies.

The reports of utilizing an appropriate strategy for 3D printing of thermosets for tribological applications and development of multi-material 3D-printed composites for non-structural applications indicated that smart thermoplastic composites matrix may be explored for tribological properties for 3D printing based structural applications (Lei et al., 2019; Kumar et al., 2023a). Therefore, as an extension to the previously reported studies, this chapter highlights the experimental investigations performed on tribological properties (based on wear resistance, friction force, and coefficient of friction) of smart thermoplastic composite of PVDF, i.e., smart programmable PVDF-graphene-Mn doped ZnO (PVDF-G-MZ) and PVDF-calcium carbonate ($CaCO_3$) (PVDF-CC).

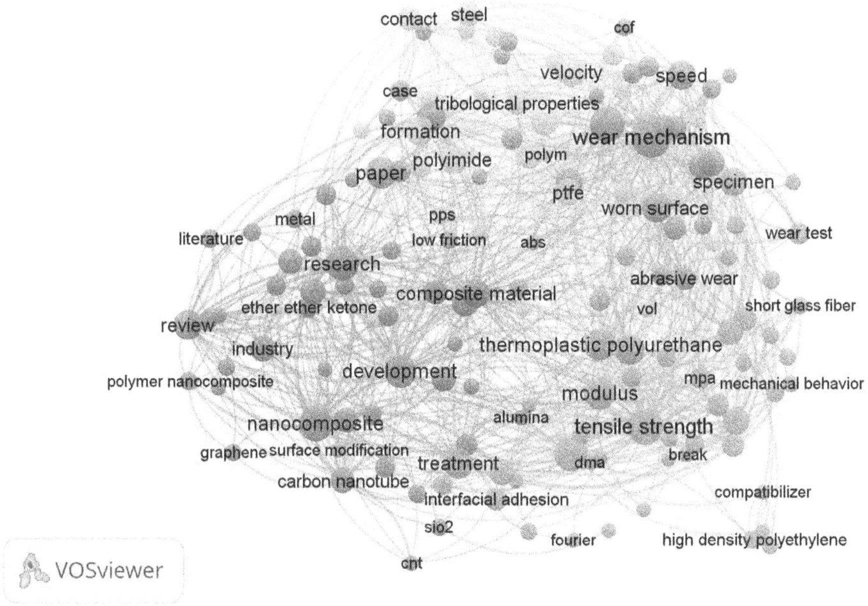

FIGURE 9.7   Web of keywords for studies reported on thermoplastic composites for tribological properties based engineering applications.

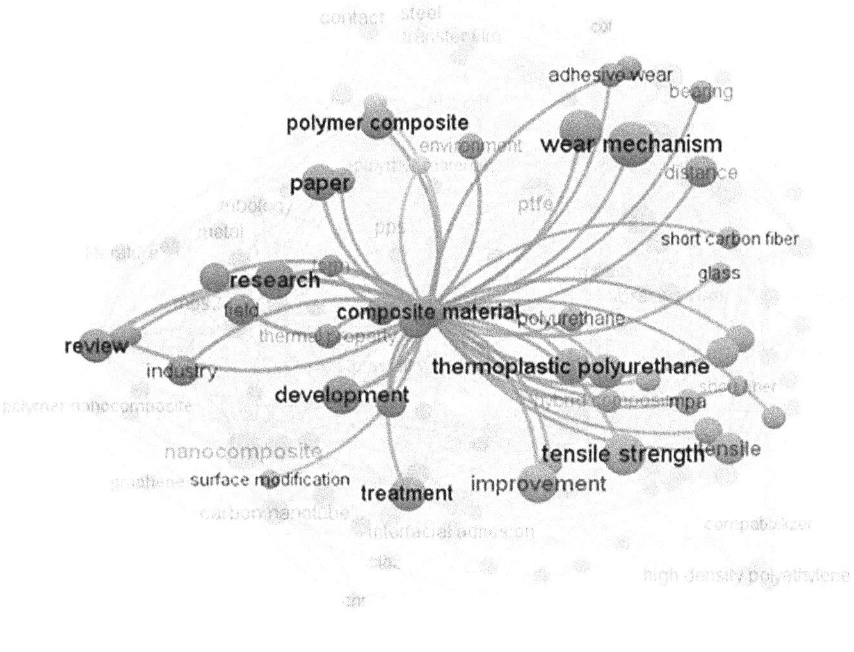

FIGURE 9.8   Studies related to composite materials based keywords as nodal point.

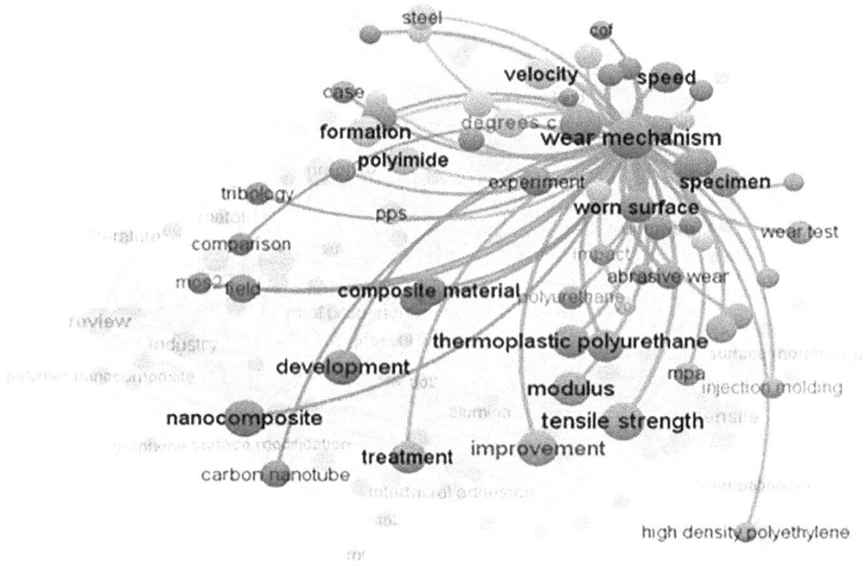

FIGURE 9.9   Studies related to wear mechanism based keywords as nodal point.

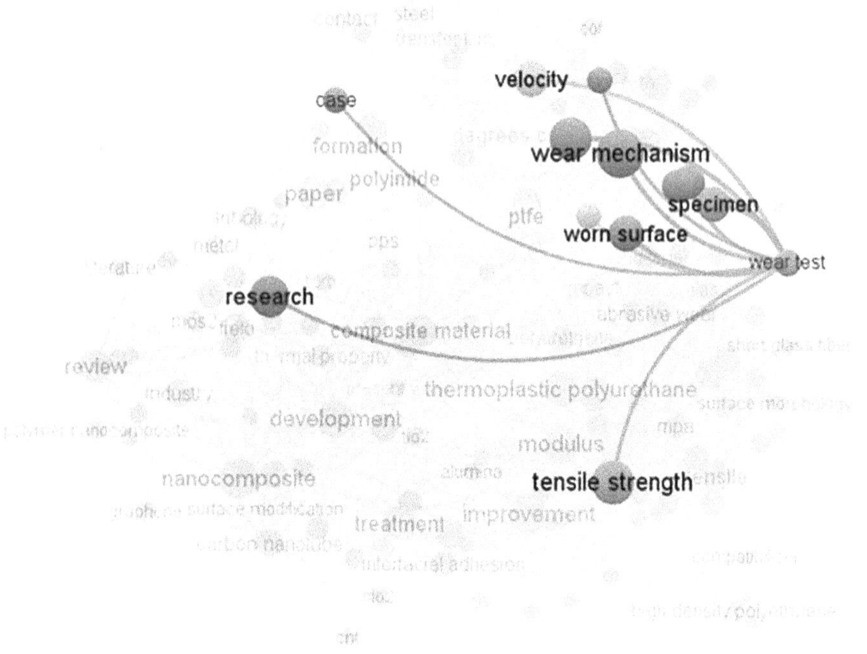

FIGURE 9.10   Studies related to wear testing based keywords as nodal point.

9.3 EXPERIMENTATION

The PVDF-G-MZ and PVDF-CC composites were prepared by mechanical blending process. The selected proportion of reinforcements such as graphene, Mn-doped ZnO, and $CaCO_3$ was obtained as per the reported literature (Kumar et al., 2023a). Therefore, PVDF-6%graphene-3%Mn doped ZnO and PVDF-6%$CaCO_3$ composites were prepared, and feedstock filaments were prepared using twin screw extruder at 220°C extrusion temperature and 120 rpm screw extrusion speed. The filaments of PVDF-G-MZ and PVDF-CC composites so prepared were used to obtain 3D-printed specimens of the composites for tribological analysis (based on pin-on-disk testing). The 8 mm diameter and 30 mm long pins of PVDF-G-MZ and PVDF-CC composites were 3D printed using fused filament fabrication (by keeping 220°C nozzle temperature, 80°C bed temperature, 40 mm/s printing speed, and rectilinear infill pattern). Three pin samples of both PVDF-G-MZ and PVDF-CC composites were prepared with varying infill densities, i.e., 100%, 75%, and 50%. The wear rate and friction force based tribological properties were investigated by wearing the 3D-printed composite pin samples against 220 grit size fine abrasive paper. The pin samples of varying infill densities were rubbed against the abrasive paper on pin-on-disk setup by considering the wear velocity of 1 m/s. Therefore, the track diameter was selected 80 mm for testing the pins along with 100 rpm disk rotation speed under 20 N applied dead weight for 300 s time duration. Further, the morphological analysis of 3D-printed worn out surfaces was also performed to investigate the effect of varying infill densities on tribological properties of PVDF-G-MZ and PVDF-CC composites for structural applications in repair and maintenance of heritage structures.

9.4 RESULTS AND DISCUSSION

The results obtained for wear rate and friction force exerted by each 3D-printed pin sample of the prepared PVDF-G-MZ and PVDF-CC composites were analyzed to ascertain the best acceptable printing conditions in terms of infill density to impart more wear resistance to the 3D-printed composite structure for desired applications. The wear and friction force observations for PVDF-G-MZ pin sample with 100% infill density are shown in Figure 9.11. Similarly, Figure 9.12 shows the observations for wear of PVDF-G-MZ sample with 75% infill density and Figure 9.13 shows the observations for wear of PVDF-G-MZ sample with 50% infill density.

The wear and friction force observations for PVDF-CC pin sample with 100% infill density are shown in Figure 9.14. Similarly, Figure 9.15 shows the observations for wear of PVDF-CC sample with 75% infill density and Figure 9.16 shows the observations for wear of PVDF-CC sample with 50% infill density.

The results obtained for wear of PVDF-G-MZ and PVDF-CC samples outlined that wear rate of the composite increased with decrease in infill density. For 100% infill density, the minimum wear of 304 μm was observed in PVDF-G-MZ and 175 μm wear was observed in PVDF-CC sample. In contrast to morphological properties,

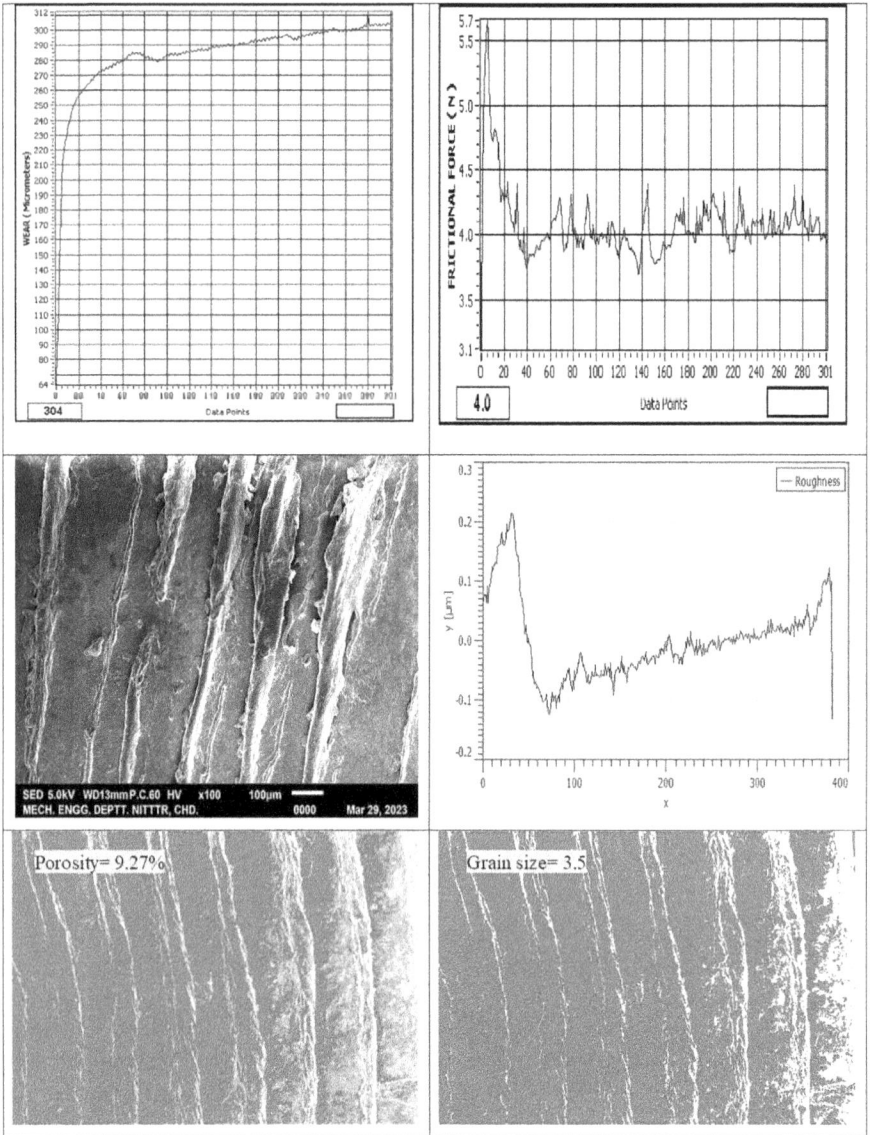

FIGURE 9.11 Tribological and morphological observations for PVDF-G-MZ pin sample with 100% infill density.

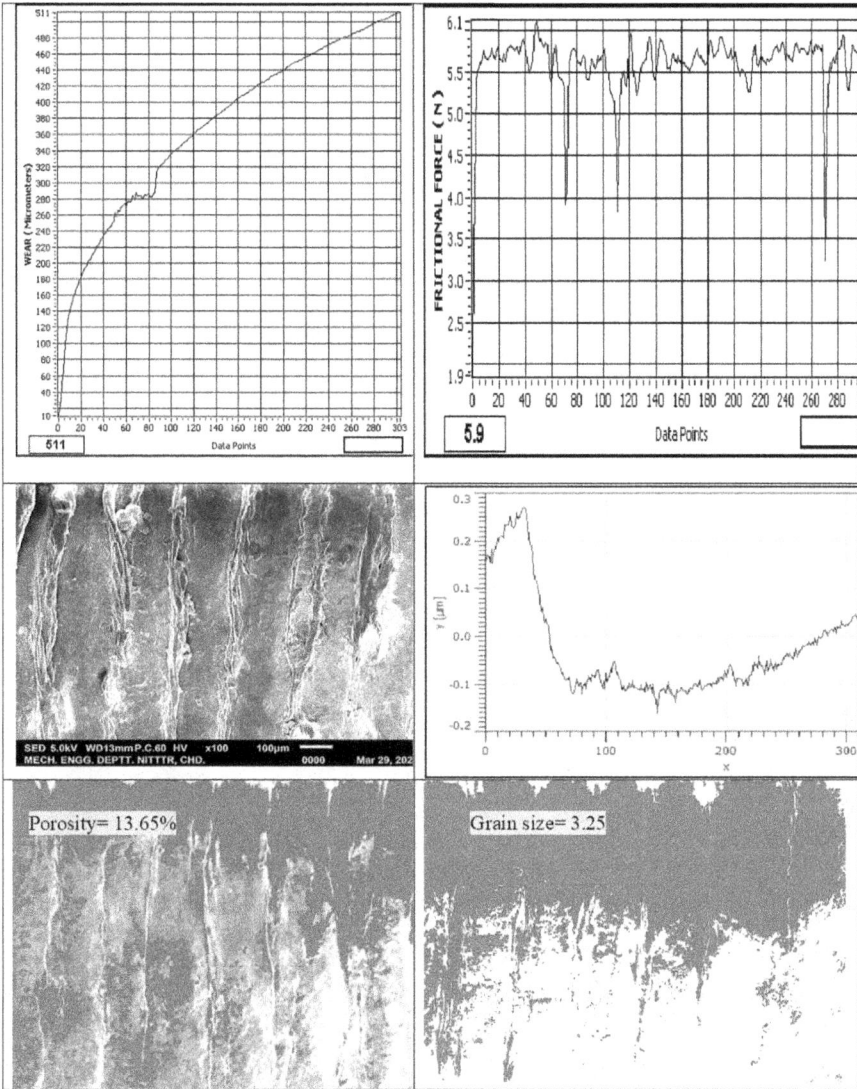

FIGURE 9.12 Tribological and morphological observations for PVDF-G-MZ pin sample with 75% infill density.

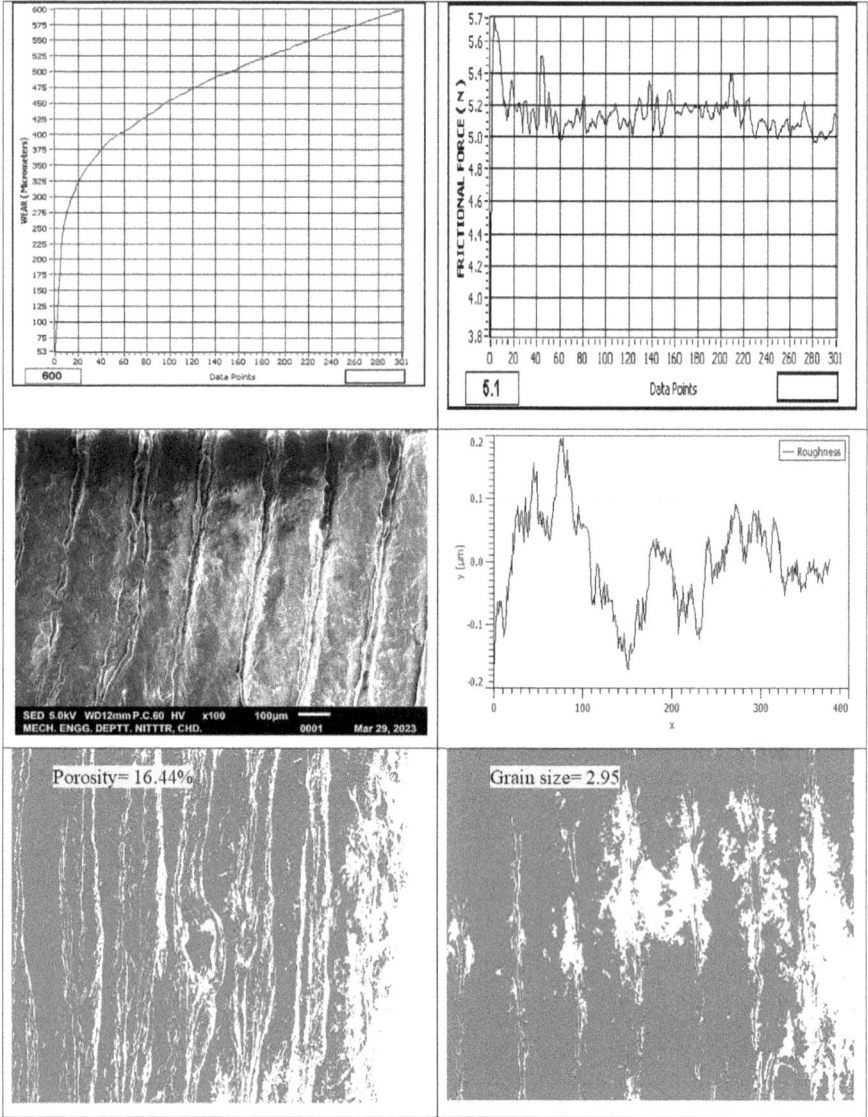

FIGURE 9.13 Tribological and morphological observations for PVDF-G-MZ pin sample with 50% infill density.

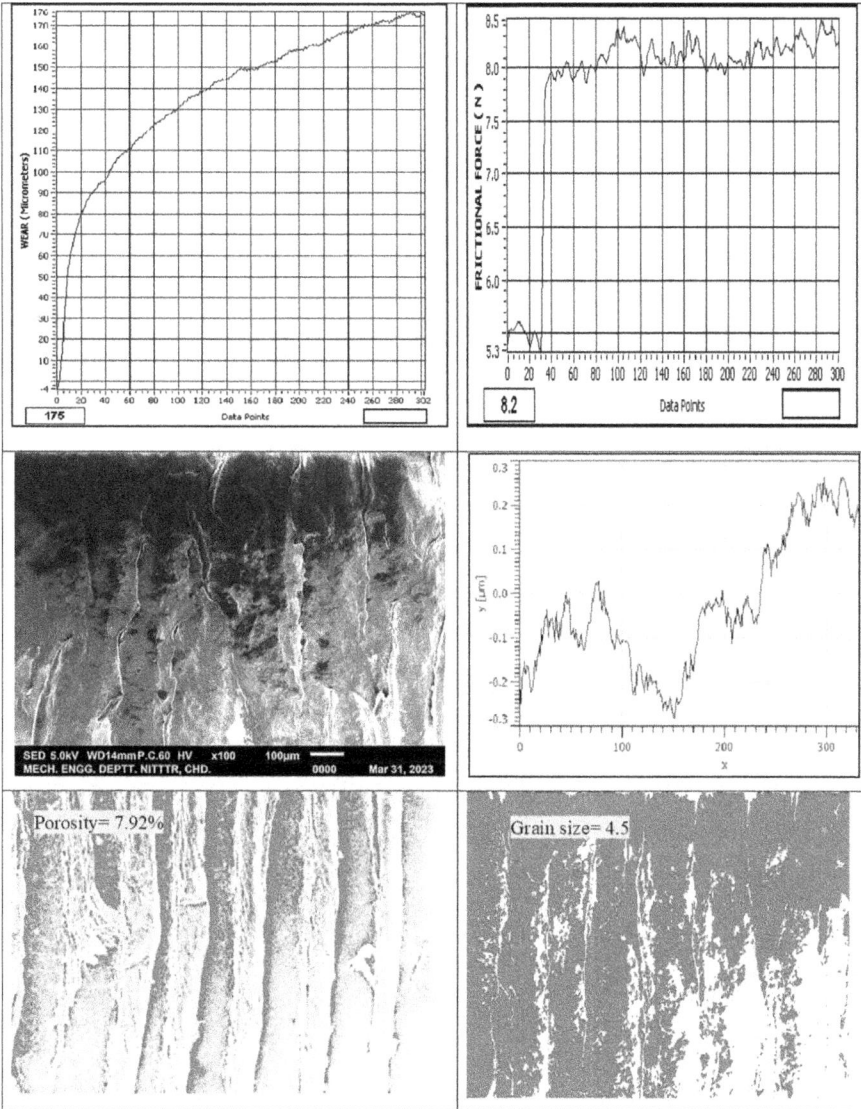

FIGURE 9.14 Tribological and morphological observations for PVDF-CC pin sample with 100% infill density.

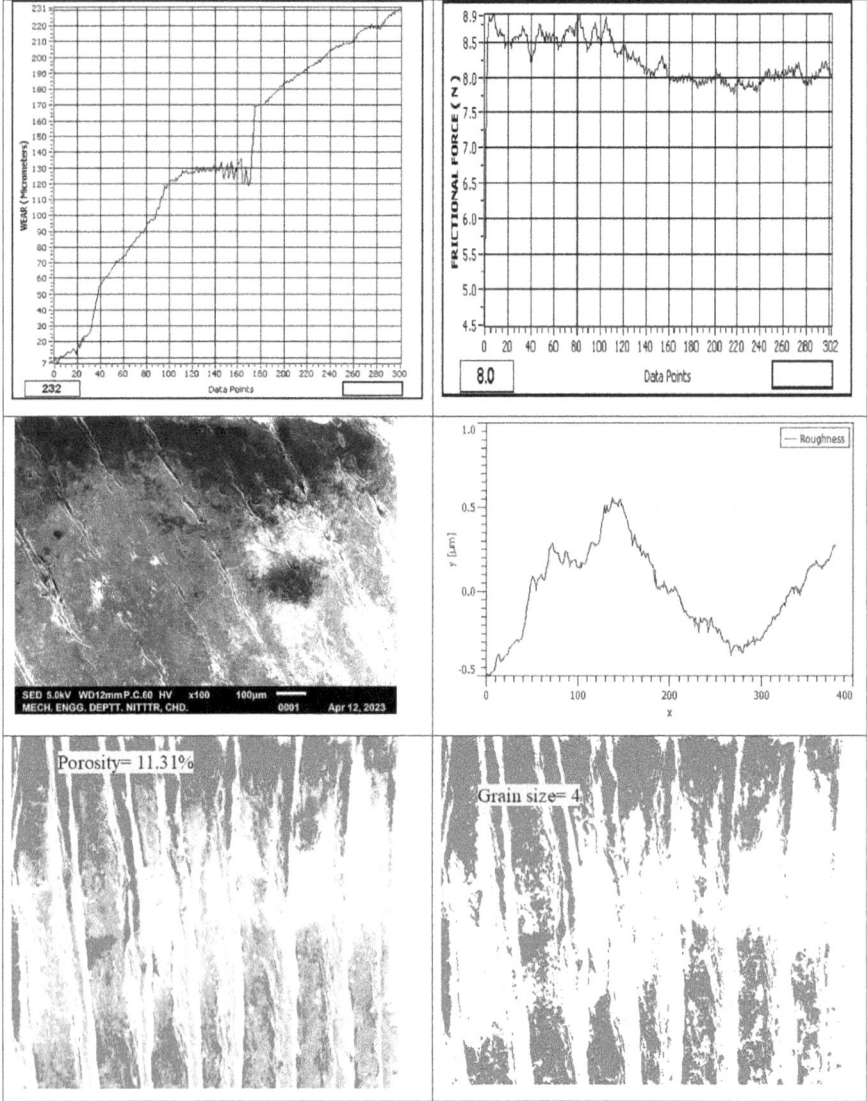

FIGURE 9.15 Tribological and morphological observations for PVDF-CC pin sample with 75% infill density.

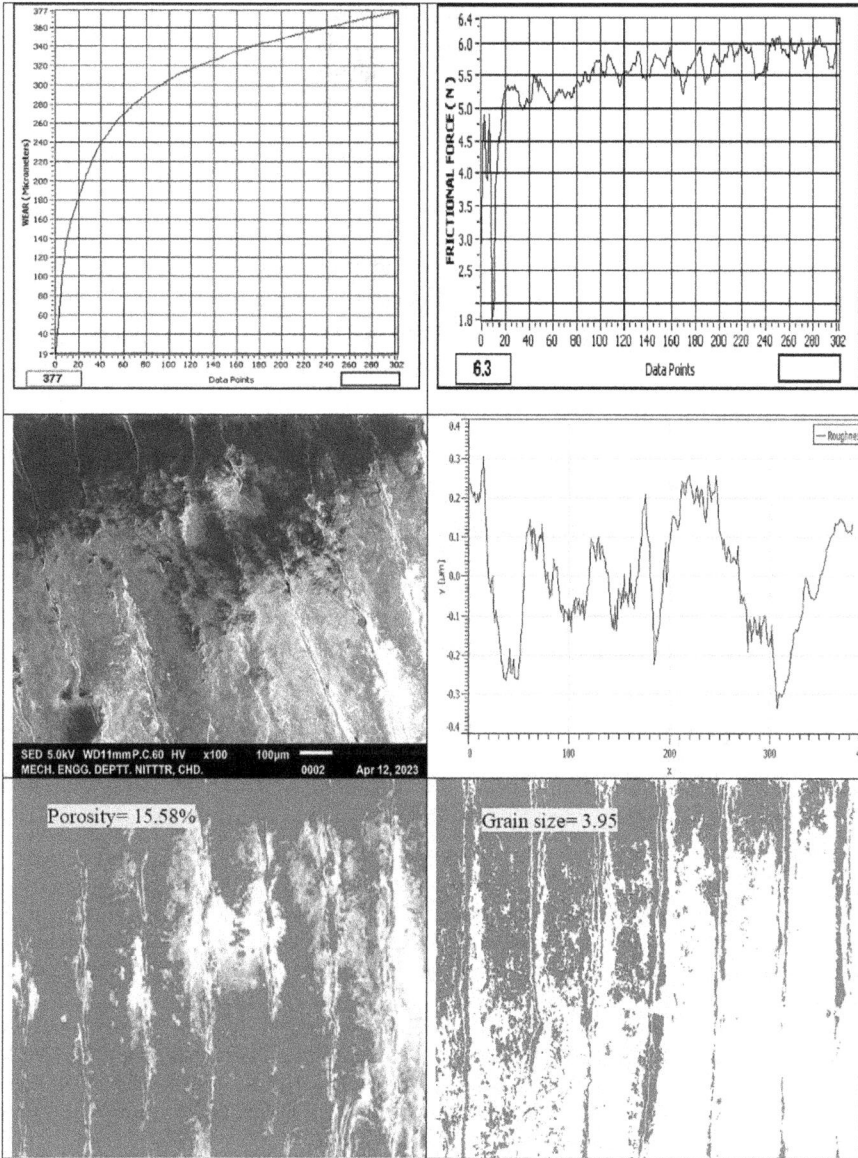

FIGURE 9.16 Tribological and morphological observations for PVDF-CC pin sample with 50% infill density.

porosity % and surface roughness (Ra) also increased with decrease in infill density. Therefore, it may be stated that 100% infill density is the best acceptable parameter to attain good wear-resistant properties in 3D-printed composite structures of PVDF-G-MZ and PVDF-CC for repair and maintenance based structural applications in heritage structures.

9.5 SUMMARY

The investigations performed on tribological and morphological properties of PVDF-G-MZ and PVDF-CC for repair and maintenance based structural applications in heritage structures highlighted that the proposed composites possess good wear-resistant properties with a minimum wear rate of 304 μm in the case of PVDF-G-MZ and 175 μm wear rate in the case of PVDF-CC. The 3D-printed composite structures of the proposed composites with 100% infill density also possess acceptable morphological properties in terms of porosity % and Ra. The minimum porosity of 7.92% and 9.27% was observed in PVDF-CC and PVDF-G-MZ composite structures with acceptable surface roughness in the range of 80–120 μm. The results of the study outlined that the PVDF-G-MZ and PVDF-CC composites possess acceptable tribological properties (i.e., high wear resistance due to less wear rate, i.e., 35 μm/min) as a result of which the same may be used successfully for structural and non-structural applications in repair and maintenance of heritage structures.

ACKNOWLEDGMENTS

The authors gratefully acknowledge the University Centre for Research and Development, Chandigarh University for their assistance and support.

REFERENCES

Ashraf, M., Gibson, I., &Rashed, M. G. (2018). Challenges and prospects of 3D printing in structural engineering. In *Proceedings of the 13th International Conference on Steel, Space and Composite Structures*, Perth, WA, Australia, Vol. *11*.

Bedarf, P., Dutto, A., Zanini, M., &Dillenburger, B. (2021). Foam 3D printing for construction: A review of applications, materials, and processes. *Automation in Construction*, *130*, 103861.

Coiai, S., Passaglia, E., Pucci, A., & Ruggeri, G. (2015). Nanocomposites based on thermoplastic polymers and functional nanofiller for sensor applications. *Materials*, *8*(6), 3377–3427.

Costa, A. P. D., Botelho, E. C., Costa, M. L., Narita, N. E., & Tarpani, J. R. (2012). A review of welding technologies for thermoplastic composites in aerospace applications. *Journal of Aerospace Technology and Management, 4*, 255–265.

Enneti, R. K., & Prough, K. C. (2019). Wear properties of sintered WC-12% Co processed via binder jet 3D printing (BJ3DP). *International Journal of Refractory Metals and Hard Materials*, *78*, 228–232.

Gao, Y., & Zhou, M. (2018). Superior mechanical behavior and fretting wear resistance of 3D-printed Inconel 625 superalloy. *Applied Sciences*, *8*(12), 2439.

Gbadeyan, O. J., Mohan, T. P., &Kanny, K. (2021). Tribological properties of 3D printed polymer composites-based friction materials. *Tribology of Polymer and Polymer Composites for Industry, 4.0*, 161–191.

Grigore, M. E. (2017). Methods of recycling, properties and applications of recycled thermoplastic polymers. *Recycling, 2*(4), 24.

Hofmann, M. (2014). 3D printing gets a boost and opportunities with polymer materials. *ACS Macro Letters, 3*(4), 382–386.

Huang, X., Fan, Y., Zhang, J., & Jiang, P. (2017). Polypropylene based thermoplastic polymers for potential recyclable HVDC cable insulation applications. *IEEE Transactions on Dielectrics and Electrical Insulation, 24*(3), 1446–1456.

Husain, M., Singh, R., & Pabla, B. S. (2023). A review on 3D printing of partially absorbable implants. *Journal of the Institution of Engineers (India): Series C, 104*, 1–20. https://doi.org/10.1007/s40032-023-00980-7

Jeon, H., Kim, Y., Yu, W. R., & Lee, J. U. (2020). Exfoliated graphene/thermoplastic elastomer nanocomposites with improved wear properties for 3D printing. *Composites Part B: Engineering, 189*, 107912.

Junk, S., & Tränkle, M. (2011). Design for additive manufacturing technologies: New applications of 3D-printing for rapid prototyping and rapid tooling. In *DS 68-5: Proceedings of the 18th International Conference on Engineering Design (ICED 11), Impacting Society through Engineering Design, Vol. 5: Design for X/Design to X*, Lyngby/Copenhagen, Denmark, 15–19 August 2011 (pp. 12–18).

Kumar, R., & Singh, R. (2023). On fracture mechanism of 3D printed nanofiber-reinforced PLA matrix. *National Academy Science Letters*, 1–5. https://doi.org/10.1007/s40009-023-01330-y

Kumar, V., Kumar, R., Singh, R., & Kumar, P. (2022a). On 3D printed biomedical sensors for non-enzymatic glucose sensing applications. *Proceedings of the Institution of Mechanical Engineers, Part H: Journal of Engineering in Medicine, 236*(8), 1057–1069.

Kumar, V., Singh, R., & Ahuja, I. P. S. (2022b). On correlation of rheological, thermal, mechanical and morphological properties of chemical assisted mechanically blended ABS-Graphene composite as tertiary recycling for 3D printing applications. *Advances in Materials and Processing Technologies, 8*(3), 2476–2495.

Kumar, V., Singh, R., & Ahuja, I. S. (2022c). On 3D printing of electro-active PVDF-graphene and Mn-doped ZnO nanoparticle-based composite as a self-healing repair solution for heritage structures. *Proceedings of the Institution of Mechanical Engineers, Part B: Journal of Engineering Manufacture, 236*(8), 1141–1154.

Kumar, V., Singh, R., & Ahuja, I. P. S. (2022d). On 4D capabilities of chemical assisted mechanical blended ABS-nano graphene composite matrix. *Materials Today: Proceedings, 48*, 952–957.

Kumar, V., Singh, R., & Ahuja, I. S. (2022e). On wear properties of mechanical blended and chemical assisted mechanical blended ABS-graphene reinforced composites. In *Encyclopedia of Materials: Plastics and Polymers,* edited by M.S.J. Hashmi (pp. 434–441). Elsevier. https://doi.org/10.1016/B978-0-12-820352-1.00109-7

Kumar, V., Singh, R., & Ahuja, I. S. (2022f). On programming of polyvinylidene fluoride-limestone composite for four-dimensional printing applications in heritage structures. *Proceedings of the Institution of Mechanical Engineers, Part L: Journal of Materials: Design and Applications, 236*(2), 319–333.

Kumar, V., Singh, R., & Ahuja, I. S. (2022g). Online health monitoring of repaired non-structural cracks with innovative 3D printed strips in heritage buildings. *Materials Letters, 327*, 133033.

Kumar, V., Singh, R., & Ahuja, I. S. (2022h). 3D printed innovative customized solution for regulating weathering effect on heritage structures. *Materials Letters, 324*, 132717.

Kumar, V., Singh, R., & Ahuja, I. S. (2023a). Multi-material printing of PVDF composites: A customized solution for maintenance of heritage structures. *Proceedings of the Institution of Mechanical Engineers, Part L: Journal of Materials: Design and Applications, 237*(3), 554–564.

Kumar, R., Singh, R., Kumar, V., Kumar, P., Prakesh, C., & Singh, S. (2021). Characterization of in-house-developed Mn-ZnO-reinforced polyethylene: A sustainable approach for developing fused filament fabrication-based filament. *Journal of Materials Engineering and Performance*, *30*(7), 5368–5382.

Kumar, V., Singh, R., Kumar, R., Ranjan, N., & Hussain, M. (2023b). ABS-PLA-Al composite for digital twinning of heritage buildings. *Materials Letters*, *346*, 134536.

Lei, D., Yang, Y., Liu, Z., Chen, S., Song, B., Shen, A., & You, Z. (2019). A general strategy of 3D printing thermosets for diverse applications. *Materials Horizons*, *6*(2), 394–404.

Mahieux, C. A., & Reifsnider, K. L. (2002) Property modeling across transition temperatures in polymers: Application to thermoplastic systems. *Journal of Materials Science*, *37*, 911–920.

Melgoza, E. L., Vallicrosa, G., Serenó, L., Ciurana, J., & Rodríguez, C. A. (2014). Rapid tooling using 3D printing system for manufacturing of customized tracheal stent. *Rapid Prototyping Journal*, *20*(1), 2–12.

Norani, M. N. M., Abdullah, M. I. H. C., Abdollah, M. F. B., Amiruddin, H., Ramli, F. R., Tamaldin, N., & Malaysia, U. T. (2021). Mechanical and tribological properties of FFF 3D-printed polymers: A brief review. *Journal of Tribology*, *29*, 11–30.

Olesik, P., Godzierz, M., & Kozioł, M. (2019). Preliminary characterization of novel LDPE-based wear-resistant composite suitable for FDM 3D printing. *Materials*, *12*(16), 2520.

Prause, E., Hey, J., Beuer, F., & Schmidt, F. (2022). Wear resistance of 3D-printed materials: A systematic review. *Dentistry Review*, *2*(2), 100051.

Rudnik, M., Hanon, M. M., Szot, W., Beck, K., Gogolewski, D., Zmarzły, P., &Kozior, T. (2022). Tribological properties of medical material (MED610) used in 3D printing PJM technology. *Tehničkivjesnik*, *29*(4), 1100–1108.

Singh, M., Kumar, S., Singh, R., Kumar, R., & Kumar, V. (2022). On shear resistance of almond skin reinforced PLA composite matrix-based scaffold using cancellous screw. *Advances in Materials and Processing Technologies*, *8*(2), 2361–2384.

Singh, S., Ramakrishna, S., & Singh, R. (2017). Material issues in additive manufacturing: A review. *Journal of Manufacturing Processes*, *25*, 185–200.

Studer, V., Pepin, A., & Chen, Y. (2002). *Nanoembossing of thermoplastic polymers for microfluidic applications*. *Applied Physics Letters*, *80*(19), 3614–3616.

Tay, Y. W. D., Panda, B., Paul, S. C., Noor Mohamed, N. A., Tan, M. J., & Leong, K. F. (2017). 3D printing trends in building and construction industry: A review. *Virtual and Physical Prototyping*, *12*(3), 261–276.

Valino, A. D., Dizon, J. R. C., Espera Jr, A. H., Chen, Q., Messman, J., &Advincula, R. C. (2019). Advances in 3D printing of thermoplastic polymer composites and nanocomposites. *Progress in Polymer Science*, *98*, 101162.

Velu, R., Raspall, F., & Singamneni, S. (2019). 3D printing technologies and composite materials for structural applications. In *Green Composites for Automotive Applications*, edited by G. Koronis & A. Silva (pp. 171–196). International Design Centre, Singapore University of Technology and Design, Singapore. Woodhead Publishing.

Wu, G., Xie, P., Yang, H., Dang, K., Xu, Y., Sain, M., & Yang, W. (2021). A review of thermoplastic polymer foams for functional applications. *Journal of Materials Science*, *56*, 11579–11604.

You, X., Zhang, Q., Yang, J., & Dong, S. (2023). Review on 3D-printed graphene-reinforced composites for structural applications. *Composites Part A: Applied Science and Manufacturing*, *167*, 107420.

10 Effect of Surface Texturing on Tribological Behavior of Additively Manufactured Parts

Alireza Hajialimohammadi
Semnan University

Rashi Tyagi
Chandigarh University

10.1 INTRODUCTION

Additive manufacturing (AM) has rapidly become a major method of manufacturing for industrial applications (Ranjan et al., 2023a). Due to the involved layer-by-layer deposition, the amount of waste material is limited, drastically reducing overall cost and conserving resources, unlike subtractive methods of manufacturing (Ralls et al., 2021; Shah et al., 2022; Tyagi et al., 2023, Tyagi & Tripathi 2023). This technology is being used in a wide variety of industries such as aerospace, mechanical, food, architectural, medical, automotive, and marine industries with the use of all different types of materials such as polymers, metal, ceramic, and concrete (Luo et al., 2022). Among these materials, the focus of this chapter is on the polymers and metals. There are various methods for AM on polymers, and the most common approaches that attract more attention in recent years are Extrusion-based additive manufacturing (EAM), powder bed fusion (PBF), and digital light processing (DLP). Many subcategories are also derived from these three categories (Ranjan et al., 2023b; Orgeldinger et al., 2023). For metals, all methods can be classified into two main categories: powder bed fusion (PBF) and direct energy deposition (DED) with subcategories of direct metal laser sintering (DMLS), selective laser sintering (SLS), selective laser melting (SLM), laser powder bed fusion (LPBF), electron beam powder bed fusion (EBPBF), and electron beam melting (EBM) that are grouped for PBF. Electron beam direct energy deposition (EBDED), laser wire direct energy deposition (LWDED), laser powder bed direct energy deposition (LPDED) or laser additive manufacturing (LAM), and wire-arc additive manufacturing (WAAM) approaches are also categorized for DED category (Orgeldinger et al., 2023). Creating specific surface texture or producing special composites to improve the wear resistance of mechanical parts is not a new idea, and in many applications, this concept has been utilized for decades (Lu & Wood, 2020; Saligheh et al., 2020). For additively manufactured parts, regardless of their material (polymer or metal), there are many supplementary methods that can be used to modify the surface before applying in the corresponding abrasion applications. Common post-processing methods that are used for surface modification of parts produced by 3D printing methods are summarized by the researchers (Tyagi et al., 2023; Pruncu et al., 2022)

Although surface texture commonly represented by surface roughness parameters (Coblas et al., 2015), some special textures are produced with advanced and additive manufacturing methods that have effect on the tribological properties of surfaces (Townsend et al., 2016). Advanced manufacturing methods like electrochemical machining (ECM), electrical discharge machining (EDM), laser surface texturing (LST), and maskless chemical etching method (MECT) have been used for specific texture implementation in mechanical parts (Laki et al., 2022; Townsend et al., 2016). In this book chapter, the effect of surface textures on the tribological properties of additively manufactured metal and polymer parts in recent research works is investigated. Metal and polymers are studied separately, and various surface textures and their effects on the wear and abrasion properties are explained in detail for 3D-printed parts.

10.2 TRIBOLOGICAL BEHAVIOR OF POLYMER PARTS

Polymer-based materials (PBM) with self-lubricating characteristics, lower cost processability, and synthetic compatibility are being widely used in many applications (Coblas et al., 2015). For functional effectiveness of this material, it is crucial to investigate the tribological properties to develop an applied 3D-printed material.

Tribological properties of 3D texture shape, such as a truncated cone structure compared with two 2.5D equivalent structures, cylinders with the height and base diameter of the cone (2.5D-base) and rods with the height and tip diameter of the cone (2.5D-tip), have been investigated by Afshar-Mohajer and Zou (2020) for 3D micro-textures Fabricated via Two-Photon Lithography and Replica Molding. The produced textures reported that the wear rate of 2.5D-base texture is higher than the cone one and the height reduction of 2.5D-base texture is increasing with increasing the wear test normal load. The effect of 3D texture on the wear characteristics of Z-ABS® material was investigated (Sınmazçelik et al., 2020). It is concluded that the use of macro-textured surfaces produced by the AM method, specifically designed for particle erosion direction, remarkably reduces particle erosion. The formation of microscale geometric textures on the surface of the part subject to particle erosion without altering the macro geometry can significantly reduce the damage caused by particle erosion as depicted in Figure 10.1 (Sınmazçelik et al., 2020).

Luo et al. (2022) investigated the effect of two surface textures on samples produced by FDM technology: (a) convex square and (b) convex triangle. The samples' materials were SCFRN and nylon. The surface texture that is produced by FDM is shown in Figure 10.2.

The effect of surface tribology and material type on the average COF was observed. For most test conditions, the average COF of convex triangular texture is lower than

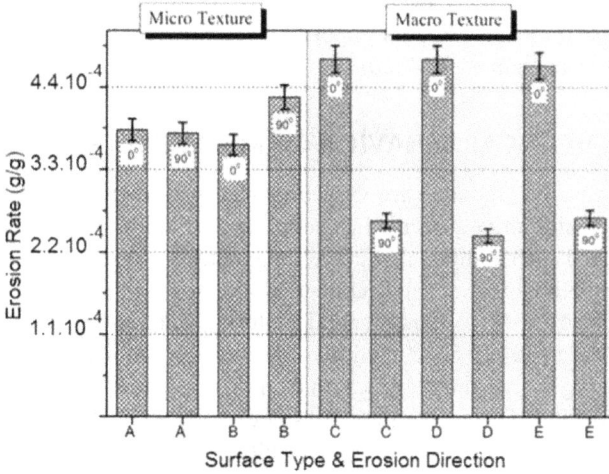

FIGURE 10.1   Wear rate results for different ABS samples produced by 3D printing with various surface texture designs (Sınmazçelik et al., 2020).

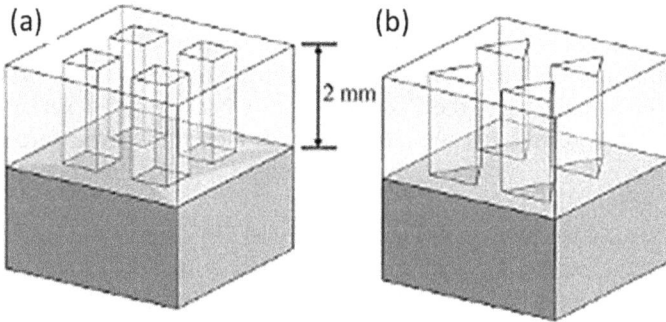

FIGURE 10.2   Convex square (a) and convex triangle (b) surface textures (Luo et al., 2022).

convex squared texture, with the fully printed texture having the maximum COF. The specific wear rate is measured for various test conditions. The same trend that has been detected for the COF can be detected for wear rate, with the convex triangular texture providing the minimum wear rate. Surface texture of 3D-printed parts can be developed by changing the printing orientation. Dangnan et al. (2020) investigated the friction and wear characteristics of ABS and Verogray materials produced with Polyjet additive manufacturing method and parallel and perpendicular printing orientations. Average COF values for different orientations and material in three testing load values of 1, 5, and 10 Newton, and for all cases, the COF in the parallel direction is lower than that in the perpendicular direction. Results of specific wear rate for different orientations and materials for two different trends are observed for ABS and Verogray materials. Unlike the Verogray, for ABS material, the wear rate in the perpendicular direction is lower than that in the parallel direction, indicating that the material specification is also an important factor in determining the tribological characteristics (Dangnan et al., 2020).

10.3 TRIBOLOGICAL BEHAVIOR OF METAL PARTS

Wear and corrosion resistance are very important parameters of metals that are being used in automotive and medical application, and the parts like implants that are produced with additive manufacturing methods should possess mentioned characteristics (Khun et al., 2018). Khun et al. investigated tribological characteristics of Ti-6Al-4V as a new biomedical material that was produced with EBM 3D printing Process. They concluded that the additively manufactured Ti-6Al-4V samples have higher wear resistance than the commercial Ti-6Al-4V material (Khun et al., 2018). Friction characteristics of different stainless steel (AISI 316 L) surface textures including flat, gecko's fibrils, dimples, pyramids, mushrooms, mesh, inclined brush, and brush were investigated by Holovenko et al. (2018). The SLM 3D printing method has been utilized for producing the samples. Various textures were produced with 3D printing. The flat surface was the most unstable and had the highest COF. Results also revealed that the inclined brush structure provided the most stable and lowest COF. Results also proved that the surface texture is a very important factor affecting the friction characteristics of the additively manufactured metal parts. They also analyzed the size and shape of wear debris separating from each surface texture during the pin-on-disk wear test to have better understanding of the COF behavior for each specified surface pattern. They have also reported that the size of texture elements on the part surface is limited by the capabilities of current SLM method, and it is not possible to have very small texture elements.

The tribological behavior of AM and cast Inconel 718 samples was investigated, as well as the effect of surface topology on the wear characteristics of the samples (Basha & Sankar, 2023). The microstructures for two AM and cast samples have been compared. As shown in this figure, AM part is characterized with the rapid solidification that limited the macro segregation (Basha & Sankar, 2023). Microstructure of the wear scars for AM samples after wear test for different counter-body materials (Basha & Sankar, 2023). Tribo-oxide film observed in some positions for casted sample is compared to the AM sample, In the case of silicon carbide counterpart, the shallow abrasive marks are shown. It is concluded that boron carbide counter-body exhibited the lowest COF due to the presence of boric acid, which acted as a lubricant, while the silicon carbide counter-body had the minimum specific wear-rate (Basha & Sankar, 2023).

Dimple density as a texture micro-feature has been developed on the surface of Ti6Al4V titanium alloy by laser, and the wear properties of various designed textures are investigated by Gaikwad et al. (2022). Schematic view of the samples with various dimple densities is indicated in Figure 10.3. Wear test results are illustrated in Figure 10.4. As shown in this figure, increasing dimple density could decrease wear rate and the dimple density can determine the tribological behavior of Ti6Al4V material produced by 3D printing method (Gaikwad et al., 2022).

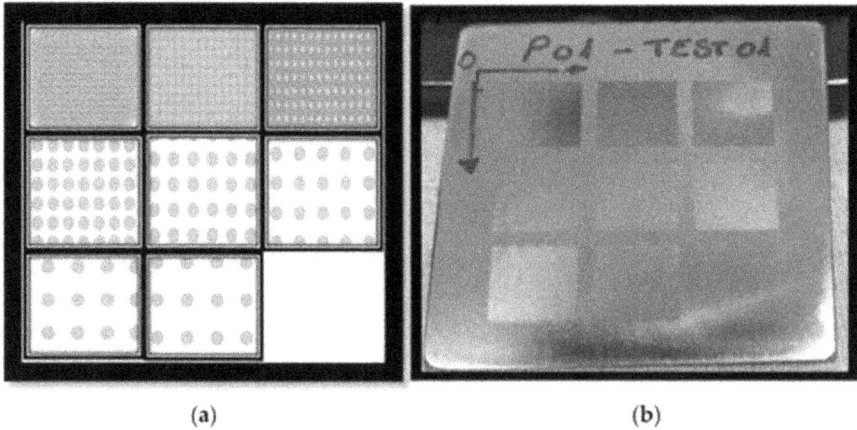

FIGURE 10.3   (a) Schematic representation of dimple density variation and (b) actual sample (Gaikwad et al., 2022).

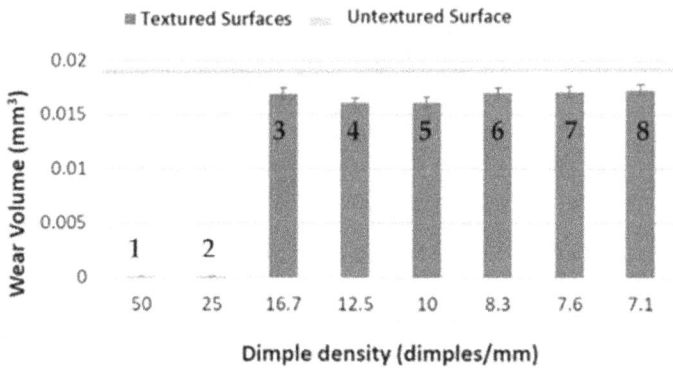

FIGURE 10.4   Wear volume of samples with various dimple densities compared with reference sample (untextured) (Gaikwad et al., 2022).

10.4   CONCLUSION

In this book chapter, the effect of using various surface patterns on the frictional and wear rate of additively manufactured metal and polymer materials has been investigated. For polymers, it is concluded that various surface textures can be utilized including cylindrical and cone bump, printing orientation, and squared and triangular convex textures. The wear rate of 2.5D-base texture is higher than the cone one and the height reduction of 2.5D-base texture is increasing with increasing wear test normal load, the friction of convex triangular texture is lower than convex squared texture, and the COF in the parallel direction is lower than that in the perpendicular direction. For metal additively manufactured parts, the effects of many surface texture features have been investigated including dimples, pyramids, mushrooms, and brush. Results indicated that these feature have considerable effect on the friction and wear of the part surface.

ACKNOWLEDGMENTS

The authors are grateful to Semnan University, Iran for their assistance and support.

REFERENCES

Afshar-Mohajer, M., & Zou, M. (2020). Multi-scale in situ tribological studies of surfaces with 3D textures fabricated via two-photon lithography and replica molding. *Advanced Materials Interfaces*, *7*(13). https://doi.org/10.1002/admi.202000299.

Basha, M. M., & Sankar, M. R. (2023). Experimental tribological study on additive manufactured Inconel 718 features against the hard carbide counter bodies. *Journal of Tribology, 145*(12), 1–11.Coblas, D. G., Fatu, A., Maoui, A., & Hajjam, M. (2015). Manufacturing textured surfaces: State of art and recent developments. *Proceedings of the Institution of Mechanical Engineers, Part J: Journal of Engineering Tribology*, *229*(1). https://doi.org/10.1177/1350650114542242.

Dangnan, F., Espejo, C., Liskiewicz, T., Gester, M., & Neville, A. (2020). Friction and wear of additive manufactured polymers in dry contact. *Journal of Manufacturing Processes*, *59*. https://doi.org/10.1016/j.jmapro.2020.09.051.

Gaikwad, A., Vázquez-Martínez, J. M., Salguero, J., & Iglesias, P. (2022). Tribological properties of Ti6Al4V titanium textured surfaces created by laser: Effect of dimple density. *Lubricants*, *10*(7). https://doi.org/10.3390/lubricants10070138.

Holovenko, Y., Antonov, M., Kollo, L., & Hussainova, I. (2018). Friction studies of metal surfaces with various 3D printed patterns tested in dry sliding conditions. *Proceedings of the Institution of Mechanical Engineers, Part J: Journal of Engineering Tribology*, *232*(1). https://doi.org/10.1177/1350650117738920.

Khun, N. W., Toh, W. Q., Tan, X. P., Liu, E., & Tor, S. B. (2018). Tribological properties of three-dimensionally printed Ti-6Al-4V material via electron beam melting process tested against 100Cr6 steel without and with Hank's solution. *Journal of Tribology*, *140*(6). https://doi.org/10.1115/1.4040158.

Laki, G., Nagy, A. L., Rohde-Brandenburger, J., & Hanula, B. (2022). A review on friction reduction by laser textured surfaces in internal combustion engines. *Tribology Online*, *17*(4). https://doi.org/10.2474/trol.17.318.

Lu, P., & Wood, R. J. K. (2020). Tribological performance of surface texturing in mechanical applications-a review. *Surface Topography: Metrology and Properties*, *8*(4). https://doi.org/10.1088/2051-672X/abb6d0.

Luo, M., Huang, S., Man, Z., Cairney, J. M., & Chang, L. (2022). Tribological behaviour of fused deposition modelling printed short carbon fibre reinforced nylon composites with surface textures under dry and water lubricated conditions. *Friction*, *10*(12). https://doi.org/10.1007/s40544-021-0574-5.Orgeldinger, C., Seynstahl, A., Rosnitschek, T., & Tremmel, S. (2023). Surface properties and tribological behavior of additively manufactured components: A systematic review. *Lubricants*, *11*(6). https://doi.org/10.3390/lubricants11060257.

Pruncu, C. I., Aherwar, A., & Gorb, S. (2022). *Tribology and Surface Engineering for Industrial Applications*, CRC Press.

Ralls, A. M., Kumar, P., & Menezes, P. L. (2021). Tribological properties of additive manufactured materials for energy applications: A review. *Processes*, *9*(1). https://doi.org/10.3390/pr9010031.

Ranjan, N., Tyagi, R., Kumar, R., & Babbar, A. (2023a). 3D printing applications of thermoresponsive functional materials: A review. *Advances in Materials and Processing Technologies*. https://www.tandfonline.com/doi/abs/10.1080/2374068X.2023.2205669.

Ranjan, N., Tyagi, R., Kumar, R., & Kumar, V. (2023b). On fabrication of acrylonitrile butadiene styrene-zirconium oxide composite feedstock for 3D printing-based rapid tooling applications. *Journal of Thermoplastic Composite Materials, 0*(0). https://doi.org/10.1177/08927057231311863.

Saligheh, A., Hajialimohammadi, A., & Abedini, V. (2020). Cutting forces and tool wear investigation for face milling of bimetallic composite parts made of aluminum and cast iron alloys. *International Journal of Engineering Transactions C: Aspects*, *33*(6). https://doi.org/10.5829/ije.2020.33.06c.12.

Shah, R., Pai, N., Rosenkranz, A., Shirvani, K., & Marian, M. (2022). Tribological behavior of additively manufactured metal components. *Journal of Manufacturing and Materials Processing*, *6*(6). https://doi.org/10.3390/jmmp6060138.

Sınmazçelik, T., Fidan, S., & Ürgün, S. (2020). Effects of 3D printed surface texture on erosive wear. *Tribology International*, *144*. https://doi.org/10.1016/j.triboint.2019.106110.

Townsend, A., Senin, N., Blunt, L., Leach, R. K., & Taylor, J. S. (2016). Surface texture metrology for metal additive manufacturing: A review. *Precision Engineering*, *46*. https://doi.org/10.1016/j.precisioneng.2016.06.001.

Tyagi, R., Singh, G., Kuma, R., Kumar, V., & Singh, S. (2023). 3D-printed sandwiched acrylonitrile butadiene styrene/carbon fiber composites: Investigating mechanical, morphological, and fractural properties. *Journal of Materials Engineering and Performance*. https://link.springer.com/article/10.1007/s11665-023-08292-8.

Tyagi, R., & Tripathi, A. (2023). Coating/cladding based post-processing in additive manufacturing. In *Handbook of Post-Processing in Additive Manufacturing: Requirements, Theories, and Methods*, edited by Gurminder Singh, Ranvijay Kumar, Kamalpreet Sandhu, Eujin Pei, & Sunpreet Singh, p. 127, CRC Press.

11 Trends of Tribology in Biomedical Application of Additively Manufactured Parts

Pratik Kumar Shaw, Suryank Dwivedi, and Amit Rai Dixit
Indian Institute of Technology (ISM, Dhanbad)

Rashi Tyagi
Chandigarh University

11.1 INTRODUCTION

Tribology, which deals with friction, wear, corrosion, and lubrication, contributes to meeting the increasing demands of sophisticated technological advancements as humanity progresses. It leads to enhanced durability of parts and improvements in the biomedical industry (Alam, 2021). When it comes to biomedical applications, there are several critical concerns that need to be addressed in tribology, including outstanding biocompatibility, high strength, and elastic modulus similar to that of bones, high corrosion resistance, and high wear resistance (Kaur et al., 2019). To tackle these challenges, additive manufacturing (AM) has gained significant traction in the biomedical industry. It is utilized for fabricating customized prosthetics, dental implants, manufactured organs and tissues, anatomical models, and pharmaceutical products (Ramakrishna et al., n.d.; Rezvani Ghomi et al., 2019). This technology leverages various biomaterials such as metals, powders, liquids, ceramics, polymers, and living cells to create intricate structures possessing exceptional tribological properties. These properties cannot be achieved through conventional manufacturing methods (Moghadasi et al., 2022; Wasyłeczko et al., 2023).

11.2 ADDITIVE MANUFACTURING TECHNIQUES AND TRIBOLOGY TESTS FOR BIOMEDICAL COMPONENTS

According to ASTM, AM is a process that involves the deposition of materials layer by layer to create an object based on CAD data, contrasting with the subtractive manufacturing technique (Tyagi et al., 2023; Ranjan et al., 2023). Figure 11.1 shows flow diagram of the AM process (Malik et al., 2022).

DOI: 10.1201/9781003400523-11

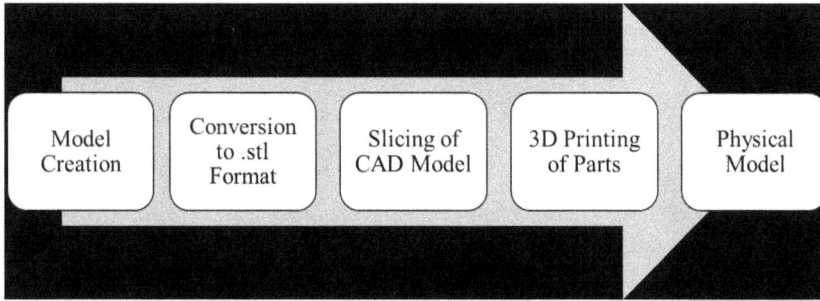

FIGURE 11.1   Flow diagram of the additive manufacturing process (Malik et al., 2022).

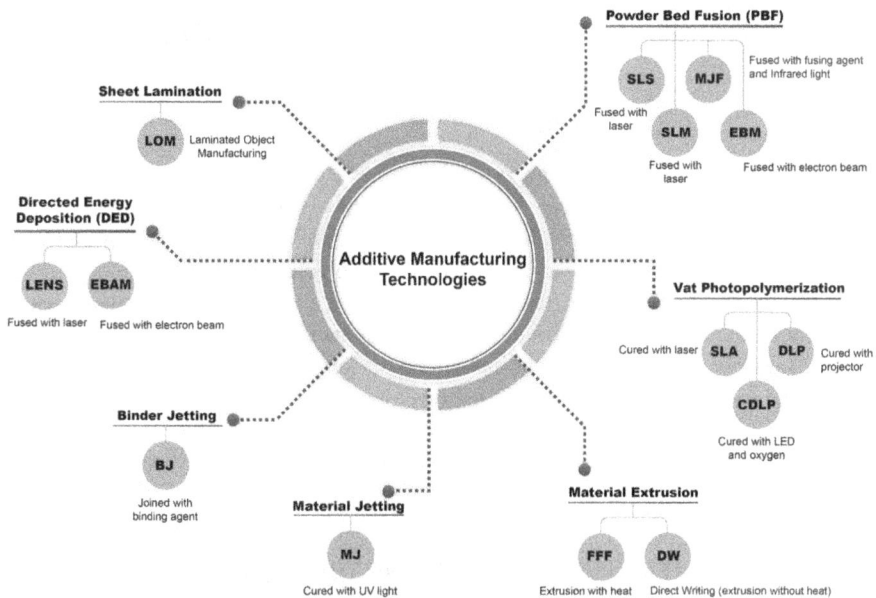

FIGURE 11.2   Classification of additive manufacturing technologies (Rafiee et al., 2020).

AM can be categorized based on the method of product formation and the type of materials employed. The classification based on the product formation methodology includes material jetting, binder jetting, extrusion, sheet lamination, and energy deposition. According to the ISO/ASTM 529000-2015 standard, AM technologies can be classified into seven different categories, and examples of AM processes are given in Figure 11.2 (Rafiee et al., 2020).

In order to evaluate the tribological properties of additively manufactured parts, mainly two types of tribometers are used: (a) Linear Reciprocating Tribometer and (b) Rotating Tribometer. A schematic of these tribometers is shown in Figure 11.3. A reciprocating tribometer is purposefully crafted to replicate the reciprocal sliding motion that occurs between two surfaces to determine the wear rate and friction.

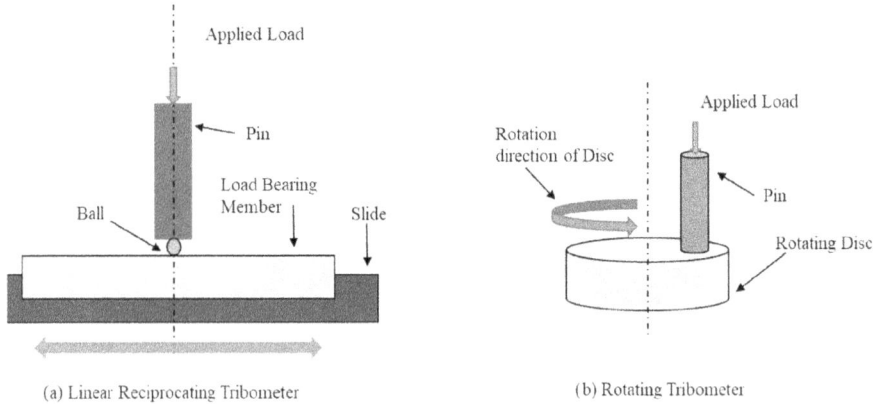

(a) Linear Reciprocating Tribometer   (b) Rotating Tribometer

FIGURE 11.3 Schematic of types of tribometer: (a) linear reciprocating tribometer and (b) rotating tribometer.

Its significance lies in its relevance to biomedical applications, particularly in the case where implants may undergo back-and-forth movements, as seen in hip or knee joint replacements. While a rotating tribometer is employed to investigate the friction and wear characteristics of materials when subjected to rotational motion, it is necessary to conduct such a tribology test to simulate a real-world scenario to ensure highly accurate material performance. The use of such tribological data can help in material selection, lubrication conditions, and design optimization for better performance and longevity of implants and medical devices.

These tests help to ensure the reliability, performance, and biocompatibility of the parts. A few tribology tests have been listed below for the AM techniques and biomedical applications in Table 11.1.

11.3 PARAMETERS INFLUENCING THE TRIBOLOGICAL PROPERTIES OF AM PARTS FOR BIOMEDICAL APPLICATION

The tribological properties of AM parts are influenced by several parameters. Understanding and controlling these factors are essential for ensuring the reliable performance and long-term functionality of AM parts in biomedical settings. Here are some key parameters that influence the tribological properties of AM parts for biomedical applications:

11.3.1 MATERIAL SECTION

Different AM materials, such as metals, polymers, ceramics, and composites, exhibit varying tribological behaviors. The choice of material can significantly impact the wear resistance, friction coefficients, and biocompatibility of AM biomedical parts. Figure 11.4 shows different types of biomaterials and their applications (Table 11.2).

TABLE 11.1

Additive Manufacturing Techniques with Tribological Tests in the Biomedical Application

Additive Manufacturing Technique	Biomedical Application	Tribological Test
Selective laser sintering (SLS)	Custom implants	Wear and friction analysis (Bertolotti et al., 2022)
Fused deposition modeling (FDM)	Tissue engineering scaffolds	Coefficient of friction (Gao et al., 2023)
Stereolithography (SLA)	Dental restorations	Surface roughness (Dikova et al., 2018)
Electron beam melting (EBM)	Orthopedic implants	Scratch resistance (Wu et al., 2023)
Binder jetting	Drug delivery systems	Wear rate (Akmal et al., 2018)
Direct ink writing (DIW)	Bioprinting of organs	Friction coefficient (Wan et al., 2020)
Material jetting	Hearing aid manufacturing	Tribocorrosion resistance (Evans et al., 2022)

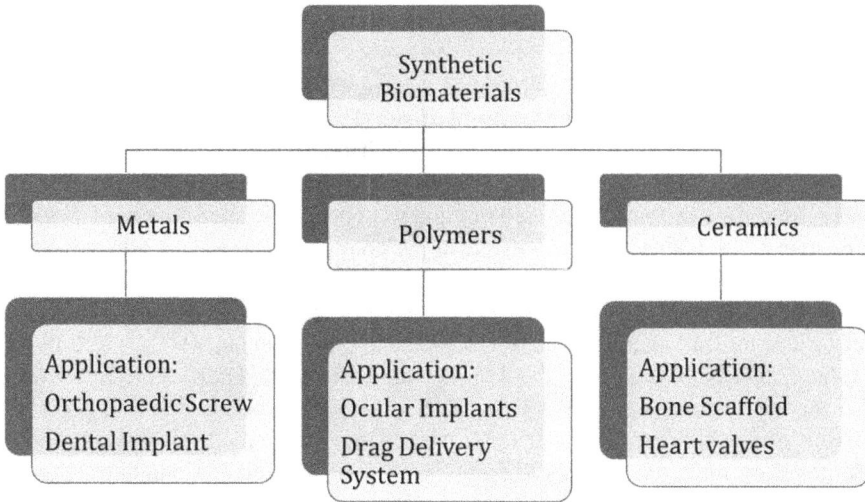

FIGURE 11.4 Different types of biomaterials and their application.

11.3.1.1 Titanium Alloys

Titanium and its alloys, such as Ti-6Al-4V, exhibit excellent tribological properties due to their low friction coefficient and high wear resistance. The material's biocompatibility makes it ideal for load-bearing implants and joint replacements. In a comparative study of the wear resistance between conventional Ti6Al4V and

TABLE 11.2
FDA-Cleared 3D-Printed Material and Its Application in the Biomedical Industry (Singh & Ramakrishna, 2017)

Material	Application
CP-Ti	Screw and abutment
Ti-6A1-4V	Artificial valve, stent, bone fixation
Ti-6Al-7Nb	Crowns, knee joint, hip joint
Ti-5Al-2.5Fe	Spinal implant
Ti-15 Zr-4Nb-2Ta-0.2Pd	Crown, bridges, dentures, implants
Ti-29Nb-13Ta-4.6Zr	Crown, bridges, dentures, implants
83%–87%Ti-13%–17%Zr (Roxolid)	Crown, bridges, dentures
316L	Knee joint, hip joint, surgical tools, screw
Co-Cr-Mo, Co-Ni-Cr-Mo	Artificial valve, plates, bolts, crowns, knee joint, hip joint
NiTi	Catheters, stents
PMMA, PE, PEEK	Dental bridges, articular cartilage, hip joint femoral surface, knee joint bearing surface, scaffolds
$SiO_2/CaO/Na_2O/P_2O_5$	Bones, dental implants, orthopedic implants
Zirconia	Porous implants, dental implants
Al_2O_3	Dental implants
$Ca_5(PO_4)_3(OH)$	Implant coating material

EBM-printed Ti6Al4V samples, the EBM-printed Ti6Al4V samples exhibited superior wear resistance due to their higher hardness (Hao et al., 2016).

The COF of the commercial and 3D-printed Ti64 samples, tested both with and without Hank's solution, is presented in Figure 11.5a as the number of laps. Throughout the entire sliding process under both dry and wet conditions, the 3D-printed TiAl4V sample exhibits higher COF compared to the commercial TiAl4V sample, despite both samples displaying stable friction during the wear tests. The fluctuations observed in the friction coefficients of both TiAl4V samples in relation to the number of laps indicate the presence of the stick-slip phenomenon, which is a result of the hard steel balls sliding on the TiAl4V samples. Figure 11.5b clearly illustrates that the application of Hank's solution leads to lower friction coefficients for both TiAl4V samples during the wear tests. Also, beta-type biomedical titanium alloys are preferred materials for medical implants owing to their low modulus, superb biocompatibility, high corrosion resistance, and superior strength when compared to stainless steel and Co-based alloys.

In Figure 11.5c and d, the hardness with depth and the specific wear rates of both the commercial and 3D-printed Ti64 samples are presented, with and without Hank's solution. This observation confirms that the 3D-printed Ti64 samples exhibit lower wear. Furthermore, even under wet conditions, the 3D-printed Ti64 samples consistently display a slightly lower average specific wear rate as compared to the commercial Ti64 samples, which have a high wear rate. Figure 11.5 demonstrates that the application of Hank's solution during sliding noticeably reduces wear in both the commercial and 3D-printed Ti64 samples (Khun et al., 2018).

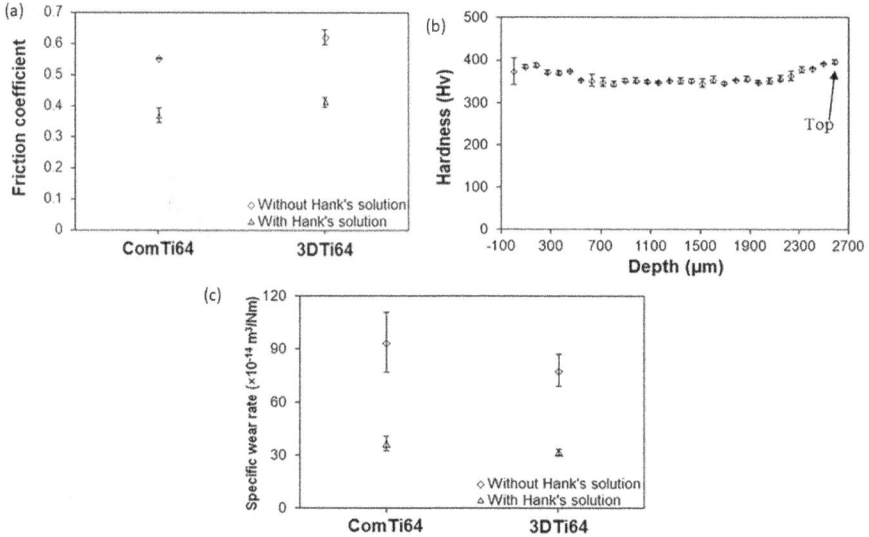

FIGURE 11.5 (a) COF of commercial and 3D-printed Ti64 samples tested, (b) hardness with depth, and (c) the specific wear rates of cast and 3D-printed Ti64 samples were tested under the same conditions (Khun et al., 2018).

11.3.1.2   Stainless Steel

Stainless steel alloys, like 316L SS, offer good tribological performance, combining low friction with adequate wear resistance. They find applications in orthopedic implants and dental components. In a study of ball-on-disk type wear analysis, it has been seen that the LB-PBF 316L SS has a low wear rate as compared to the CM 316L SS in the dry condition, under 1, 5, and 10 N of the applied load level. This is due to the higher hardness of LB-PBF than that of CM material, and it also happens due to the difference in roughness between the samples. Figure 11.6 shows the comparative study of COF and wear rate between LB-PBF and CM samples (Upadhyay & Kumar, 2020). In comparison to hot pressing and conventionally manufactured samples, the 316L stainless steel specimens created using SLM technology demonstrated higher wear resistance (Hashim, 2023).

11.3.1.3   Cobalt-Chromium Alloys

Cobalt-chromium (Co-Cr) alloys, such as Co-Cr-Mo, demonstrate exceptional wear resistance, making them suitable for knee and hip replacements. They have a comparatively low coefficient of friction, reducing the risk of wear debris-induced inflammation. In the ball-on-disk tribology test under dry conditions, cast and SLM Co-Cr alloy samples were tested. The experiment utilized a 5 N load and a 200 rpm rotational speed. The SLM sample demonstrated lower plastic deformation and exhibited significantly higher wear resistance compared to the Casting sample, which reflects that SLM samples are more suitable for implant manufacturing (Fu et al., 2022). The CAST specimens show considerable fluctuations in the friction coefficient, with an average value of 0.50. On the other hand, the SLM specimens maintain a relatively constant friction coefficient, hovering around 0.46, which is slightly lower

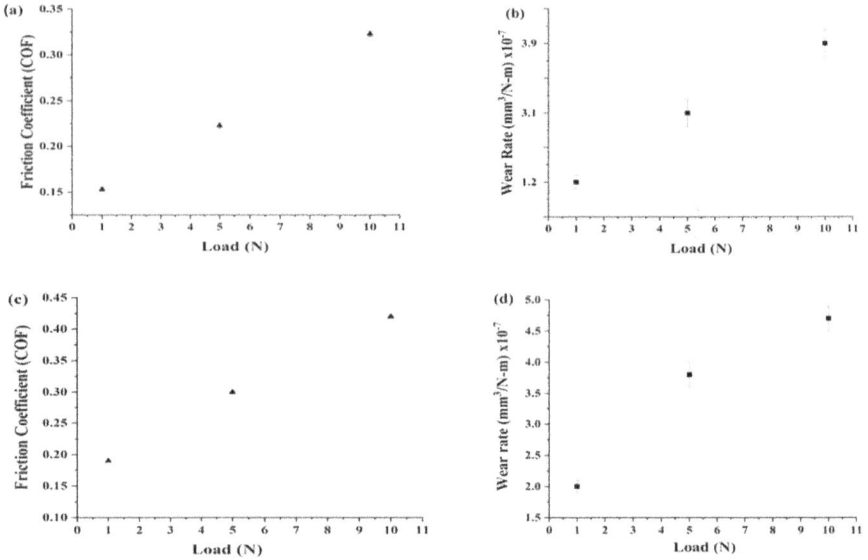

FIGURE 11.6 (a) and (b) COF and wear rate of LB-PBF; (c) and (d) the same parameters for a conventionally manufactured sample (Upadhyay & Kumar, 2020).

than the cast sample. The wear marks determine the dominant wear mechanisms for the CAST and SLM alloys, as illustrated in Figure 11.7. The worn surfaces of all tested specimens revealed the presence of parallel grooves, plastic deformation, and surface fatigue in the sliding direction. These observations suggest that fatigue and abrasive wear mechanisms were active during the testing conditions. For the CAST specimens, the worn surfaces displayed extensive partial exfoliation and metal wear debris. On the other hand, the grooves on the worn surface of SLM specimens were formed by the ploughing action and microcutting, and they appeared to be noticeably shallower compared to the grooves observed on the CAST specimens. Furthermore, a lower plastic deformation has been reported for the SLM sample.

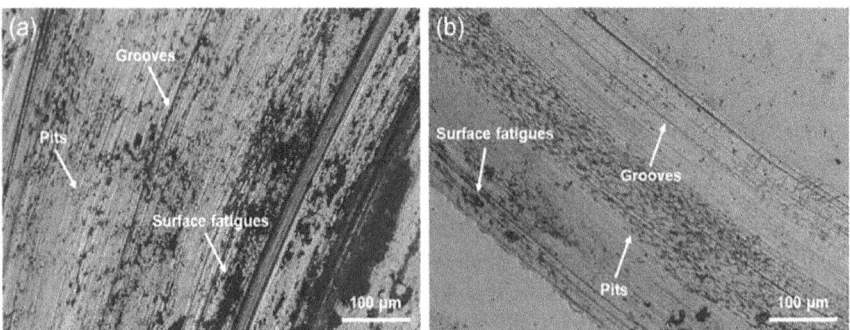

FIGURE 11.7 Metallurgical-microscopy wear scars images of (a) CAST and (b) SLM samples (Fu et al., 2022).

11.3.1.4   Polymeric Materials

Biocompatible polymers, such as polyether ether ketone (PEEK), polyethylene (PE), and polyurethane (PU), exhibit varying tribological properties. PEEK, for instance, has excellent wear resistance, making it suitable for spinal implants and articulating joints. Also, the COF of PEEK materials can be reduced by introducing carbon fibers in polymer structure, which also reduces the wear rate (Singh et al., 2019). The tribological performance of two distinct PEEK-based polymer coatings has been investigated using a ball-on-disk machine operating in sliding mode at varying load conditions. Subsequently, the findings were meticulously compared with those of the conventional metal coating, COF and wear rates are low for the PEEK due to low Hertzian pressure and temperature at the contact zone (Massocchi et al., 2021).

11.3.1.5   Bioactive and Biodegradable Materials

Hydroxyapatite (HA) coatings on metallic substrates improve tribological properties, mainly in bone implants. Biodegradable materials like polylactic acid (PLA) and polyglycolic acid are used in temporary implants and scaffolds, and their tribological performance is tailored based on the specific application (Bartolo et al., 2012). These types of materials are also useful for drug delivery systems.

11.3.1.6   Nanocomposite and Hybrid Materials

Nanocomposite materials, such as graphene-reinforced polymers or ceramics, can enhance wear resistance and reduce friction. Hybrid materials combining metals and polymers offer a balance of mechanical and tribological properties for specific biomedical applications (Ahmad et al., 2010; Raj et al., 2022a,b; Raj and Dixit, 2022). A PLA and naturally extracted corn cob composite formed using the FDM technique show a low COF and 10% higher wear resistance as compared to PLA. Also, PLA-CC composite has a higher load carrying capacity as PLA. Figures 11.8 and 11.9 represent the COF and specific wear rate at different percentages of CC in the PLA-CC composite.

COF increases with the increase in the applied load due to heat generated during friction, which softens the composite and increases the area of contact between the rubbed surfaces. Also, such heat softens the composite and increases the shear resistance, which increases the wear rate at higher applied loads. The lowest COF and wear rate are reported for PLA-CC 10 (Fouly et al., 2022).

11.3.2   Surface Modification Techniques

Surface modification methods seek to modify the surface characteristics of AM components while preserving their internal properties. The objective is to enhance friction and wear resistance, thereby improving overall performance and minimizing the risk of failure in rigorous biomedical applications. Various surface modification approaches have been investigated to tackle the tribological issues encountered in AM-produced biomedical components.

11.3.2.1   Coatings for Wear Resistance

These coatings are carefully applied to the surface of the substrate material to enhance its performance and prolong its service life. In the realm of biomedical applications, these coatings find frequent utilization on implants, surgical instruments, and various

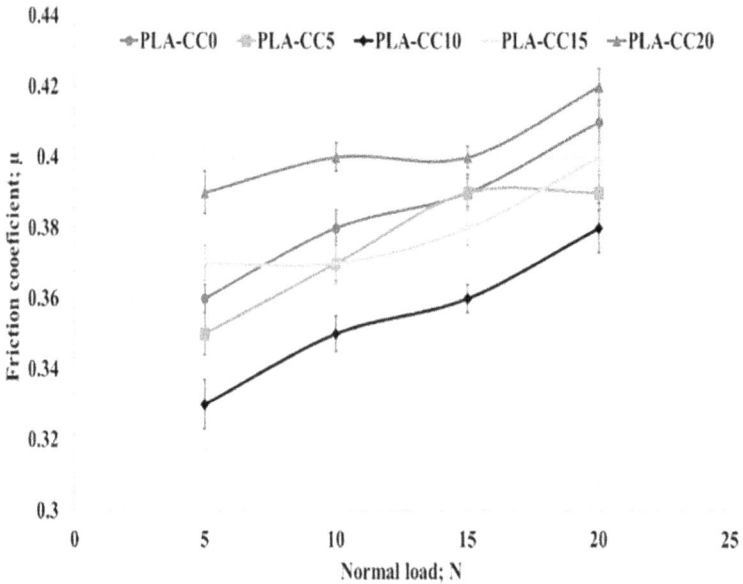

FIGURE 11.8 COF of PLA-CC at different percentages of CC under varying loads (Fouly et al., 2022).

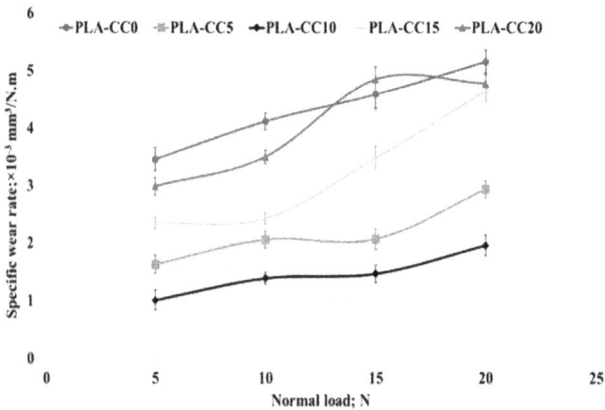

FIGURE 11.9 Specific wear rate of PLA-CC at different percentages of CC under varying loads (Fouly et al., 2022).

medical devices to augment their durability and biocompatibility. DLC coatings are thin amorphous carbon films known for their exceptional hardness, low friction, and high wear resistance. These coatings are widely used to enhance the tribological properties of AM biomedical parts, including implants and surgical instruments. By reducing the coefficient of friction between the part and contacting surfaces, DLC coatings effectively minimize wear and the risk of abrasive damage. Moreover, the hardness of DLC coatings provides protection against abrasive wear and surface deformation, resulting in improved longevity for AM biomedical components (Marian et al., 2023). HA coatings are bioactive materials employed to improve the tribological and bio-compatible properties of AM biomedical parts. HA, a natural component of bone, promotes osseointegration when applied as a coating on metallic AM implants. This process enhances load transfer between the implant and the surrounding bone, leading to reduced wear and improved stability. Furthermore, the bioactivity of HA coatings encourages tissue growth, which contributes to a more stable implant and lowers long-term wear rates (Akram et al., 2023). Bioactive glass coatings are also utilized in AM biomedical parts to enhance tribological properties and encourage tissue integration. The unique composition of bioactive glass fosters the formation of a biologically active hydroxyapatite layer when in contact with body fluids. This hydroxyapatite layer facilitates osseointegration and minimizes wear between the implant and surrounding tissues (Ramezani & Ripin, 2023). Additionally, bioactive glass coatings can be employed in drug delivery systems to locally release therapeutic agents, aiding in healing and reducing inflammation (Negut & Ristoscu, 2023).

A layered Ti6Al4V/316L substrate is produced using the LB-PBF process, and a coating of TiO2 ceramic layer is formed on the layered Ti6Al4V/316L substrate using the anodic oxidation method at different coating times (15, 30, and 60 minutes). COF and wear rate are measured using the reciprocating tribo tester, and the results of all the samples are depicted in Figure 11.10. The wear rate for the untreated 316L is much higher, while the A-60 sample has the lowest wear rate among all the samples. This is due to the TiO_2 ceramic layer formed on the specimen, which increases the load carrying capacity and hardness of the material while reducing plastic deformation (Turalıoğlu et al., 2021).

Plasma oxidation has been carried out on an L-PBF manufactured sample of pure titanium to enhance the tribological properties. Plasma oxidation treatments were conducted for 1 hour and 4 hours at 650°C and 750°C in a 100% O2 environment, respectively. After plasma oxidation, the hard oxide layer that develops reduces severe adhesive wear by improving shear resistance and providing a chemically inert structure for wear prevention. Figure 11.11 represents the hardness, COF, and wear rate of

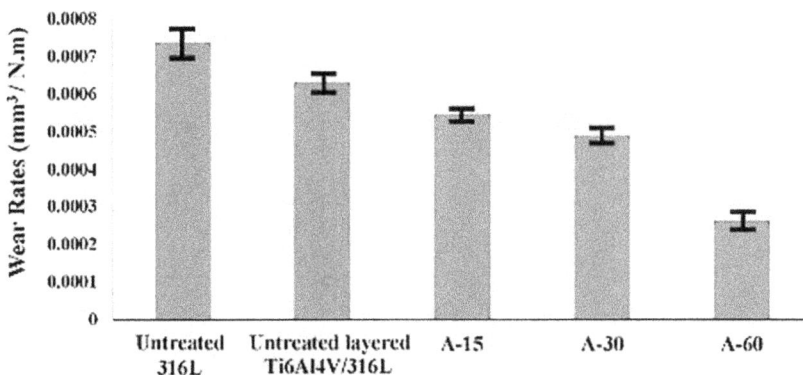

FIGURE 11.10 Wear rate of all samples (Turalıoğlu et al., 2021).

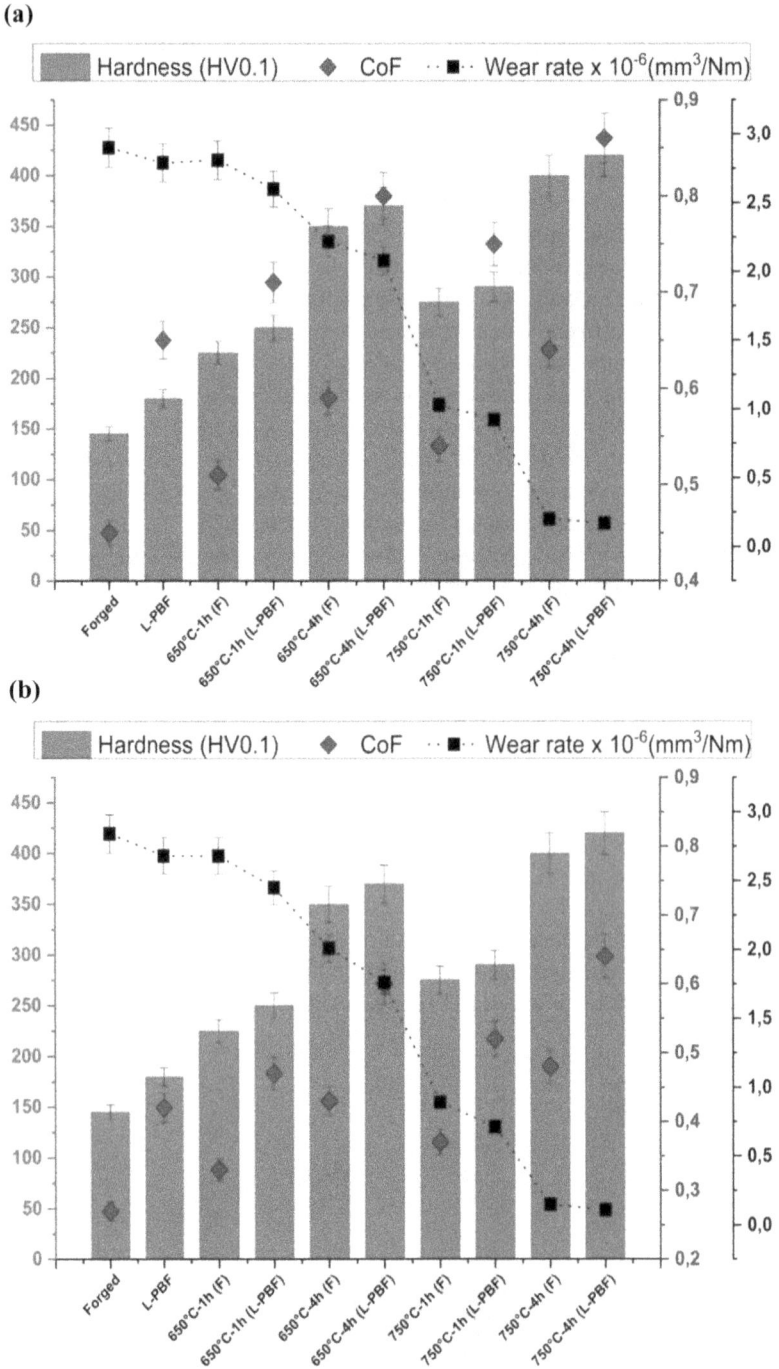

FIGURE 11.11 Hardness, COF, and wear rate of forged and L-PBF specimens under (a) dry and (b) SBF conditions (Kovacı, 2019).

forged and L-PBF specimens in (a) dry and (b) SBF conditions. Forged Cp-Ti samples that hadn't been treated had a higher rate of wear than any other samples, and untreated L-PBF Cp-Ti samples had a higher rate of wear than L-PBF/plasma-oxidized samples. Therefore, the increased hardness of both the plasma oxidation process and the L-PBF process can be stated to have reduced wear rates (Kovacı, 2019).

As discussed for all coatings for wear and corrosion resistance, some other types of coating, like TiN coating, silver coating, and zirconia coating, can also be employed to enhance the tribological properties of AM parts for biomedical devices. Because all these coatings have excellent biocompatibility and high wear and corrosion resistance.

11.3.2.2 Surface Texturing for Friction Reduction

Surface texture is another method to reduce friction by reducing the contact area. This can be produced by micromachining, laser surface texturing, and micro- and nanosurface generation using AM itself.

11.3.2.2.1 Laser Surface Texturing

It is another texturing method employed to reduce friction in AM biomedical parts. Laser surface texturing creates precise micro- or nano-sized surface patterns, effectively reducing the effective contact area and modifying the frictional behavior at the interface. During the laser ablation process, a laser beam is directly focused onto the work surface, causing the material to be removed through heating. This results in the melting and vaporization of the work material in the irradiated zone. As a consequence, selective material is removed, and the surface topography undergoes modification. This results in lower friction coefficients, making laser surface texturing ideal for prosthetics, joint replacements, and other load-bearing AM biomedical components. Figure 11.12 shows the schematic of principle of the laser surface texturing (LST) (Riveiro et al., 2018).

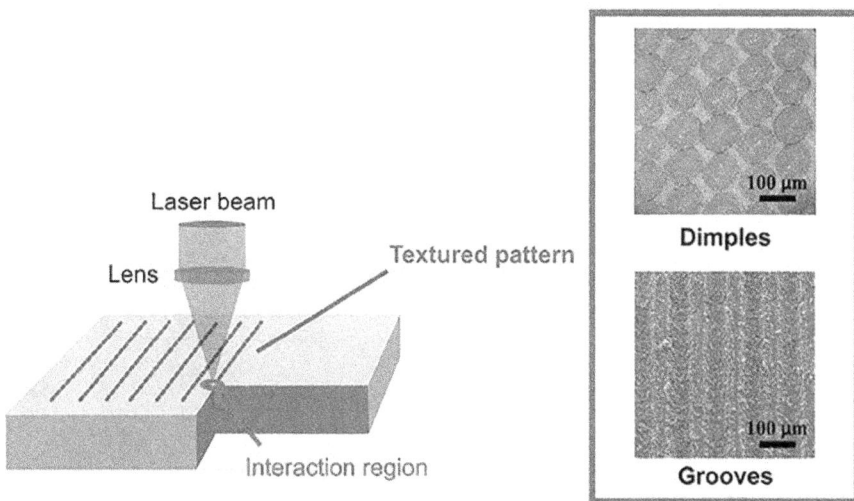

FIGURE 11.12   Schematic of principle of the laser surface texturing (LST) (Riveiro et al., 2018).

11.3.2.2.2 Micro/Nanosurface Patterns

Micro/nanosurface patterns generated through AM processes are also beneficial for improving the tribological properties of biomedical parts. These patterns retain lubricants, minimize surface adhesion, and promote fluid retention, which enhances lubrication and reduces friction-induced wear (Dwivedi et al., 2023). Incorporating tailored surface patterns in AM implants and prosthetics further improves wear resistance and reduces wear debris generation. COF and wear rate are significantly improved for the additively manufactured pattern through the SLM technique as compared to simple casted samples due to the smaller contact area (Dwivedi et al., 2022a). Figure 11.13 shows the generation of the AM texture of SS 316l using the SLM technique. The specific wear rate and COF of different types of patterns under lubricating and dry conditions, respectively, where grooves (parallel) have low COF and wear rate for dry conditions, and dimple patterns show low COF and wear rate for lubricating conditions as compared with cast samples. Cellular structures can provide lightweight components for implants and prostheses. In this regard, the S1 to S5 cellular structure of Ti6Al4V is produced by SLM, as shown in Figure 11.14. Sliding ball-on-plate tribology tests were performed with PBS fluid at 37°C to mimic the human body environment. As a result, high wear resistance was reported with an increase in the lower open cell size due to lower contact stress. A SEM micrograph of worn surfaces from S1 to S5 is shown in Figure 11.14. From the figure, it may be concluded that all the Ti6Al4V samples exhibited plastic deformation as a result of the abrasive wear test highlighted in figures a3, b3, c3, d3, and e3 (Bartolomeu et al., 2017).

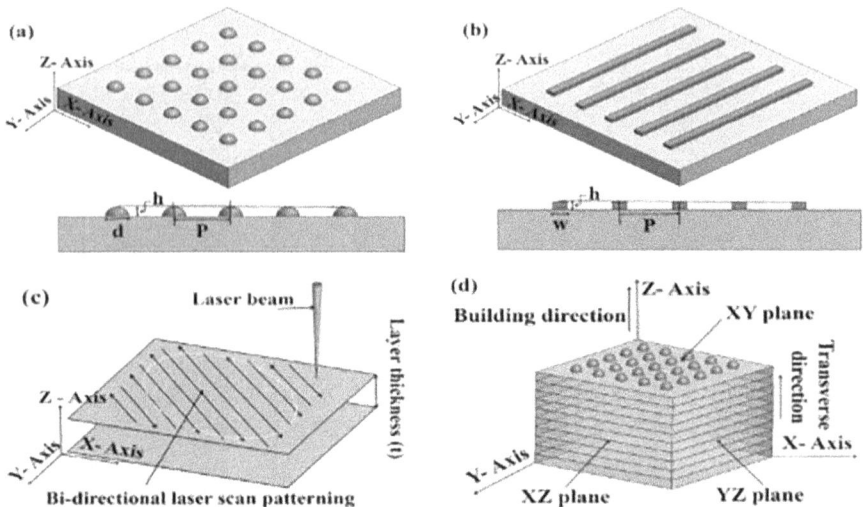

FIGURE 11.13 Isometric view of (a) dimple textured and (b) groove textured surfaces of SS 316L; (c and d) laser scanning; bi-directional rotation scan and build direction (BD); and transverse direction (TD), respectively (Dwivedi et al., 2022a).

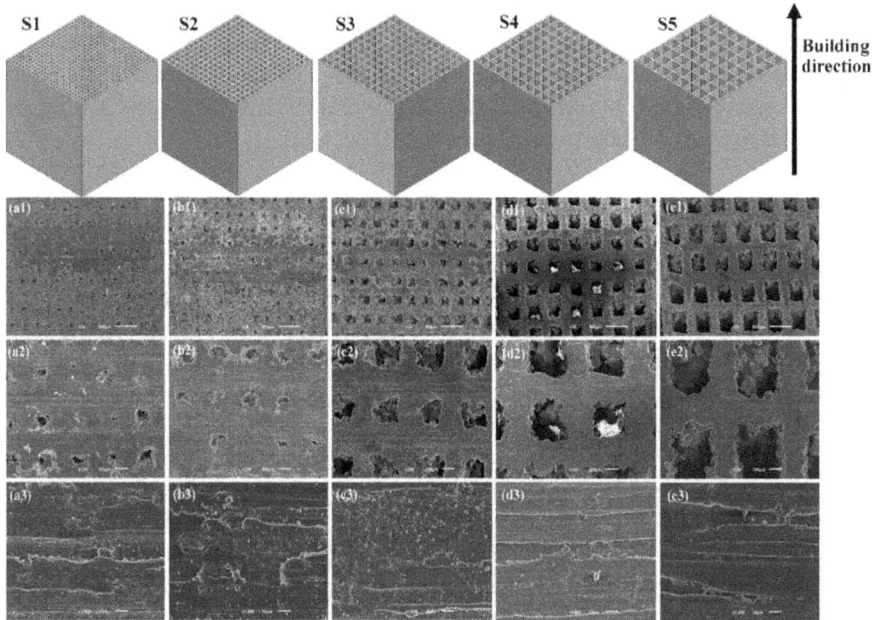

FIGURE 11.14 SEM micrograph of worn surfaces of S1–S5 (Bartolomeu et al., 2017).

11.3.2.3 Lubrication Solutions for Tribology Improvement

Lubrication is a critical factor in improving the tribological performance of AM parts used in biomedical applications. To address the challenges of wear and friction, various lubrication solutions have been developed, drawing inspiration from nature and advanced material science.

Synovial fluid (Mimicking Lubricant), a natural lubricant found in joints, offers low friction and wear protection. Synovial fluid-mimicking lubricants aim to replicate its composition and lubricating properties. Typically comprising water, hyaluronic acid, and lubricin-like molecules, these lubricants form a boundary layer on the contacting surfaces of AM biomedical parts, reducing friction and wear between components (Dwivedi et al., 2022b). The incorporation of synovial fluid-mimicking lubricants in AM implants and prosthetics enhances joint functionality and increases longevity (Lawson et al., 2021).

Self-lubricating hydrogels are designed to release lubricating molecules in response to mechanical forces, such as shear stress. These hydrogels can be incorporated into AM biomedical parts to provide an inherent lubrication mechanism that activates during movement or load-bearing activities (Porte et al., 2019). Self-lubricating hydrogels offer sustained lubrication, reducing friction and wear over extended periods, thereby enhancing the performance and lifespan of AM implants and prosthetics.

Injectable lubricating hydrogels offer a versatile lubrication solution for AM biomedical parts, allowing precise delivery and conforming to complex geometries. These hydrogels can be injected into joint spaces or integrated into porous scaffolds for tissue engineering. Injectable lubricating hydrogels reduce friction and wear in the joint space, providing pain relief and improving the functionality of AM implants. Injectable hydrogels have garnered significant attention from biomaterials scientists due to their potential applications in cartilage and bone tissue engineering. These hydrogels offer a minimally invasive alternative to implantation surgery as they can be administered through injections. Moreover, they possess the unique capability to take on any desired shape, allowing them to adapt to irregular defects and provide a suitable match (Liu et al., 2017).

11.4   IN VITRO WEAR STUDY OF AM PARTS

In vitro wear tests are controlled in laboratory environments; this involves the utilization of simulated biological fluids and wear testers, which imitate the conditions encountered by biomedical components inside the human body. Typically, these studies are carried out during the initial phases of development to evaluate the wear and corrosion characteristics of additively manufactured biomedical components before advancing to more intricate and costly in vivo studies.

As per ASTM G 99-5 standards, all wear tests are carried out employing a pin-on-disk wear test machine, utilizing freshly prepared SBF (simulated bodily fluid) maintained at a constant temperature of $37 \pm 1°C$. Table 11.3 presents the SBF chemical composition for a volume of 1 L (Reger et al., 2023).

However, the corrosion performance of biometallic materials and their alloys, intended for use in bio implant applications, normally uses Hank's balanced salt solutions (HBSS). These solutions are commonly chosen to simulate the corrosive conditions within living organisms' bodies. HBSS is considered a more aggressive environment than artificial blood plasma, mainly due to its higher chloride concentration (Reger et al., 2023).

In vitro, corrosion tests of a developed Ti-Ta-Nb–Mo–Zr high entropy alloy by plasma arc AM technique have been performed, and the decrease in corrosion potential along with the increase in corrosion current density raise the corrosion rate for the high pH value of the SBF solution. This suggests that the Ti-Ta-Nb-Mo-Zr HEA (High Entropy Alloy) developed demonstrates superior corrosion resistance at lower pH values of the SBF solution. It shows excellent biocompatibility in terms of corrosion resistance and is highly suitable for knee implant applications (Kumar et al., 2023).

TABLE 11.3
Simulated Bodily Fluid Chemical Composition for a Volume of 1 L

Order	Reagent	Amount
1	NaCl (Sodium chloride)	7.996 g
2	$NaHCO_3$ (Sodium bicarbonate)	0.350 g
3	KCl (Potassium chloride)	0.224 g
4	$K_2HPO_4 \cdot 3H_2O$ (Potassium dihydrogen phosphate)	0.228 g
5	$MgCl_2 \cdot 6H_2O$ (Magnesium chloride hexahydrate)	0.305 g
6	1 M HCl	39 mL
7	$CaCl_2$ (Calcium chloride dihydrate)	0.278 g
8	Na_2SO_4 (Sodium sulfate anhydrous)	0.071 g
9	$(CH_2OH)_3CNH_2$ (Tris hydrochloride)	6.057 g
10	1 M HCl	0–5 mL

Teeth surface wear is normally performed by a two-body contact wear test, which demonstrates the consequence of direct interaction between the surface of the teeth being tested and the opposing teeth (Suwannaroop et al., 2011). The load exerted on the teeth prior to testing serves as a means of simulating the effects of aging under oral conditions. It was observed that subjecting the samples to 250,000 cycles of loading is approximately equivalent to one year of wear experienced by natural teeth in the mouth. In order to ensure reliable in vitro outcomes that closely resemble clinical wear, the wear test employed standardized values for the applied force and sliding movement within the chewing simulator (DeLong & Douglas, 1991). This standardization helped achieve nearly comparable results between the laboratory simulation and real-world wear cases. Form lab resin, 3D-printed (using the DLP technique) denture teeth show excellent wear resistance as compared to conventionally prefabricated denture teeth (Mohamed et al., 2023). Also, the teeth printed by ASIGA and NextDent show low wear resistance as compared to Form Lab denture teeth due to a different fabrication method. Figure 11.15 shows SEM images of the wear pattern of denture teeth printed on different systems (Gad et al., 2023).

So, conducting wear studies in a controlled laboratory environment (in vitro) has its drawbacks, as it fails to completely mimic the intricate and ever-changing conditions present within the human body. Therefore, the shift towards conducting studies within a living organism (in vivo) becomes imperative to authenticate the results and ascertain the long-term effectiveness of the biomedical components.

In vivo wear studies involve the examination of additively manufactured parts within living organisms or human participants. These investigations take place in carefully regulated clinical environments, enabling the observation of the components' wear characteristics under actual physiological conditions for an extended duration. Undertaking in vivo wear studies helps in real-life validation, biocompatibility assessment, long-term performance, and patient-specific response (Bandyopadhyay et al., 2023). Even though they are important, conducting in vivo studies is more intricate, requires more time, and incurs higher costs when compared to in vitro studies. Moreover, ethical factors and the necessity for obtaining regulatory approvals further contribute to the difficulties involved in performing in vivo wear testing.

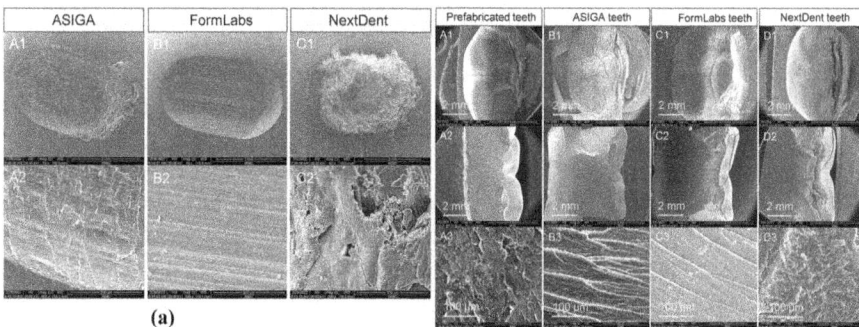

FIGURE 11.15 (a) SEM images of the denture teeth wear pattern at high (A2, B2, C2) and low (A1, B1, C1) magnification. (b) SEM images of fractured teeth.

11.5 FUTURE PERSPECTIVES AND CHALLENGES

Integrating tribological analysis into the design process of additively manufactured biomedical components is highly essential for optimizing their performance and reliability. By considering tribological factors from the early stages of design, engineers can select suitable materials, surface treatments, and lubrication methods to minimize wear and friction as discussed earlier. Creating design frameworks that incorporate tribological considerations will lead to improved functionality, reduced failure rates, and enhanced patient outcomes for biomedical applications.

The development of multi-material 3D printing technologies holds tremendous potential for fabricating complex biomedical systems with enhanced functionality. By combining different materials with distinct mechanical properties, researchers can create patient-specific implants, drug delivery devices, and tissue constructs that closely mimic the native tissues' behavior. However, this area faces challenges related to material compatibility, process optimization, and ensuring seamless integration between different materials within the printed structure.

As additively manufactured biomedical components are increasingly used in clinical settings, it becomes crucial to perform long-term biocompatibility and wear assessments. Biocompatibility studies need to evaluate how these components interact with the body over extended periods, identifying potential adverse reactions or degradation issues. Long-term wear assessment helps ensure that the materials and designs chosen maintain their functionality and structural integrity over the device's expected lifespan.

Regulatory bodies have an impact on ensuring the safety and efficacy of medical devices. As additive manufacturing continues to advance in the biomedical sector, regulatory agencies face the challenge of developing appropriate guidelines and standards specific to AM technologies. Addressing issues related to material traceability, process validation, and quality control will be crucial to gaining regulatory approval for the medical industry.

Standardization is essential for ensuring the consistency, quality, and reproducibility of additively manufactured biomedical components. Establishing industry-wide standards for material properties, testing methods, and performance evaluation will foster trust in AM technologies and facilitate widespread adoption in clinical settings.

11.6 CONCLUSION

The AM process possesses the capability to produce intricate parts in comparison to other manufacturing methods. This feature makes it particularly suitable for applications in the medical industry, where customized products are required. A thorough review of the literature demonstrates that the performance of AM biomedical devices is significantly influenced by their tribological properties. By adjusting the process parameters during the manufacturing of these parts, their tribological properties can be notably improved. Proper material selection, surface modification, and lubrication techniques have been applied to AM parts, resulting in z lower wear rate and COF compared to conventionally manufactured components. This substantiates the superior performance and durability of AM parts. Furthermore, an in vitro wear study indicates enhanced functionality and biocompatibility of AM parts when subjected to body fluids.

REFERENCES

Ahmad, I., Kennedy, A., & Zhu, Y. Q. (2010). Wear resistant properties of multi-walled carbon nanotubes reinforced Al2O3 nanocomposites. *Wear*, *269*(1), 71–78. https://doi.org/10.1016/j.wear.2010.03.009.

Akmal, J. S., Salmi, M., Mäkitie, A., Björkstrand, R., & Partanen, J. (2018). Implementation of industrial additive manufacturing: Intelligent implants and drug delivery systems. *Journal of Functional Biomaterials*, *9*(3), 41. https://doi.org/10.3390/jfb9030041.

Akram, W., Zahid, R., Usama, R. M., AlQahtani, S. A., Dahshan, M., Basit, M. A., & Yasir, M. (2023). Enhancement of antibacterial properties, surface morphology and in vitro bioactivity of hydroxyapatite-zinc oxide nanocomposite coating by electrophoretic deposition technique. *Bioengineering*, *10*(6), 693. https://doi.org/10.3390/bioengineering10060693.

Alam, Md. S. (2021). Tribology in recent biomedical engineering: a review. *Material Science & Engineering International Journal*, *5*(4), 103–109. https://doi.org/10.15406/mseij.2021.05.00165.

Bandyopadhyay, A., Mitra, I., Goodman, S. B., Kumar, M., & Bose, S. (2023). Improving biocompatibility for next generation of metallic implants. *Progress in Materials Science*, *133*, 101053. https://doi.org/10.1016/j.pmatsci.2022.101053.

Bartolo, P., Kruth, J.-P., Silva, J., Levy, G., Malshe, A., Rajurkar, K., Mitsuishi, M., Ciurana, J., & Leu, M. (2012). Biomedical production of implants by additive electro-chemical and physical processes. *CIRP Annals*, *61*(2), 635–655. https://doi.org/https://doi.org/10.1016/j.cirp.2012.05.005.

Bartolomeu, F., Sampaio, M., Carvalho, O., Pinto, E., Alves, N., Gomes, J. R., Silva, F. S., & Miranda, G. (2017). Tribological behavior of Ti6Al4V cellular structures produced by selective laser melting. *Journal of the Mechanical Behavior of Biomedical Materials*, *69*, 128–134. https://doi.org/10.1016/j.jmbbm.2017.01.004.

Bertolotti, L., Sharma, A., & Parekh, S. G. (2022). Electron beam manufacturing 3D printed titanium with titanium nitride coating shows favorable overall performance when compared to selective laser sintering cobalt chrome in patient specific talus implants. *National Journal of Clinical Orthopaedics*, *6*(4), 22–27. https://doi.org/10.33545/orthor.2022.v6.i4a.378.

DeLong, R., & Douglas, W. H. (1991). An artificial oral environment for testing dental materials. *IEEE Transactions on Biomedical Engineering*, *38*(4), 339–345. https://doi.org/10.1109/10.133228.

Dikova, T. D., Dzhendov, D. A., Ivanov, D., & Bliznakova, K. (2018). Dimensional accuracy and surface roughness of polymeric dental bridges produced by different 3D printing processes, *Archives of Materials Science and Engineering, 94*(2), 65–75. https://www.rapidshape.de/.

Dwivedi, S., Dixit, A. R., Das, A. K., & Adamczuk, K. (2022a). Additive texturing of metallic implant surfaces for improved wetting and biotribological performance. *Journal of Materials Research and Technology*, *20*, 2650–2667. https://doi.org/10.1016/j.jmrt.2022.08.029.

Dwivedi, S., Dixit, A. R., Das, A. K., & Nag, A. (2023). A novel additive texturing of stainless steel *316*L through binder jetting additive manufacturing. *International Journal of Precision Engineering and Manufacturing-Green Technology*, pp. 1–9. https://doi.org/10.1007/s40684-023-00508-5.

Dwivedi, S., Rai Dixit, A., & Kumar Das, A. (2022b). Wetting behavior of selective laser melted (SLM) bio-medical grade stainless steel *316*L. *Materials Today: Proceedings*, *56*, 46–50. https://doi.org/10.1016/j.matpr.2021.12.046.

Evans, D., Rahman, M. H., Heintzen, M., Welty, J., Leslie, J., Hall, K., & Menezes, P. L. (2022). 4-Additively manufactured functionally graded metallic materials. In P. Kumar, M. Misra, & P. L. Menezes (Eds.), *Tribology of Additively Manufactured Materials* (pp. 107–136). Elsevier. https://doi.org/10.1016/B978-0-12-821328-5.00004-4.

Fouly, A., Assaifan, A. K., Alnaser, I. A., Hussein, O. A., & Abdo, H. S. (2022). Evaluating the mechanical and tribological properties of 3D printed polylactic-acid (PLA) green-composite for artificial implant: Hip joint case study. *Polymers, 14*(23), 5299. https://doi.org/10.3390/polym14235299.

Fu, W., Liu, S., Jiao, J., Xie, Z., Huang, X., Lu, Y., Liu, H., Hu, S., Zuo, E., Kou, N., & Ma, G. (2022). Wear resistance and biocompatibility of Co-Cr dental alloys fabricated with CAST and SLM techniques. *Materials, 15*(9), 3263. https://doi.org/10.3390/ma15093263.

Gad, M. M., Alalawi, H., Akhtar, S., Al-Ghamdi, R., Alghamdi, R., Al-Jefri, A., & Al-Qarni, F. D. (2023). Strength and wear behavior of three-dimensional printed and prefabricated denture teeth: An in vitro comparative analysis. *European Journal of Dentistry, 17*(04), 1248–1256. https://doi.org/10.1055/s-0042-1759885.

Gao, J., Li, M., Cheng, J., Liu, X., Liu, Z., Liu, J., & Tang, P. (2023). 3D-printed GelMA/PEGDA/F127DA scaffolds for bone regeneration. *Journal of Functional Biomaterials, 14*(2), 96. https://doi.org/10.3390/jfb14020096.

Hao, Y.-L., Li, S.-J., & Yang, R. (2016). Biomedical titanium alloys and their additive manufacturing. *Rare Metals, 35*(9), 661–671. https://doi.org/10.1007/s12598-016-0793-5.

Hashim, S. (2023). Tribological Behavior of 316L Stainless Steel Manufactured by Selective Laser Melting, in *4th International Conference on Architectural & Civil Engineering Sciences* (pp. 142–146).

Kaur, S., Ghadirinejad, K., & Oskouei, R. H. (2019). An overview on the tribological performance of titanium alloys with surface modifications for biomedical applications. *Lubricants, 7*(8), 65. https://doi.org/10.3390/lubricants7080065.

Khun, N. W., Toh, W. Q., Tan, X. P., Liu, E., & Tor, S. B. (2018). Tribological properties of three-dimensionally printed Ti-6Al-4V material via electron beam melting process tested against 100Cr6 steel without and with Hank's solution. *Journal of Tribology, 140*(6), 061606. https://doi.org/10.1115/1.4040158.

Kovacı, H. (2019). Comparison of the microstructural, mechanical and wear properties of plasma oxidized Cp-Ti prepared by laser powder bed fusion additive manufacturing and forging processes. *Surface and Coatings Technology, 374*, 987–996. https://doi.org/10.1016/j.surfcoat.2019.06.095.

Kumar, P., Jain, N. K., Jaiswal, S., & Gupta, S. (2023). Development of Ti-Ta-Nb-Mo-Zr high entropy alloy by μ-plasma arc additive manufacturing process for knee implant applications and its biocompatibility evaluation. *Journal of Materials Research and Technology, 22*, 541–555. https://doi.org/10.1016/j.jmrt.2022.11.167.

Lawson, T. B., Mäkelä, J. T. A., Klein, T., Snyder, B. D., & Grinstaff, M. W. (2021). Nanotechnology and osteoarthritis. Part 2: Opportunities for advanced devices and therapeutics. *Journal of Orthopaedic Research, 39*(3), 473–484. https://doi.org/10.1002/jor.24842.

Liu, M., Zeng, X., Ma, C., Yi, H., Ali, Z., Mou, X., Li, S., Deng, Y., & He, N. (2017). Injectable hydrogels for cartilage and bone tissue engineering. *Bone Research, 5*(1), 17014. https://doi.org/10.1038/boneres.2017.14.

Malik, A., Rouf, S., Ul Haq, M. I., Raina, A., Valerga Puerta, A. P., Sagbas, B., & Ruggiero, A. (2022). Tribo-corrosive behavior of additive manufactured parts for orthopaedic applications. *Journal of Orthopaedics, 34*, 49–60. https://doi.org/10.1016/j.jor.2022.08.006.

Marian, M., Zambrano, D. F., Rothammer, B., Waltenberger, V., Boidi, G., Krapf, A., Merle, B., Stampfl, J., Rosenkranz, A., Gachot, C., & Grützmacher, P. G. (2023). Combining multi-scale surface texturing and DLC coatings for improved tribological performance of 3D printed polymers. *Surface and Coatings Technology, 466*, 129682. https://doi.org/10.1016/j.surfcoat.2023.129682.

Massocchi, D., Riboni, G., Lecis, N., Chatterton, S., & Pennacchi, P. (2021). Tribological characterization of polyether ether ketone (PEEK) polymers produced by additive manufacturing for hydrodynamic bearing application. *Lubricants, 9*(11), 112. https://doi.org/10.3390/lubricants9110112.

Moghadasi, K., Mohd Isa, M. S., Ariffin, M. A., Mohd jamil, M. Z., Raja, S., Wu, B., Yamani, M., Bin Muhamad, M. R., Yusof, F., Jamaludin, M. F., Ab Karim, M. S. B., Abdul Razak, B. B., & Yusoff, N. B. (2022). A review on biomedical implant materials and the effect of friction stir based techniques on their mechanical and tribological properties. *Journal of Materials Research and Technology*, *17*, 1054–1121. https://doi.org/10.1016/j.jmrt.2022.01.050.

Mohamed, A., Takaichi, A., Kajima, Y., Takahashi, H., & Wakabayashi, N. (2023). Bond strength of CAD/CAM denture teeth to a denture base resin in a milled monolithic unit. *Journal of Prosthodontic Research*, *67*(4), 610–618. https://doi.org/10.2186/jpr.JPR_D_22_00190

Negut, I., & Ristoscu, C. (2023). Bioactive glasses for soft and hard tissue healing applications—A short review. *Applied Sciences*, *13*(10), 6151. https://doi.org/10.3390/app13106151.

Porte, E., Cann, P., & Masen, M. (2019). Fluid load support does not explain tribological performance of PVA hydrogels. *Journal of the Mechanical Behavior of Biomedical Materials*, *90*, 284–294. https://doi.org/10.1016/j.jmbbm.2018.09.048.

Rafiee, M., Farahani, R. D., & Therriault, D. (2020). Multi-material 3D and 4D printing: A survey. *Advanced Science*, *7*(12), 1902307. https://doi.org/10.1002/advs.201902307.

Raj, R., & Dixit, A. R. (2022). Direct ink writing of carbon-doped polymeric composite ink: A review on its requirements and applications. *3D Printing and Additive Manufacturing*, *10*(4), 828–854. https://doi.org/10.1089/3dp.2021.0209.

Raj, R., Dixit, A. R., Łukaszewski, K., Wichniarek, R., Rybarczyk, J., Kuczko, W., & Górski, F. (2022a). Numerical and experimental mechanical analysis of additively manufactured ankle–foot orthoses. *Materials*, *15*(17), 6130. https://doi.org/10.3390/ma15176130.

Raj, R., Dixit, A. R., Singh, S. S., & Paul, S. (2022b). Print parameter optimization and mechanical deformation analysis of alumina-nanoparticle doped photocurable nanocomposites fabricated using vat-photopolymerization based additive technology. *Additive Manufacturing*, *60*, 103201. https://doi.org/10.1016/j.addma.2022.103201.

Ramakrishna, S., Mayer, J., Wintermantel, E., & Leong, K. W. (n.d.). Biomedical applications of polymer-composite materials: A review. *Composites Science and Technology, 61*(9), 1189–1224. www.elsevier.com/locate/compscitech.

Ramezani, M., & Ripin, Z. M. (2023). An overview of enhancing the performance of medical implants with nanocomposites. *Journal of Composites Science*, *7*(5), 199. https://doi.org/10.3390/jcs7050199.

Ranjan, N., Tyagi, R., Kumar, R., & Kumar, V. (2023). On fabrication of acrylonitrile butadiene styrene-zirconium oxide composite feedstock for 3D printing-based rapid tooling applications. *Journal of Thermoplastic Composite Materials, 0*(0), 08927057231186310. https://doi.org/10.1177/089270572311863.

Reger, N. C., Devi, K. B., Balla, V. K., & Das, M. (2023). In-vitro corrosion and wear studies of ceramic layers on additively manufactured Zr metal for implant applications. *Transactions of the Indian Institute of Metals*, *76*(7), 1949–1958. https://doi.org/10.1007/s12666-023-02893-6.

Rezvani Ghomi, E., Khalili, S., Nouri Khorasani, S., Esmaeely Neisiany, R., & Ramakrishna, S. (2019). Wound dressings: Current advances and future directions. *Journal of Applied Polymer Science*, *136*(27), 47738. https://doi.org/10.1002/app.47738.

Riveiro, A., Maçon, A. L. B., del Val, J., Comesaña, R., & Pou, J. (2018). Laser surface texturing of polymers for biomedical applications. *Frontiers in Physics, 6,* 16. https://doi.org/10.3389/fphy.2018.00016.

Singh, S., Prakash, C., & Ramakrishna, S. (2019). 3D printing of polyether-ether-ketone for biomedical applications. *European Polymer Journal*, *114*, 234–248. https://doi.org/10.1016/j.eurpolymj.2019.02.035.

Singh, S., & Ramakrishna, S. (2017). Biomedical applications of additive manufacturing: Present and future. *Current Opinion in Biomedical Engineering*, *2*, 105–115. https://doi.org/10.1016/j.cobme.2017.05.006.

Suwannaroop, P., Chaijareenont, P., Koottathape, N., Takahashi, H., & Arksornnukit, M. (2011). In vitro wear resistance, hardness and elastic modulus of artificial denture teeth. *Dental Materials Journal*, *30*(4), 461–468. https://doi.org/10.4012/dmj.2010-200.

Turalıoğlu, K., Taftalı, M., Tekdir, H., Çomaklı, O., Yazıcı, M., Yetim, T., & Yetim, A. F. (2021). The tribological and corrosion properties of anodized Ti6Al4V/316L bimetallic structures manufactured by additive manufacturing. *Surface and Coatings Technology*, *405*, 126635. https://doi.org/10.1016/j.surfcoat.2020.126635.

Tyagi, R., Singh, G., Kuma, R., Kumar, V., & Singh, S. (2023). 3D-printed sandwiched acrylonitrile butadiene styrene/carbon fiber composites: Investigating mechanical, morphological, and fractural properties. *Journal of Materials Engineering and Performance*, *0*(0), 1–14. https://link.springer.com/article/10.1007/s11665-023-08292-8.

Upadhyay, R. K., & Kumar, A. (2020). Scratch and wear resistance of additive manufactured 316L stainless steel sample fabricated by laser powder bed fusion technique. *Wear*, *458–459*, 203437. https://doi.org/10.1016/j.wear.2020.203437.

Wan, X., Luo, L., Liu, Y., & Leng, J. (2020). Direct ink writing based 4D printing of materials and their applications. *Advanced Science*, *7*(16), 2001000. https://doi.org/10.1002/advs.202001000.

Wasyłeczko, M., Remiszewska, E., Sikorska, W., Dulnik, J., & Chwojnowski, A. (2023). Scaffolds for cartilage tissue engineering from a blend of polyethersulfone and polyurethane polymers. *Molecules*, *28*(7), 3195. https://doi.org/10.3390/molecules28073195.

Wu, Y., Wang, Y., Liu, M., Shi, D., Hu, N., & Feng, W. (2023). Mechanical properties and in vivo assessment of electron beam melted porous structures for orthopedic applications. *Metals*, *13*(6), 1034. https://doi.org/10.3390/met13061034.

12 Tribological Effect of 3D Printing in Industrial Applications

Harpreet Kaur Channi
Chandigarh University

12.1 INTRODUCTION

The science and engineering of tribology studies friction, wear, and lubrication of interacting surfaces in relative motion. The Greek word tribes means rubbing, meaning "the science of rubbing." Leonardo da Vinci was the first to explain the two laws of friction regulating the motion of a rectangular block sliding across a flat surface. Still, his work remained unpublished until the 20th century (Xu et al. 2023). Due to massive industrial progress since the beginning of the 20th century, knowledge in all fields of tribology has developed greatly, necessitating a better understanding. It encompasses understanding how surfaces interact, the effects of forces on those surfaces, and the mechanisms of wear and friction that occur during their contact. Tribology is vital in mechanical engineering, materials science, automotive engineering, manufacturing, and other sectors where moving parts must function well and last. Tribology seeks to improve mechanical systems' efficiency, dependability, and lifetime by reducing friction, wear, and lubrication (Quader 2021).

Tribology has been studied since ancient times as humanity observed and dealt with friction, wear, and lubrication. Ancient civilizations had a tribological understanding. Chariot axles and wooden bearings were lubricated with animal fats, plant oils, and waxes by ancient Egyptians and Sumerians. Leonardo da Vinci contributed substantially to friction and lubrication research in the 15th and 16th centuries. His investigations of fluid flow, including water and air, illuminated lubrication concepts. The British engineer Dr. H. Peter Jost invented "tribology" in 1966. "Tribology—Lubrication and Friction," his paper highlighted the economic consequences of friction and wear and stressed the necessity for research and development (Ren et al. 2023). Tribology groups and organizations were founded in the mid-20th century to promote the study and use of tribological concepts. The Institution of Mechanical Engineers' Tribology Committee was founded in 1946 as the Tribology Group in the UK.

The Japanese Society of Tribologists was formed in 1961, too. Late 20th-century tribology became a scientific field. Lubrication mechanics, wear characterization, surface engineering, and friction reduction strategies were extensively researched. Tribology relies on the development of new lubricants and procedures (Baligidad et al. 2023).

Synthetic, solid, and additive oils like graphite and molybdenum disulphide have been widely explored and employed to minimize friction and wear. Tribology has profited from materials science and surface engineering advances. Wear-resistant materials, coatings, and surface treatments have improved tribologically stressed components' performance and durability—Automotive, aerospace, energy, Manufacturing, and healthcare use tribology. Tribology investigates friction, wear, and lubrication of moving surfaces. "The science of rubbing" is derived from the Greek term tribos, meaning rubbing. Leonardo da Vinci was the first to describe the two friction rules governing a rectangular block moving on a flat surface (Shahrubudin et al. 2019). He didn't publish till the 20th century. Since the start of the 20th century, tribology knowledge has grown in all sectors due to significant industrial advancements.

12.2 OVERVIEW OF 3D PRINTING TECHNOLOGY

3D printing, also known as additive manufacturing, is a groundbreaking technique that creates three-dimensional items layer by layer. It's a flexible, fast-changing production method popular in many sectors. 3D printing basics: 3D printing builds tangible objects from computer models layer by layer. The process typically consists of the following steps, as shown in Figure 12.1.

12.2.1 MATERIALS

3D printing works with polymers, metals, ceramics, composites, and biological materials. Different printing technologies are designed to handle specific material types. Common polymer-based filaments used in 3D printing include ABS, PLA, PETG, and nylon. Metal 3D printing employs metal powders or wire-based materials (Mitsouras and Liacouras 2017).

12.2.2 PRINTING TECHNOLOGIES

Various 3D printing technologies exist, each with its strengths and limitations. Some prominent technologies include (Shahrubudin et al. 2019):

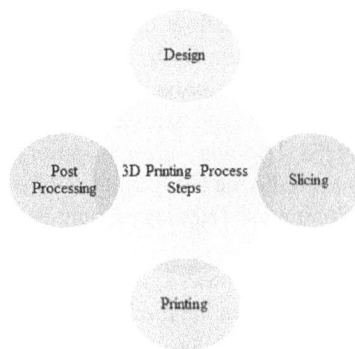

FIGURE 12.1   3D Printing process steps.

- **Fused Deposition Modelling (FDM):** This technology uses thermoplastic filaments that are melted and extruded through a nozzle to create layers. It is one of the most widely accessible and affordable 3D printing methods.
- **Stereolithography (SLA)/Digital Light Processing (DLP):** SLA and DLP technologies use liquid photopolymer resins that are solidified using light sources such as lasers or projectors. These technologies provide high-resolution prints with excellent surface detail.
- **Selective Laser Sintering (SLS):** SLS utilizes a laser to selectively fuse powdered materials, such as plastics or metals, to form each layer. It allows for the creation of complex geometries and functional parts.
- **Direct Metal Laser Sintering (DMLS)/Selective Laser Melting (SLM):** These metal 3D printing technologies use lasers to selectively melt metal powders, layer by layer, to create fully dense metal parts with high precision.

12.3 APPLICATIONS

3D printing offers several advantages over traditional manufacturing methods, such as design flexibility, cost-effective prototyping, customization, reduced waste, and on-demand manufacturing. Because of its benefits, 3D printing has applications in various industries and fields, including automotive, aerospace, healthcare, consumer goods, architecture, and education (Rudnik et al. 2022). It enables rapid prototyping, customized manufacturing, complex geometries, and on-demand production of spare parts. Rolling, turning, stamping, grinding, and polishing are tribology-based industrial operations. Most transportation techniques also use tribology in their mechanical components, and they slide or roll over at the contact between the wheels and the surfaces. Excavators, oil rigs, mine slurry pumps, and tunnel boring drills use tribology. Many industries utilize lubricants to control friction and wear (Pieterse and Nel 2016).

12.3.1 Tribological Effect of 3D Printing in Industrial Applications

Tribology has come a long way from ancient civilizations' observations to establishing dedicated research societies and applying advanced technologies in modern industries. The field continues to evolve, addressing emerging challenges and striving for improved performance, efficiency, and sustainability in mechanical systems. The tribological effect of 3D printing in industrial applications refers to the study of how materials' friction, wear, and lubrication characteristics are affected when manufactured using 3D printing techniques. Tribology plays a crucial role in determining the performance and longevity of mechanical components, especially in industrial settings where reliability and efficiency are essential. Below are the key considerations regarding the tribological effects of 3D printing in industrial applications (Ballesteros et al. 2021):

- **Material Selection:** The choice of materials for 3D printing can significantly impact the tribological properties of printed parts. Various materials, such as polymers, metals, and ceramics, can be used for 3D printing, each with its tribological characteristics. For example, certain polymer-based filaments may offer low friction and self-lubricating properties, while metal alloys can provide high strength and wear resistance.

- **Surface Finish:** The surface finish of 3D-printed parts can affect their tribological behaviour. Surface roughness from 3D printing's layer-by-layer additive manufacturing process may increase friction and wear. Polishing or surface treatment may minimize friction and improve surface quality.

- **Design Considerations:** The tribological performance of 3D-printed components depends on their design. Reduce friction and wear by optimizing shape to minimize sliding contact or adding lubrication channels. Adding functional aspects like integrated lubrication or wear-resistant components to the invention may also enhance tribological behaviour (Rudnik et al. 2022).

- **Lubrication:** Industrial lubrication reduces friction and wear. 3D Manufacturing lubricating channels or reservoirs is difficult. Multi-material printing and post-printing lubrication treatments may improve 3D-printed component lubrication.

- **Performance Validation:** Before industry use, 3D-printed items must be tested for tribological performance. This might entail friction and wear testing under appropriate operating circumstances compared to typical.

- **Anisotropy:** Depending on the printing orientation, 3D-printed items may have anisotropic mechanical and tribological characteristics. This anisotropy from layer-by-layer deposition may affect printed components' friction, wear, and lubrication. When developing industrial 3D-printed items, anisotropy must be considered.

- **Material Composites:** Combining materials or adding additives in 3D printing creates complex material composites. Lubricants, reinforcements, and wear-resistant particles may improve the tribological performance of printed components. Graphite or molybdenum disulphide may minimize friction and wear.

- **Post-Processing Techniques:** 3D-printed components may be improved tribologically by post-processing. Surface coating, polishing, and heat treatment improve wear resistance, friction, and self-lubrication.

- **Manufacturing Parameters:** Layer thickness, printing speed, and temperature affect the tribological qualities of 3D-printed components. The printed components' surface quality, mechanical strength, and tribological performance depend on optimal parameter selection. Process optimization and parameter management are studied to enhance 3D printing tribology (Friedrich 2018).

- **Real-Life Applications:** 3D printing has tribological applications in automotive, aerospace, manufacturing, and healthcare. 3D-printed vehicle components with optimized tribological qualities minimize engine friction, increase fuel economy, and boost performance. In healthcare, the customized wear-resistant prostheses may enhance patient outcomes.

- **Future Developments:** Tribological qualities of 3D-printed materials and components are being improved. Explore new materials, optimize printing techniques, and create sophisticated surface treatments to enhance 3D-printed components' wear resistance, friction reduction, and lubrication (Hanon and Zsidai 2020, Gbadeyan et al. 2021).

In industrial applications, 3D printing's tribological impact demands careful consideration of material selection, surface polish, design optimization, lubrication, and performance validation. By recognizing and resolving these aspects, 3D printing may be used to ensure components' dependable and efficient performance in industrial environments. By considering these additional factors, industries can maximize the potential of 3D printing technology while addressing tribological challenges, optimizing printed components' performance in industrial applications and manufacturing methods, and identifying opportunities for further optimization (Li et al. 2018).

12.4   INFLUENCE OF MANUFACTURING PARAMETERS ON TRIBOLOGICAL BEHAVIOUR

Tribological behaviour includes friction, wear, and lubrication. Manufacturing influences material tribology. The manufacturing parameters and their influence on tribological behaviour are listed below:

- **Surface Roughness:** Surface roughness impacts friction and wear. Interlocking and adhesion between mating surfaces enhance friction and wear on rougher surfaces. By increasing lubricant film production and minimizing surface interactions, smoother surfaces reduce friction and wear.
- **Material Hardness:** Wear resistance depends on material hardness. Harder materials wear less because they can sustain greater contact forces and distortion. Extremely hard materials may have more friction owing to adhesion (Mohamed et al. 2017).
- **Surface Treatment:** Heat treatment, coating, and surface modification may change a material's tribology. Carburizing or nitriding may harden the surface, boosting wear resistance. DLC and TiN minimize friction and wear.
- **Lubrication:** Reducing friction and wear requires lubrication. Manufacturing lubricants and methods affect tribological behaviour. Lubrication prevents direct contact between surfaces, decreasing friction and wear (Sedlaček et al. 2012).
- **Manufacturing Defects:** Manufacturing imperfections like cracks, voids, and inclusions may function as stress concentration sites, causing premature failure and wear. Controlling production procedures to reduce flaws is crucial for tribological performance (Hanon and Zsidai 2021).
- **Material Composition:** Material composition, including alloying element type and concentration, may affect tribological behaviour. Carbon or molybdenum may improve steel wear resistance.
- **Microstructure:** The grain size, phase distribution, and dislocation density affect tribological characteristics. Fine grains and well-dispersed phases increase wear resistance, whereas coarse grains or phase segregation accelerate wear (Ramola et al. 2019).
- **Manufacturing characteristics**: Manufacturing characteristics affect tribological behaviour in complicated and interrelated ways. Manufacturing methods, material selection, and application-specific lubrication procedures may optimize tribological performance.

12.5 REAL-LIFE APPLICATIONS OF 3D PRINTING AND TRIBOLOGY

Additive manufacturing, or 3D printing, is used in many sectors. Tribology—the study of friction, wear, and lubrication—opens up new possibilities and advantages. 3D printing and tribology uses are shown below (Rao et al. 2010):

12.5.1 CUSTOMIZED PROSTHETICS

3D printing enables the production of personalized prosthetic devices that precisely match the anatomical requirements of individuals. By integrating tribological principles, such as incorporating low-friction materials or designing lubrication channels, prosthetics can provide smoother movement and reduce wear on contact surfaces. Using CAD, a child limb prosthesis may be cast. 3D technology reduces cost, weight, and prosthesis replacement time. Transradial, transfemoral, transtibial, and transhumeral kinds are commonly considered (Fergason and Smith 1999).

12.5.2 TRANSRADIAL

Transradial prosthetic arms attach below the elbow. Passive devices are aesthetic. Active prostheses are twofold. Figure 12.2 shows a cable-operated prosthetic device allowing the user to regulate movement manually. Myoelectric prosthetic implants use sensors to detect upper arm muscle action and open and close the hand (Tsikandylakis et al. 2014, Koudelkova et al. 2023).

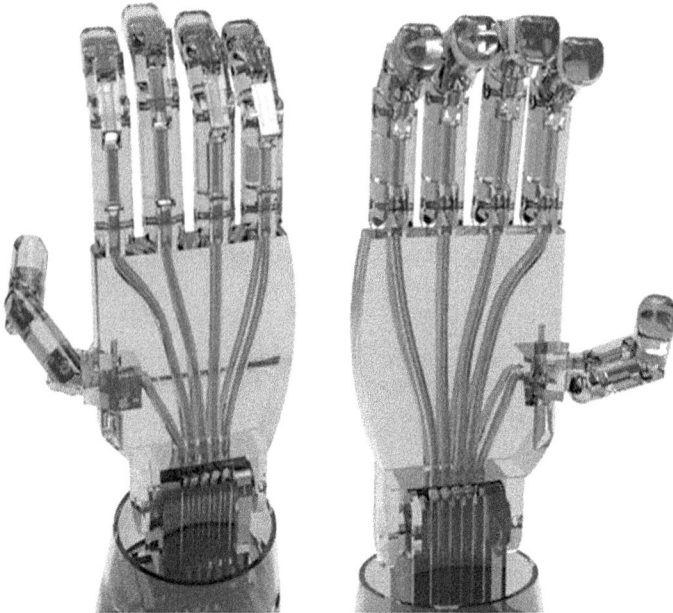

FIGURE 12.2 Transradial prosthetics limb (Koudelkova et al. 2023).

12.5.3 TRANSHUMERAL

Transhumeral prostheses attach above the elbow but below the shoulder. Figure 12.3 shows that a transhumeral prosthesis is more difficult to manoeuvre than a transradial prosthesis because it lacks an elbow. Active and passive transhumeral prosthetics exist. Modern functioning transhumeral prosthesis move using myoelectric sensors or a mix of sensors and wires (Anjum et al. 2017, Kuiken et al. 2004, Mereu et al. 2021).

12.5.4 TRANSTIBIAL

Transtibial prostheses are below-knee prosthetic legs. Figure 12.4 shows that the prosthesis's main purpose is to distribute weight and give comfort as the knee may move freely. Patients require walking therapy since transtibial prostheses don't move (Anjum et al. 2017, Selvam et al. 2021, Fardan et al. 2023).

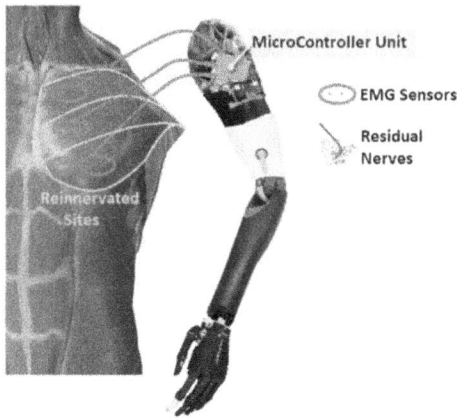

FIGURE 12.3   Transhumeral customized prosthetics (Kuiken et al. 2004, Mereu et al. 2021).

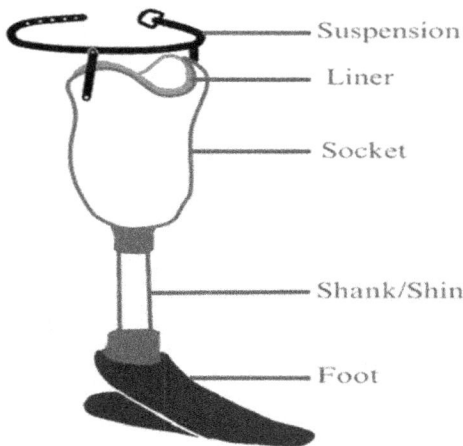

FIGURE 12.4   Customized transtibial prosthetic limb (Fardan et al. 2023).

12.5.5 TRANSFEMORAL

Transfemoral prostheses (above-knee leg replacement) may be difficult. Figure 12.5 shows how the residual limb's strength affects the prosthetic knee joint's hip mobility. After extensive rehabilitation, a transfemoral prosthesis may operate normally. Comfort and stability depend on socket fit (Brueck et al. 2009).

Partial prostheses for missing hands and feet may be active or passive, depending on the demands and budget. Due to various injuries and missing digits, busy prosthetic hands are difficult to make. A skilled medical practitioner must analyze patients to equip them with the best prosthetic hand. A body-powered arm harnessed to the shoulder is another prosthetic arm. The patient's other shoulder mobility, like a transhumeral prosthesis, controls the prosthesis. Other prosthetic implants, such as artificial eyes, fingers, and noses, are aesthetic. Technology has improved prosthetics. A new prosthesis can move more and fit better (Bättig and Schiffmann 2022, Muñoz-Vásquez et al. 2023).

12.5.6 BEARINGS AND BUSHINGS

3D printing allows for the creation of complex geometries with precise specifications. Bearings and bushings, which are essential components in various machinery, can be custom-designed using tribological considerations. This includes optimizing

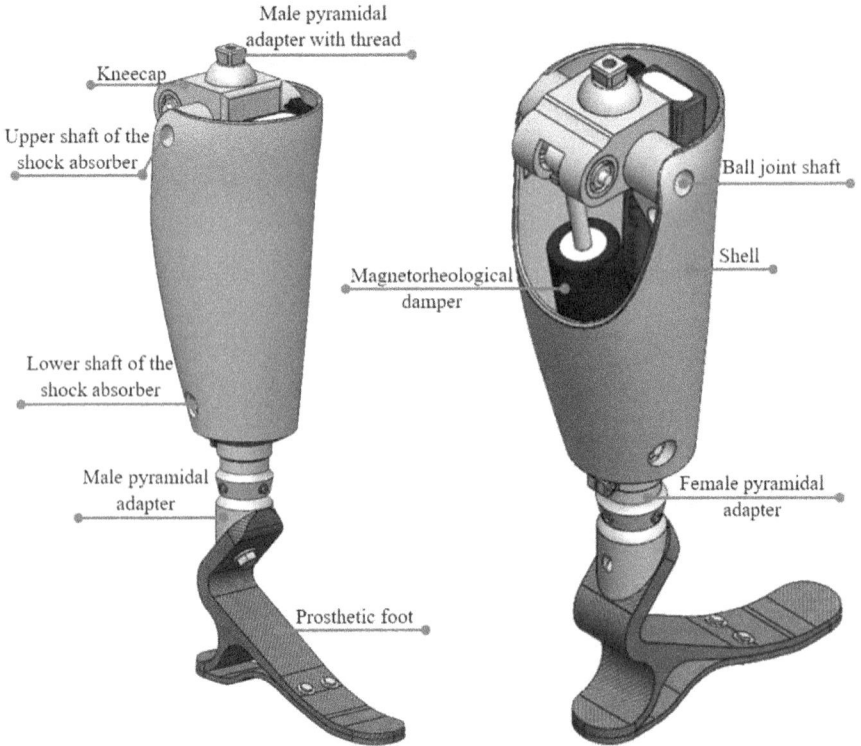

FIGURE 12.5 Transfemoral customized crafted prosthetics (Muñoz-Vásquez et al. 2023).

FIGURE 12.6   Bearing and bushes (Kočiško et al. 2023).

the internal lubrication pathways, surface finishes, and materials to reduce friction and wear. Bearings and bushings are essential for wheel applications. Without it, the wheel bore wears down quicker due to friction. Industrial revolutions transporting big loads, such as metal wheels in timber mills or manufacturing factories, need this. The process of transmitting load energy between a wheel and an axle is called bearing. Bearings reduce friction between wheels and axles. As demonstrated in Figure 12.6, bushings slide instead of rolling like most bearings (Kočiško et al. 2023). Still, bushings are bearings because they improve rotation efficiency.

12.5.7   AEROSPACE COMPONENTS

In the aerospace industry, 3D printing is utilized for manufacturing lightweight and high-strength components. By incorporating tribological knowledge into the design and material selection, critical parts such as turbine blades, bearings, and seals can be optimized to withstand high temperatures, reduce friction, and minimize wear under demanding operating conditions. Aerospace uses sophisticated technical materials and complicated geometry to decrease weight and improve performance. Aerospace components often include conformal cooling channels, thin walls, and difficult curved surfaces. 3D printing can create such features and lightweight, stable, complex constructions. This considerable design flexibility allows topological optimization of components and functional feature integration in a single part (Ciampa et al. 2018).

Material jetting can print colourful graphics with an injection moulding-like quality. These attractive models help designers grasp a part's shape and fit before making production choices. Due to its exact surface quality, 3D printing is perfect for aerodynamic testing and analysis. As seen in Figure 12.7, 3D printing is utilized to make beautiful aeronautical components, including door handles, light housings, control wheels, and whole dashboard designs (Iftekar et al. 2023).

12.5.8   TOOLING AND MOLD MANUFACTURING

3D printing rapidly produces tooling and molds used in manufacturing processes. Integrating tribological principles during the design and material selection can

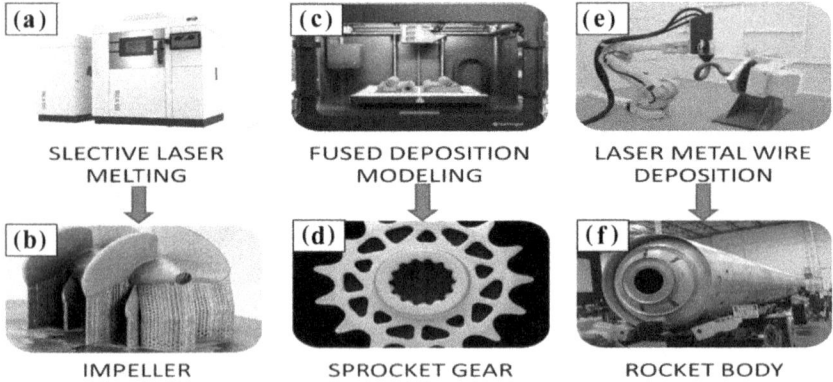

FIGURE 12.7   Aerospace components (Iftekar et al. 2023).

improve the performance and lifespan of these tools. For example, incorporating self-lubricating features or using wear-resistant materials can enhance the efficiency and durability of the devices. Injection moulding is a manufacturing process that uses a metal mold to shape molten plastic resins (Chua et al. 2015). The mold is often referred to as tooling. Mold and tooling are relatively interchangeable for the metal mold at the heart of the injection moulding process, as shown in Figure 12.8 (Hirsch 2014). Mold tooling may also describe machining the mold out of a metal block. A core and cavity is cut into the negative shape of the part. The injection moulding machines hold the mold together while injecting it with molten plastic at high pressure. The mold is opened once the plastic has cooled and the part is ejected. The process is repeated until the desired amount of features has been completed. Injection moulding is a cost-efficient way to manufacture a high volume of complex plastic parts. High-quality tooling is required to produce high-quality parts. ICOMold by Fathom uses premium components and materials for all injection moulding tooling and will work closely with you to ensure every project meets your requirements (Cole and Sherman 1995, Schoenherr et al. 2023).

12.5.9   AUTOMOTIVE APPLICATIONS

3D printing is increasingly used in the automotive industry for producing parts, prototypes, and customized components. Considering tribological factors in the design and material selection, such as optimizing surface textures, incorporating self-lubricating properties, or reducing friction between moving parts, can enhance fuel efficiency, reduce wear, and improve overall vehicle performance (Cole and Sherman 1995). Electronics and signal testing fuel these advances. Recent advances in breaking systems, traction control, and accident avoidance have been made possible by microprocessors, rapid communications, and sensor technologies. Modern automobiles utilize electronic ignition systems and hybrid architectures to lessen their consumption of conventional engine fuels. Producers perform intensive testing and inspections throughout manufacturing to assure quality and

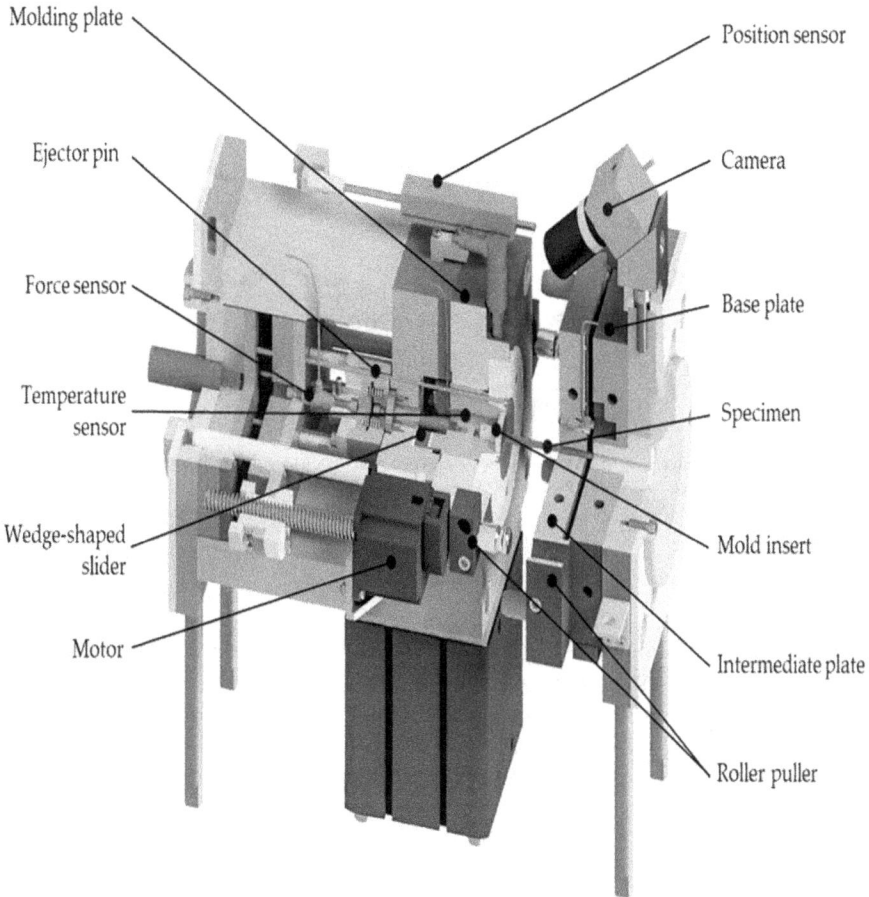

FIGURE 12.8 Tooling and mold manufacturing (Schoenherr et al. 2023).

safety. Manufacturing uses robotics to decrease costs, speed up production, and remove human error (Blawert et al. 2004).

Electronic components and systems, manufacturing methods, and testing are used in the sector. Spectrum has several digitizers and arbitrary waveform generators to handle the car industry's diverse electronic signals. PCIe, PXIe, PXI, and LXI are supported. They provide sampling speeds from 100 KS/s to 5 GS/s, bandwidths from 50 kHz to 1.5 GHz, and resolutions from 8 to 16 bits. Spectrum provides tiny, high-channel-count products for in-vehicle testing, such as vibration analysis and data recording. They're perfect for CAN communication and mechanical parameter testing systems with analogue and digital inputs and outputs.

Spectrum's digitizer NETBOX devices may be utilized with a DC power source for mobile applications. Digitizer cards provide 1–16 channels. Spectrum's StarHub technology can join up to sixteen cards to create instruments with 256 completely synchronous channels, perfect for applications with many sensors and sensor arrays.

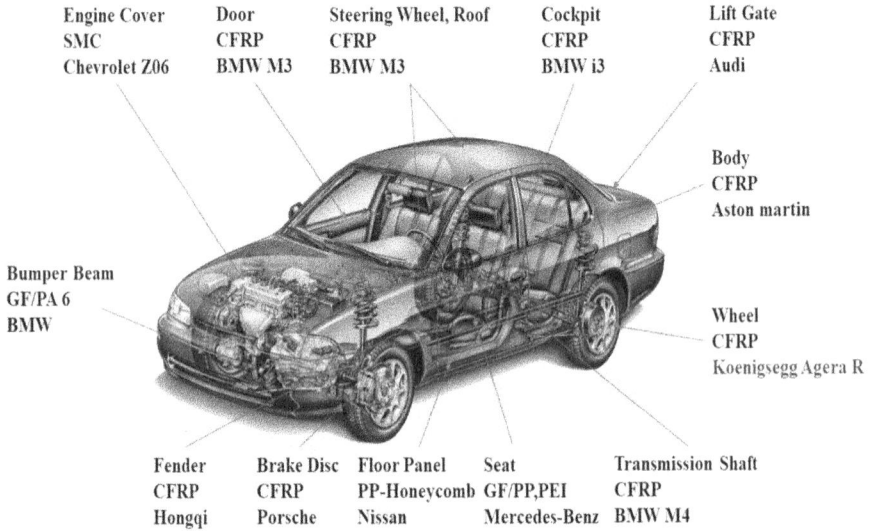

FIGURE 12.9 Automotive applications (Du et al. 2023).

Automotive applications include mechanical testing, vibration analysis, ultrasonic inspection, data logging, component testing, RADAR ranging, telematics, engine performance, ignition monitoring, CAN communication monitoring, infotainment systems (RF, audio, video, navigation), and anti-lock braking systems as shown in Figure 12.9 (Blawert et al. 2004, Du et al. 2023).

12.6 BIOMEDICAL DEVICES

3D printing has revolutionized the production of biomedical devices and implants. Integrating tribological considerations, such as using biocompatible materials with low wear rates or incorporating lubrication mechanisms, can enhance the performance and longevity of these devices, leading to improved patient outcomes. The combination of 3D printing and tribology has significantly contributed to biomedical devices, offering numerous benefits in terms of customization, functionality, and patient outcomes (Blawert et al. 2004). The tribological impact of 3D printing on biomedical devices:

- **Customization and Patient-Specific Designs:** 3D printing enables the production of personalized biomedical devices tailored to the patient's anatomy and needs. By utilizing imaging data, such as CT scans or MRIs, 3D-printed devices can be precisely designed and fabricated, ensuring optimal fit and function. This customization reduces friction, pressure points, and discomfort for patients, enhancing overall tribological performance.
- **Material Selection and Biocompatibility:** 3D printing allows using a wide range of biocompatible materials suitable for biomedical applications. The selection of materials with appropriate tribological properties, such as low friction coefficients, wear resistance, and corrosion

resistance, is crucial for the performance and longevity of biomedical devices. 3D-printed implants, prosthetics, or surgical instruments can be optimized for improved tribological behaviour, reducing wear and tissue damage (Ketabchi et al. 2023).

- **Complex Structures and Integration:** Additive manufacturing enables the creation of complex and intricate geometries, facilitating the incorporation of interior features and functional elements in biomedical devices. By integrating tribological principles during the design stage, 3D-printed devices can enhance their tribological performance by using lubrication channels, textured surfaces, or self-lubricating structures. Such features minimize friction, wear, and the need for external lubrication, improving functionality.
- **Bioactive Surface Modifications:** Surface modifications play a crucial role in enhancing the tribological behaviour of biomedical devices. 3D-printed devices can undergo surface treatments, such as coating deposition or functionalization, to improve biocompatibility and reduce friction and wear. Bioactive coatings can promote tissue integration, while low-friction coatings reduce wear and enhance the longevity of implants or prosthetics (Uddin et al. 2023).
- **Surgical Guides and Models:** 3D printing facilitates the production of surgical guides and models used for preoperative planning and practice. These guides and models aid surgeons in understanding patient-specific anatomical structures and performing procedures with precision. By considering tribological factors, such as optimizing contact surfaces and reducing friction, 3D-printed surgical guides can improve the accuracy and success of surgical interventions.
- **Drug Delivery Systems:** 3D printing enables the fabrication of intricate drug delivery systems such as implants, microneedles, or scaffolds that provide controlled release of medications or therapies. Tribological considerations, such as surface roughness or lubricant coatings, can affect drug release rates, minimize tissue damage, and optimize the functionality of these systems.

The convergence of 3D printing and tribology in biomedical devices offers immense potential for customization, improved functionality, and patient care. While advancements have been made, ongoing research and development are necessary to optimize the tribological behaviour of 3D-printed biomedical devices and ensure their long-term performance and compatibility with biological systems.

12.7   ROBOTICS AND MECHANISMS

3D printing is utilized in the production of mechanical components and mechanisms. By considering tribological factors, such as selecting materials with low friction coefficients, optimizing mating surfaces, or incorporating lubrication systems, improving these systems' efficiency, reliability, and lifespan is possible. Robotics and Mechanisms promote the research and development of technologies that allow intelligent, goal-oriented systems and new tools to monitor, operate, and control systems. These applications demonstrate 3D printing–tribology synergy. 3D printing and friction, wear, and lubrication may create unique solutions that meet particular needs and improve performance in numerous sectors (Wang et al. 2023).

12.7.1 ANISOTROPY IN 3D-PRINTED COMPONENTS AND ITS TRIBOLOGICAL EFFECTS

Anisotropy in 3D-printed components refers to the directional dependence of material properties and characteristics resulting from the layer-by-layer additive manufacturing process. This anisotropy can have significant tribological effects on the performance and behaviour of the printed components (Khosravani et al. 2022). Anisotropy and its tribological impact are discussed below:

- **Mechanical Strength:** 3D-printed components typically exhibit lower mechanical strength in the build direction (vertical direction) than in-plane direction (horizontal direction). This anisotropic mechanical behaviour can affect the load-carrying capacity of the component, potentially leading to increased wear and reduced tribological performance, particularly in applications with high loads or sliding contact.
- **Surface Roughness:** Anisotropic surface roughness is a common characteristic of 3D-printed components due to the layer-by-layer deposition process. The roughness can vary between the build and in-plane directions, affecting the frictional behaviour and wear rates. Higher roughness in the build direction may increase friction and wear compared to the smoother in-plane movement.
- **Layer Adhesion and Delamination:** Anisotropy in layer adhesion strength can impact the tribological behaviour of 3D-printed components. Weak interlayer bonding can lead to delamination or separation under mechanical stresses, increasing friction, wear, and reduced component integrity.
- **Material Orientation and Microstructure:** The orientation of material deposition in 3D printing affects the microstructure of the printed component, leading to variations in material properties. This anisotropic microstructure can influence tribological properties in different directions, such as hardness, elastic modulus, and wear resistance. For example, materials printed with aligned fibres or reinforcing particles in specific orientations may exhibit different wear rates and frictional behaviour depending on the sliding movement (Ji et al. 2019).
- **Post-Processing Techniques:** Post-processing techniques, such as heat treatment or surface finishing, can mitigate the negative effects of anisotropy on tribological behaviour. Heat treatment can improve interlayer bonding and enhance material properties in the build direction, reducing anisotropy. Surface finishing processes, such as polishing or coating, can help reduce surface roughness and improve tribological performance by promoting smoother sliding contact (Hanon et al. 2020).

It is essential to consider anisotropy and its tribological effects during designing and applying 3D-printed components. Understanding the specific tribological requirements and considering factors such as load direction, contact surfaces, and lubrication can aid in optimizing the performance of anisotropic features. Additionally, ongoing research and advancements in additive manufacturing processes, materials, and post-processing techniques address anisotropy-related challenges, aiming to improve the overall tribological behaviour of 3D-printed components (Zhang et al. 2020, Dhakal et al. 2023).

12.8 FUTURE DIRECTIONS AND RESEARCH OPPORTUNITIES OF TRIBOLOGICAL EFFECT OF 3D PRINTING

The tribological effect of 3D printing is an active area of research, and several future directions and research opportunities hold significant potential (Roy and Mukhopadhyay 2021).

- **Material Development:** Further research is needed to develop new materials specifically tailored for tribological applications in 3D printing. This includes developing materials with enhanced wear resistance, low friction coefficients, and improved self-lubricating properties. Investigating the influence of material composition, reinforcement additives, and surface treatments on the tribological behaviour of 3D-printed components can lead to material selection and optimization advancements (Pant et al. 2020).
- **Surface Engineering and Post-Processing Techniques:** Surface engineering is crucial in optimizing the tribological performance of 3D-printed components. Exploring post-processing techniques, such as polishing, coating, or surface texturing, can improve surface finish, reduce friction, and enhance wear resistance. Additionally, investigating novel surface modification methods, such as laser surface texturing or electrochemical treatments, can offer new opportunities for tailoring surface properties and enhancing tribological behaviour.
- **Design Optimization and Topology Optimization:** Research efforts can be directed towards developing design methodologies and optimization algorithms that consider tribological factors in the design process of 3D-printed components. This includes topology optimization techniques that can generate optimal designs by considering material distribution, load paths, and tribological requirements. Such approaches can lead to the creation of lightweight and efficient components with improved tribological performance.
- **Multi-Material and Hybrid Printing:** Exploring multi-material 3D printing techniques and hybrid printing approaches can provide opportunities to create components with tailored tribological properties. Researchers can achieve enhanced performance by combining materials with different tribological characteristics or incorporating inserts, fibres, or coatings into 3D-printed structures, such as improved wear resistance, reduced friction, and customized functionality.

- **Simulation and Modelling:** The development of accurate simulation and modelling techniques for predicting and analyzing the tribological behaviour of 3D-printed components is crucial. This includes modelling the contact mechanics, lubrication, and wear processes specific to additive manufacturing. Such simulations can aid in the virtual design and optimization of components, reducing the need for extensive experimental testing.
- **In situ, Monitoring and Control:** Real-time monitoring and controlling tribological behaviour during 3D printing offers quality assurance and optimization opportunities. Integration of sensors, such as temperature sensors or force sensors, into 3D printers can enable the detection of anomalies or variations in tribological behaviour during the printing process. This information can then be used to adjust printing parameters or identify potential issues, ensuring consistent and optimized tribological performance (Holovenko et al. 2018).
- **Application-Specific Studies:** Conducting tribological studies focused on specific applications, such as aerospace, automotive, or medical devices, can provide valuable insights into the performance requirements and challenges in those domains. Investigating the tribological behaviour of 3D-printed components under different operating conditions, loads, temperatures, or lubrication regimes specific to these applications can lead to tailored solutions and advancements (Pan et al. 2023).

Overall, future research on the tribological effect of 3D printing lies in material development, surface engineering, design optimization, modelling, and application-specific studies. By addressing these areas, researchers can unlock the full potential of 3D printing technology in achieving enhanced tribological performance, opening up new opportunities for various industries and applications.

12.9   GLOBAL STATUS OF TRIBOLOGICAL EFFECT OF 3D PRINTING IN RESEARCH

Bibliometric analysis, a scientific computer-assisted review procedure, may identify essential research and authors by reviewing all relevant publications. Books, journals, and other magazines, particularly those with scientific content, may be analyzed using bibliometric techniques. The area of library and information science relies heavily on bibliometric techniques. Bibliometrics and scientometrics, the study of scientific metrics and indicators, are so intertwined that they are sometimes considered synonymous terms (Tambunan et al. 2023). The study of friction, wear, and lubrication between rubbing surfaces (tribology) is one of the many areas that has taken notice of 3D printing, often known as additive manufacturing. The requirement for new materials and the quantity of trash transported to landfills are reduced. Therefore, 3D printing is aiding in lessening the damage the industrial industry does to the natural world. 3D printing has also contributed to the decline in manufacturing costs. The section below discusses the evolution of research in tribology and 3D printing on different parameters, such as research sub-area-wise, year-wise (2018–2023), country-wise, keywords used, document type, and funding sponsors, as shown in Table 12.1.

TABLE 12.1

Steps for Collecting Information on the Scopus Database

Search Steps	Query on Scopus	Description	Documents Count
1	Tittle-Abs-Key	Tribology effect	21,167
2	And (Limit-To-Subarea)	Material Science	19,810
		Engineering	
		Biochemistry, Genetics, and Molecular Biology	
		Medicine	
		Multidisciplinary	
		Pharmacology, Toxicology, and Pharmaceutics	
		Health profession	
3	And (Limit-To Publication year)	2018–2023	14,374
4	And (Limit-To-Document type)	Conference	2,011
		Book chapter	
		Short survey	
5	And (Exclude Funding sponsor)	Science and Engineering Research Board	2,000

Figure 12.10 shows the most used keyword in the research between 2018 and 2023 in tribology. The most widely used keywords are tribology, friction, wear of materials, wear of resistance, friction coefficients, tribological properties, and scanning electron microscopy. The treemap shows the percentage usage of a particular keyword, as depicted in Figure 12.11. Similarly, Figure 12.12 shows the keyword use trend over the past 10 years.

To estimate the exponentially declining returns of searching for references in scientific publications, Bradford's law was originally stated by Samuel C. Bradford in 1934. The number of journals in each group will be proportional to 1:n:n²; if the journals in a field are sorted by article, count into three groups, each containing around one-third of all reports. The concept may be stated in many different ways (Palaniyappan et al. 2023). Figure 12.13 shows Bradford's law, which indicates the number of articles published by the top four sources in this case.

Using the frequency scale of 1–14, China to the USA collaboration frequency of research is 14, China to the United Kingdom is 11, China to Australia is 6, Algeria to Saudi Arabia is 3, China to Japan is 5, Germany to France is 5, Italy to Spain is 5, South Africa to Nigeria is 5, Germany to Spain is 4, South Africa to Botswana is 4, and the rest of the countries has frequency 3–1 as depicted in Figure 12.14. Several articles published by different countries are shown in Figure 12.15. India ranks 1st, China ranks 2nd, Germany ranks 3rd, the USA ranks 4th, Malaysia ranks 5th, Poland

FIGURE 12.10 Most used keywords.

ranks 6th, and the United Kingdom ranks 7th. Other countries have less than 25% of the published papers.

A thematic narrative is an ongoing series of topics that spans many periods. There is a group of words that compose each issue. This study's thematic evolution chart is divided into five sub-periods, i.e., 2010–2012, 2013–2015, 2016–2018, 2019–2020, and 2021–2023, based on different sub-themes. Three-field plot analysis explains the relationship between three other pieces of information. In this context, three-field plots were used to visualize the relationship between the study's source, author, and keywords variables, as shown in Figures 12.16 and 12.17, respectively.

12.10 CONCLUSION

Finally, 3D printing's tribological influence in industrial applications affects several industries. The design, material selection, and manufacture of 3D-printed components may benefit from tribological concepts. Complex geometries and bespoke designs optimized industrial component performance using 3D printing. 3D-printed parts may be more efficient, durable, and friction-free by addressing tribological aspects, including surface roughness, lubrication, and wear resistance. Customized components for particular applications may be made using 3D printing. 3D-printed parts may be adjusted for performance and wear by including tribological concerns

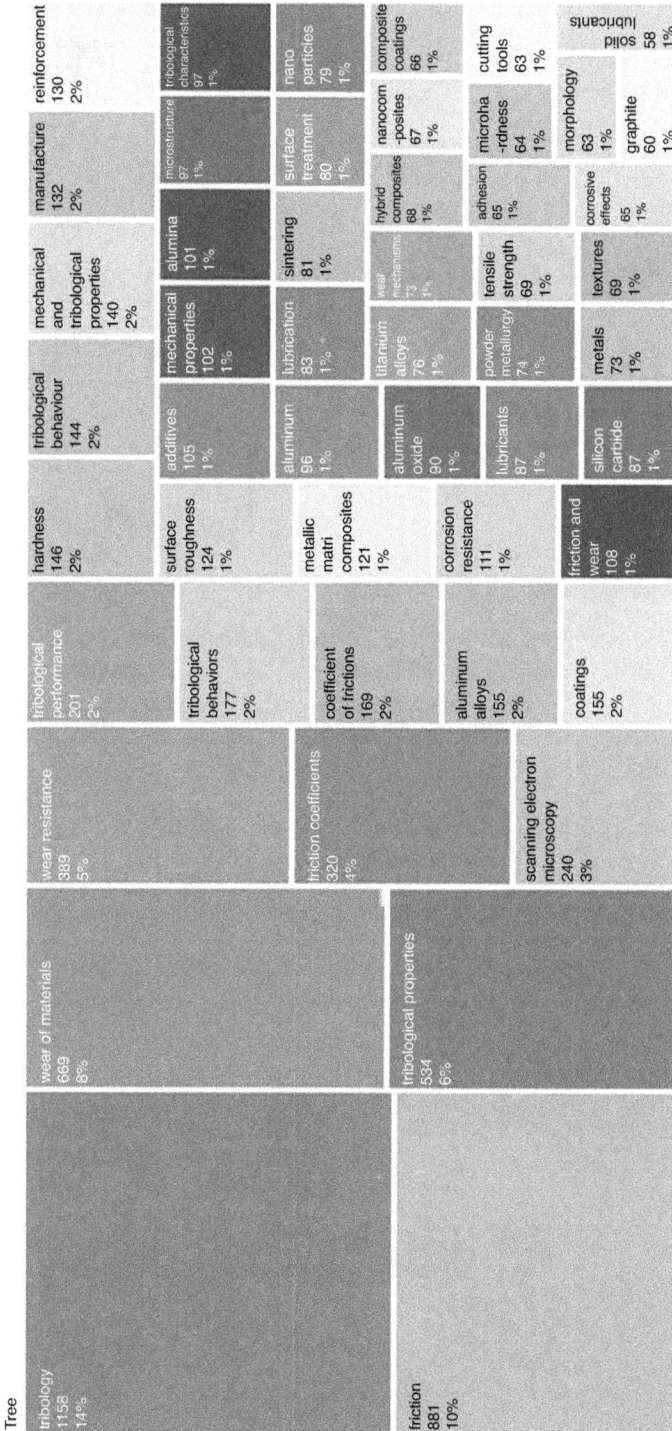

FIGURE 12.11 Treemap.

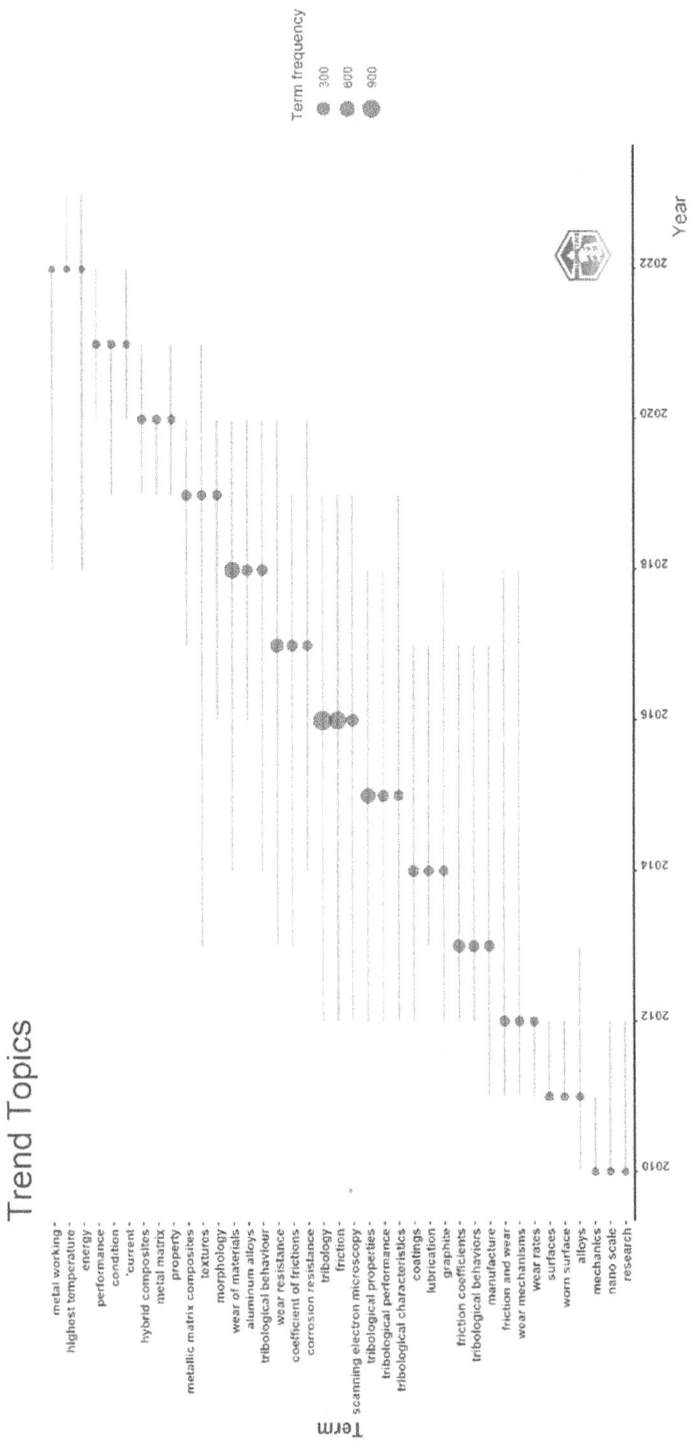

FIGURE 12.12   Trend topics.

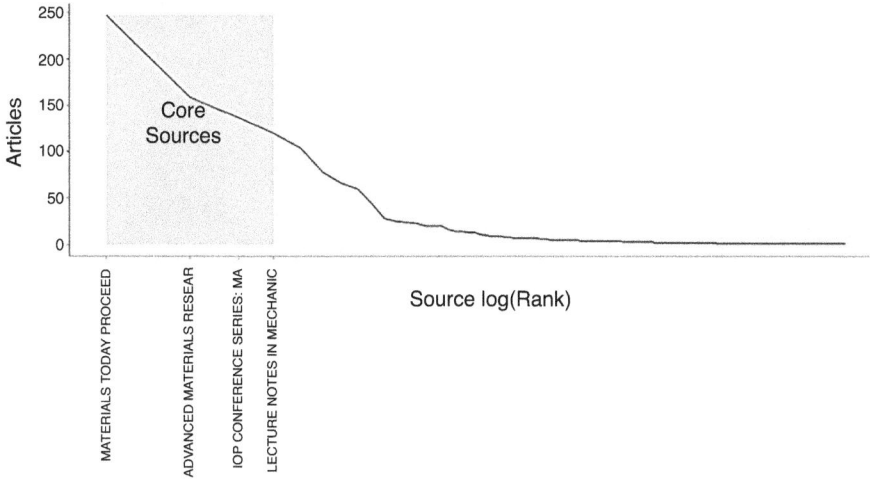

FIGURE 12.13 Bradford law.

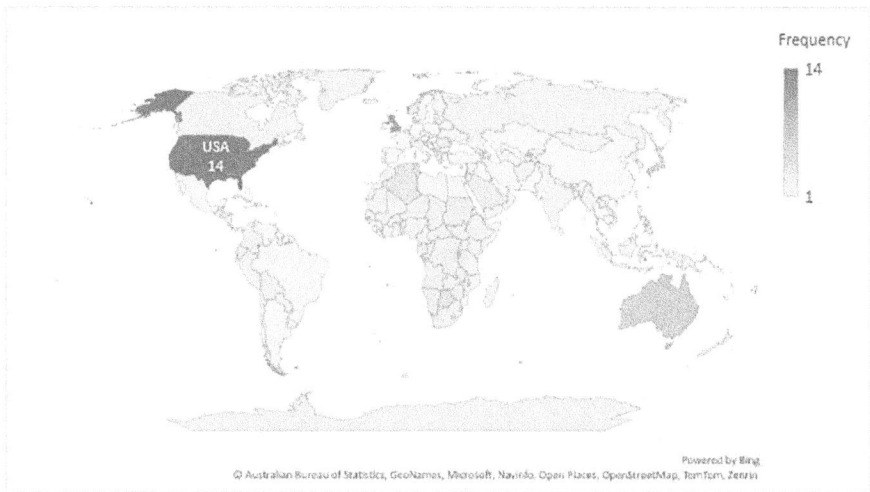

FIGURE 12.14 Country collaboration map.

like material selection and surface quality optimization. Rapid prototyping with 3D printing allows design revisions and idea testing. During prototyping, tribological considerations may reveal friction, wear, and lubrication concerns early in development, allowing for essential corrections and improvements. 3D printing may use several materials with different tribological qualities. 3D-printed components may increase frictional properties, wear resistance, and performance by carefully choosing and optimizing materials depending on application requirements. Industrial applications may benefit from 3D printing. 3D printing may expedite manufacturing operations, eliminate material waste, and shorten production cycles by removing conventional machining and assembly methods.

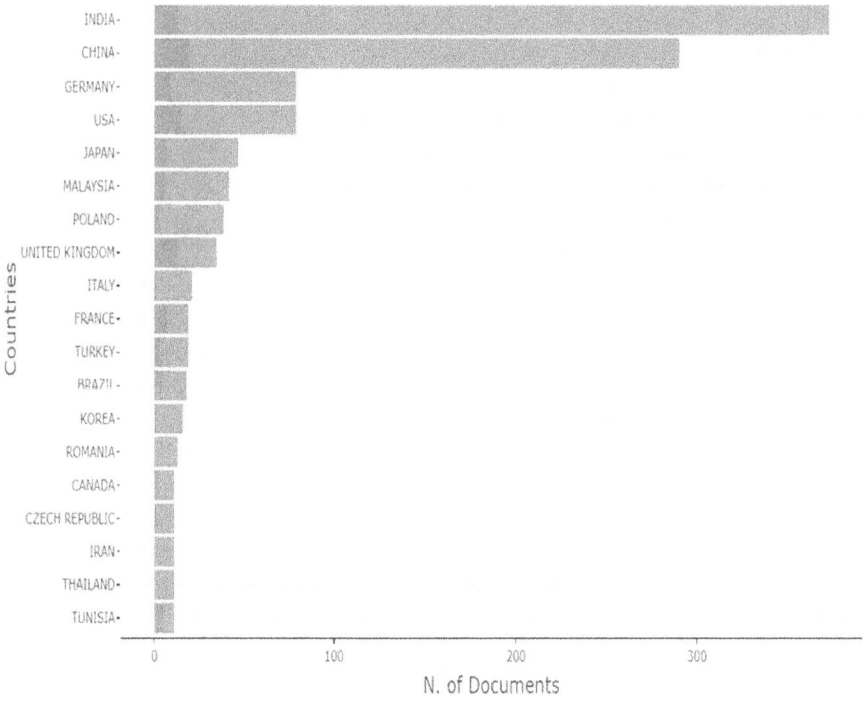

FIGURE 12.15 No documents published by different countries.

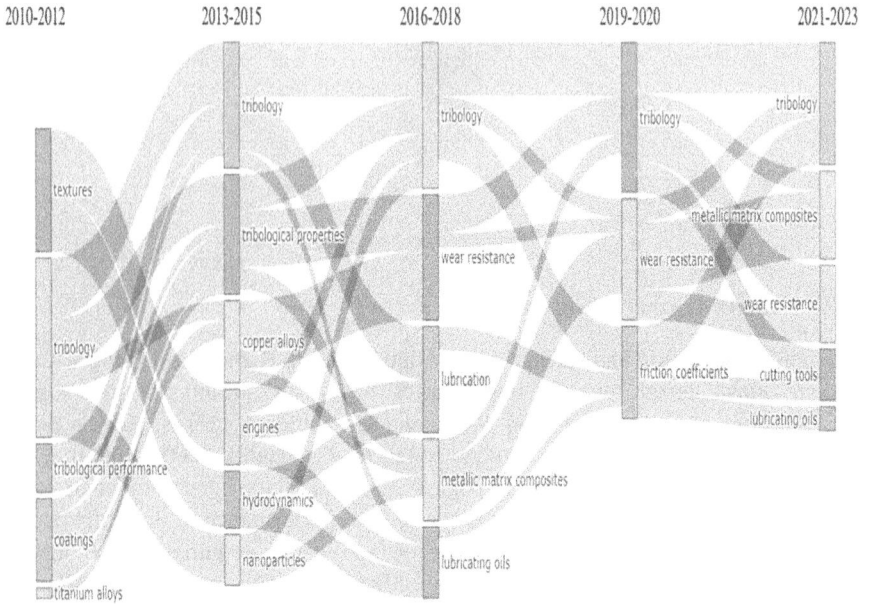

FIGURE 12.16 Thematic evolution.

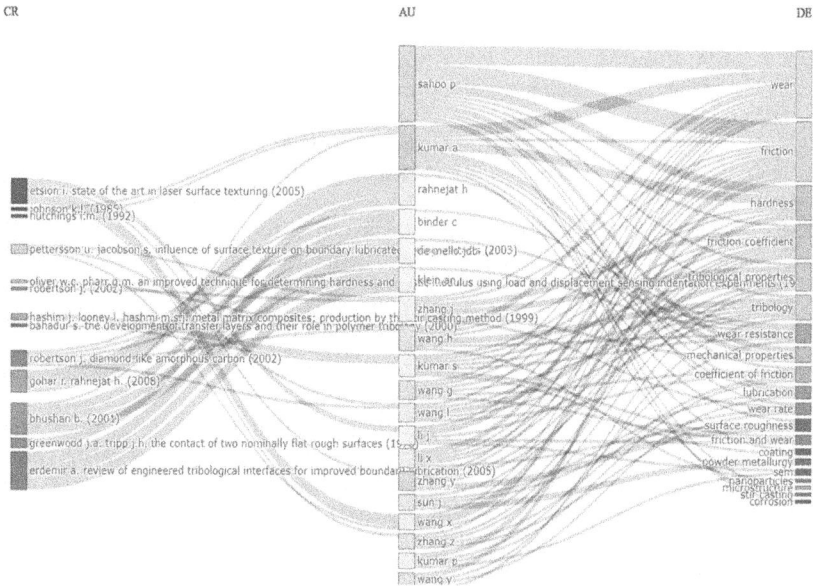

FIGURE 12.17 Three field plot.

Tribology, which examines friction, wear, and lubrication of interacting surfaces, has drawn attention to 3D printing, also known as additive manufacturing. This decreases the requirement for virgin materials and landfill garbage. Thus, 3D printing is reducing Manufacturing's environmental effect. Additionally, 3D printing lowers manufacturing costs. The section below discusses the evolution of research in tribology and 3D printing on different parameters such as research sub-area, year-wise (2018–2023), country-wise, keywords, document type, and funding sponsors. The most widely used keywords are tribology, friction, wear of materials, wear of resistance, friction coefficients, tribological properties, and scanning electron microscopy. China to the USA collaboration frequency of research is 14, the highest among all collaborations. This study's thematic evolution chart is divided into five sub-periods, i.e., 2010–2012, 2013–2015, 2016–2018, 2019–2020, and 2021–2023, based on different sub-themes.

While 3D printing may improve tribological performance, it also has drawbacks. The final components' tribological behaviour may be affected by surface polish, material qualities, and 3D printing constraints. Therefore, careful evaluation of application requirements, comprehensive testing, and continual improvement are needed to maximize 3D printing's tribological performance in industrial applications. Using tribological concepts to design and manufacture industrial 3D-printed components improves performance, durability, customization, and cost. Further study and development in this subject will open up more opportunities for enhancing tribological behaviour in 3D-printed industrial components.

ACKNOWLEDGEMENTS

The authors express gratitude to Chandigarh University for their assistance and support.

REFERENCES

Anjum, I., M. A. Khan, M. Aadil, A. Faraz, M. Farooqui and A. Hashmi (2017). "Transradial vs. transfemoral approach in cardiac catheterization: A literature review." *Cureus* 9(6). https://doi.org/10.7759/cureus.1309

Baligidad, S. M., T. Arunkumar, G. Thodda and K. Elangovan (2023). "Fabrication of HAp/ rGO nanocomposite coating on PEEK: Tribological performance study." *Surfaces and Interfaces* 38: 102865.

Ballesteros, L. M., E. Zuluaga, P. Cuervo, J. S. Rudas and A. Toro (2021). "Tribological behavior of polymeric 3D-printed surfaces with deterministic patterns inspired in snake skin morphology." *Surface Topography: Metrology and Properties* 9(1): 014002.

Bättig, P. and J. Schiffmann (2022). "Unstable tilting motion of flexibly supported gas bearing bushings." *Mechanical Systems and Signal Processing* 162: 107981.

Blawert, C., N. Hort and K. Kainer (2004). "Automotive applications of magnesium and its alloys." *Transactions of the Indian Institute of Metals* 57(4): 397–408.

Brueck, M., D. Bandorski, W. Kramer, M. Wieczorek, R. Höltgen and H. Tillmanns (2009). "A randomized comparison of transradial versus transfemoral approach for coronary angiography and angioplasty." *JACC: Cardiovascular Interventions* 2(11): 1047–1054.

Chua, C. K., K. F. Leong and Z. H. Liu (2015). Rapid tooling in manufacturing. In *Handbook of Manufacturing Engineering and Technology*, edited by Andrew Y. C. Nee. Springer: 2525–2549.

Ciampa, F., P. Mahmoodi, F. Pinto and M. Meo (2018). "Recent advances in active infrared thermography for non-destructive testing of aerospace components." *Sensors* 18(2): 609.

Cole, G. and A. Sherman (1995). "Light weight materials for automotive applications." *Materials Characterization* 35(1): 3–9.

Dhakal, N., X. Wang, C. Espejo, A. Morina and N. Emami (2023). "Impact of processing defects on microstructure, surface quality, and tribological performance in 3D printed polymers." *Journal of Materials Research and Technology* 23: 1252–1272.

Du, B., Q. Li, C. Zheng, S. Wang, C. Gao and L. Chen (2023). "Application of lightweight structure in automobile bumper beam: A review." *Materials* 16(3): 967.

Fardan, M. F., B. W. Lenggana, U. Ubaidillah, S.-B. Choi, D. D. Susilo and S. Z. Khan (2023). "Revolutionizing prosthetic design with auxetic metamaterials and structures: A review of mechanical properties and limitations." *Micromachines* 14(6): 1165.

Fergason, J. and D. G. Smith (1999). "Socket considerations for the patient with a transtibial amputation." *Clinical Orthopaedics and Related Research* 361: 76–84.

Friedrich, K. (2018). "Polymer composites for tribological applications." *Advanced Industrial and Engineering Polymer Research* 1(1): 3–39.

Gbadeyan, O. J., T. Mohan and K. Kanny (2021). "Tribological properties of 3D printed polymer composites-based friction materials." *Tribology of Polymer and Polymer Composites for Industry* 4.0: 161–191.

Hanon, M. M., Y. Alshammas and L. Zsidai (2020). "Effect of print orientation and bronze existence on tribological and mechanical properties of 3D-printed bronze/PLA composite." *The International Journal of Advanced Manufacturing Technology* 108: 553–570.

Hanon, M. M. and L. Zsidai (2020). Tribological and mechanical properties investigation of 3D printed polymers using DLP technique. *AIP Conference Proceedings*, AIP Publishing.

Hanon, M. M. and L. Zsidai (2021). "Comprehending the role of process parameters and filament color on the structure and tribological performance of 3D printed PLA." *Journal of Materials Research and Technology* 15: 647–660.

Hirsch, J. (2014). "Recent development in aluminium for automotive applications." *Transactions of Nonferrous Metals Society of China* 24(7): 1995–2002.

Holovenko, Y., M. Antonov, L. Kollo and I. Hussainova (2018). "Friction studies of metal surfaces with various 3D printed patterns tested in dry sliding conditions." *Proceedings of the Institution of Mechanical Engineers, Part J: Journal of Engineering Tribology* 232(1): 43–53.

Iftekar, S. F., A. Aabid, A. Amir and M. Baig (2023). "Advancements and limitations in 3D printing materials and technologies: A critical review." *Polymers* 15(11): 2519.

Ji, Z., C. Yan, S. Ma, S. Gorb, X. Jia, B. Yu, X. Wang and F. Zhou (2019). "3D printing of bioinspired topographically oriented surfaces with frictional anisotropy for directional driving." *Tribology International* 132: 99–107.

Ketabchi, M. R., S. M. Soltani and A. Chan (2023). "Synthesis of a new biocomposite for fertiliser coating: Assessment of biodegradabilityand thermal stability." *Environmental Science and Pollution Research International* 30: 93722–93730

Khosravani, M. R., S. Rezaei, H. Ruan and T. Reinicke (2022). "Fracture behavior of aniso-tropic 3D-printed parts: Experiments and numerical simulations." *Journal of Materials Research and Technology* 19: 1260–1270.

Kočiško, M., P. Baron and D. Paulišin (2023). "Research of dynamic characteristics of bearing reducers of the twinspin class in the start-up phase and in the initial operating hours." *Machines* 11(6): 595.

Koudelkova, Z., A. Mizera, M. Karhankova, V. Mach, P. Stoklasek, M. Krupciak, J. Minarcik and R. Jasek (2023). "Verification of finger positioning accuracy of an affordable tran-sradial prosthesis." *Designs* 7(1): 14.

Kuiken, T. A., G. A. Dumanian, R. D. Lipschutz, L. A. Miller and K. Stubblefield (2004). "The use of targeted muscle reinnervation for improved myoelectric prosthesis control in a bilateral shoulder disarticulation amputee." *Prosthetics and Orthotics International* 28(3): 245–253.

Li, H., M. Ramezani, M. Li, C. Ma and J. Wang (2018). "Effect of process parameters on tribo-logical performance of 316L stainless steel parts fabricated by selective laser melting." *Manufacturing Letters* 16: 36–39.

Mereu, F., F. Leone, C. Gentile, F. Cordella, E. Gruppioni and L. Zollo (2021). "Control strate-gies and performance assessment of upper-limb TMR prostheses: A review." *Sensors* 21(6): 1953.

Mitsouras, D. and P. C. Liacouras (2017). "3D printing technologies." In *3D Printing in Medicine: A Practical Guide for Medical Professionals*, edited by Frank J. Rybicki and Gerald T. Grant. Springer: 5–22.

Mohamed, O. A., S. H. Masood, J. L. Bhowmik and A. E. Somers (2017). "Investigation on the tribological behavior and wear mechanism of parts processed by fused deposition addi-tive manufacturing process." *Journal of Manufacturing Processes* 29: 149–159.

Muñoz-Vásquez, S., Z. A. Mora-Pérez, P. A. Ospina-Henao, C. H. Valencia-Niño, M. Becker and J. G. Díaz-Rodríguez (2023). "Finite element analysis in the balancing phase for an open source transfemoral prosthesis with magneto-rheological damper." *Inventions* 8(1): 36.

Palaniyappan, S., N. K. Sivakumar, M. Bodaghi, M. Kumar and M. Rahaman (2023). "A feasibility study of various joining techniques for three-dimensional printed polylactic acid and wood-reinforced polylactic acid biocomposite." *Proceedings of the Institution of Mechanical Engineers, Part L: Journal of Materials: Design and Applications* 14644207231189956. https://doi.org/10.1177/14644207231189956

Pan, L., Z. Xu and M. Skare (2023). "Sustainable business model innovation literature: A bib-liometrics analysis." *Review of Managerial Science* 17(3): 757–785.

Pant, M., R. M. Singari, P. K. Arora, G. Moona and H. Kumar (2020). "Wear assessment of 3-D printed parts of PLA (polylactic acid) using Taguchi design and Artificial Neural Network (ANN) technique." *Materials Research Express* 7(11): 115307.

Pieterse, F. and A. L. Nel (2016). The advantages of 3D printing in undergraduate mechanical engineering research. *2016 IEEE Global Engineering Education Conference (EDUCON)*, IEEE.

Quader, R. (2021). *Moisture Sensitivity of PLA/PBS Blends during Ultrasonic Welding and Fused Deposition Modeling*, North Dakota State University.

Ramola, M., V. Yadav and R. Jain (2019). "On the adoption of additive manufacturing in healthcare: A literature review." *Journal of Manufacturing Technology Management* 30(1): 48–69.

Rao, S. V., M. G. Cohen, D. E. Kandzari, O. F. Bertrand and I. C. Gilchrist (2010). "The transradial approach to percutaneous coronary intervention: Historical perspective, current concepts, and future directions." *Journal of the American College of Cardiology* 55(20): 2187–2195.

Ren, Y., H. Wu, J. Du, B. Liu, X. Wang, Z. Jiao, Y. Tian and I. Baker (2023). "Effect of laser scanning speeds on microstructure, tribological and corrosion behavior of Ti-23Nb alloys produced by laser metal deposition." *Materials Characterization* 197: 112647.

Roy, R. and A. Mukhopadhyay (2021). "Tribological studies of 3D printed ABS and PLA plastic parts." *Materials Today: Proceedings* 41: 856–862.

Rudnik, M., M. M. Hanon, W. Szot, K. Beck, D. Gogolewski, P. Zmarzły and T. Kozior (2022). "Tribological properties of medical material (MED610) used in 3D printing PJM technology." *Tehnički vjesnik* 29(4): 1100–1108.

Schoenherr, M., H. Ruehl, T. Guenther, A. Zimmermann and B. Gundelsweiler (2023). "Adhesion-induced demolding forces of hard coated microstructures measured with a novel injection molding tool." *Polymers* 15(5): 1285.

Sedlaček, M., B. Podgornik and J. Vižintin (2012). "Correlation between standard roughness parameters skewness and kurtosis and tribological behaviour of contact surfaces." *Tribology International* 48: 102–112.

Selvam, P. S., M. Sandhiya, K. Chandrasekaran, D. H. Rubella and S. Karthikeyan (2021). *Prosthetics for Lower Limb Amputation. Prosthetics and Orthotics*, IntechOpen.

Shahrubudin, N., T. C. Lee and R. Ramlan (2019). "An overview on 3D printing technology: Technological, materials, and applications." *Procedia Manufacturing* 35: 1286–1296.

Tambunan, H. B., N. W. Priambodo, J. Hartono, I. A. Aditya and M. Triani (2023). "Research trends on microgrid systems: A bibliometric network analysis." *International Journal of Electrical & Computer Engineering* 13(3): 2529.

Tsikandylakis, G., Ö. Berlin and R. Brånemark (2014). "Implant survival, adverse events, and bone remodeling of osseointegrated percutaneous implants for transhumeral amputees." *Clinical Orthopaedics and Related Research* 472: 2947–2956.

Uddin, M. S., M. I. Khan, S. B. Shafiq, S. Sadik, M. S. Rana, K. M. S. Bhuiya, S. A. Udoy and M. K. H. Rafi (2023). "Structural analysis and material selection for biocompatible cantilever beam in soft robotic nanomanipulator." *BIBECHANA* 20(2): 183–189.

Wang, C., J. Cai, G. Cheng, J. Wang and D. Tang (2023). "Numerical investigations of tribological characteristics of biomimetic-textured surfaces." *Sustainability* 15(17): 13054.

Xu, J., N. Liu, F. Zhang, J. Du, C. Zheng, X. Gao and K. Liu (2023). "Frictional behaviors of 3D-printed polylactic acid components with spiral-groove surface textures under oil lubrication." *Journal of Tribology* 145(1): 011803.

Zhang, Y., C. Purssell, K. Mao and S. Leigh (2020). "A physical investigation of wear and thermal characteristics of 3D printed nylon spur gears." *Tribology International* 141: 105953.

13 Emerging Applications of 3D-Printed Parts with Enhanced Tribological Properties

Ratnesh Raj, Annada Prasad Moharana,
Vishal Kumar
Indian Institute of Technology (Indian School of Mines)

Rashi Tyagi
Chandigarh University

Amit Rai Dixit
Indian Institute of Technology (Indian School of Mines)

13.1 INTRODUCTION

3D printing, also referred to as additive manufacturing (AM), is a manufacturing process that involves the creation of three-dimensional objects through the layer-by-layer addition of material, guided by a digital model or design (Tyagi et al., 2023; Ranjan et al., 2023a; Phadke et al., 2022). It is an innovative manufacturing technique that enables the production of intricate and customized objects with complex geometries (Tyagi & Tripathi, 2023 ; Ranjan et al., 2023b; Raj et al., 2022). The use of 3D-printed parts in industrial applications introduces specific tribological considerations related to friction, wear, and lubrication. The tribological performance of these components is influenced by factors such as material properties, surface finish, and the layering process employed during 3D printing (Dwivedi et al., 2022). Tribology is basically a scientific discipline that explores the interaction between surfaces when they are in relative motion, focusing on the consequences of friction, wear, and lubrication. It involves the comprehensive understanding, analysis, and management of friction, lubrication, and wear phenomena that arise from the contact and sliding or rolling of two surfaces (Zhang, 2013).

Friction behavior in 3D-printed parts varies depending on factors like material composition, surface roughness, and part design (Dwivedi et al., 2022). The interaction between these parts and their mating surfaces results in varying friction coefficients and wear rates. Understanding and optimizing the frictional characteristics of 3D-printed components is crucial to ensure proper functionality and minimize

DOI: 10.1201/9781003400523-13

energy losses in industrial settings (Hanon et al., 2019). Considering wear resistance is vital when utilizing 3D-printed parts in industrial applications. The choice of printing materials, such as polymers, metals, or composites, impacts the wear resistance of these components. Factors like surface finish, applied load, and operating conditions affect the wear rate and mechanisms experienced by 3D-printed parts (Hu et al., 2019). Evaluating wear behavior and durability is essential to ensure reliability and longevity in industrial environments. Effective lubrication is essential to minimize friction and wear in industrial applications involving 3D-printed parts (Keshavamurthy et al., 2021). The surface characteristics and material composition of these components influence their lubrication requirements. Selecting appropriate lubricants and optimizing lubrication strategies can enhance performance and reduce the risk of premature wear or failure (Neville et al., 2007).

A comprehensive understanding of the tribological effects of 3D-printed parts in industrial applications necessitates thorough testing, analysis, and material selection. Researchers and engineers are exploring various techniques, including surface treatments (Chabak et al., 2021), post-processing methods (Hanon and Zsidai, 2020), and material modifications (Fouly et al., 2022), to improve the tribological properties of 3D-printed components. By optimizing these factors, the industrial usability of 3D-printed parts can be enhanced, enabling reliable and efficient operation across diverse applications.

This chapter offers an introduction to the core principles of tribology, which involve the examination of friction, wear, and lubrication. It also provides an exploration of diverse AM techniques, an area that is continually evolving. Additionally, the chapter emphasizes the importance of tribological studies specifically in relation to 3D-printed parts. Understanding the tribological characteristics of these components is essential to ensure their dependable and efficient operation. The chapter further discusses the extensive range of applications for tribology across various industries, including Classical and Open Systems Tribology, Biotribology, Nanotribology, Tribotronics, and Aerospace Tribology. These specialized branches of tribology address the distinctive challenges and demands of their respective industries, thereby facilitating advancements in materials, coatings, lubricants, and design strategies to enhance overall performance and durability.

13.2 FRICTION AND WEAR

Friction and wear are predominantly influenced by the properties of the sliding surfaces (Zhang, 2013). The intricate nature of these occurrences introduces difficulties in accurately predicting and comprehending them. This challenge arises from the need to analyze various surface attributes, including microstructural properties, the existence of organic compounds and oxides, moisture content, geometric deviations, and potential contaminants absorbed from the ambient environment. Hence, when two surfaces make contact, their distinctive characteristics significantly influence the interaction. This encompasses mechanical aspects along with stress–strain dynamics within the sliding region, frequently accompanied by the formation of physical or chemical bonds (Gropper et al., 2016). The estimation of contact stresses can be approximated by considering the notion of idealized smooth surfaces, assuming the absence of geometrical inconsistencies. However, achieving genuinely smooth surfaces at the molecular scale poses difficulties. The connection between contact stresses and deformations can be deduced through theoretical examination, as originally established by Hertz for linear elastic materials. This analysis is relevant when the two bodies are in frictional or elastic contact, presuming that the radius of the contact body is substantially greater than the dimensions of the contact area (Purushothaman and Thankachan, 2014). There are typically three types of surface contacts: (a) Elastic contact, (b) Viscoelastic contact, and (c) Plastic contact (Kildashti et al., 2023). Elastic contact can be categorized into two main types: conformal and nonconformal contacts. Conformal contact arises when mating surfaces closely match, as seen in scenarios like a bearing sliding on shafts or between a wire and tool during drawing processes. In contrast, nonconformal contacts involve point or line contacts (Liu et al., 2023). In the context of polymers, the deformation response is shaped by plasticity, elasticity, and viscoelastic mechanisms. In polymer plates, viscosity has a significant impact, particularly when holding the amorphous phase at temperatures above the glass transition temperature (Deshmukh et al., 2020). Energy loss is associated with viscoelastic loading and unloading processes, leading to temperature generation due to the material's lower thermal conductivity. Plastic deformation, however, remains stable. Both plastic and viscoelastic processes are temperature-dependent, with their intensity increasing as the temperature rises. When the material exhibits ductile behavior, the applied contact load induces plastic deformation (Nie et al., 2021). At the same time, the stress equivalent at the critical point impacts the uniaxial yield stress of the material. In such scenarios, the material transitions from an elastic phase to an elastic-plastic state.

Friction forces can be classified as either beneficial or unfavorable. Friction is indispensable for enabling the use of vehicle tires on roads, facilitating walking on surfaces, and enabling the act of picking up objects. In specific contexts such as clutches, vehicle brakes, and power transmission mechanisms like belt drives, friction is deliberately augmented. Conversely, in scenarios involving moving and rotating parts like seals and bearings, friction is considered undesirable. Elevated levels of friction lead to amplified material degradation (wear rate) and energy dissipation. As a result, endeavors are undertaken to diminish friction in such operational conditions (De Castro et al., 2022).

Friction is the resistance force that opposes the relative movement of two surfaces in contact, often within a fluid medium. As these sliding surfaces undergo relative motion, the frictional interaction between them transforms kinetic energy into thermal energy or heat. The coefficient of friction is commonly used to quantify friction, denoted by a dimensionless scalar value (μ). It represents the ratio of frictional force (F) between the surfaces to the applied force (F_n) on them, as described in the following equation:

$$F = \mu(F_n) \tag{13.1}$$

In the investigation of sliding scenarios, it becomes crucial to observe and analyze frictional characteristics. Variations in friction, coupled with wear data, offer valuable information for modeling and comprehending the fundamental mechanisms involved. Typically, friction is divided into two primary categories: (a) static friction, which pertains to the resistance encountered when two objects are at rest relative to each other, and (b) dynamic friction, which emerges when the interacting objects undergo relative motion (Faghihnejad and Zeng, 2013).

Normally, the interaction between tiny projections on metallic surfaces is primarily characterized by plastic behavior. Metals like cobalt, titanium, and magnesium, possessing a hexagonal closed packed (hcp) crystal lattice configuration, display a coefficient of friction around 0.5 when in contact with similar metals during sliding movements (Ma et al., 2013). Within ceramics, the interaction at the microscopic projections usually combines both elastic and plastic responses. When the surface irregularities are minimal, the interaction might lean towards being entirely elastic (Yin and Komvopoulos, 2010). Nevertheless, when confronted with greater surface roughness, the transition to a plastic state becomes more likely. In cases of elastic contacts, the coefficient of friction usually remains consistent irrespective of the applied normal load. Certain polymer composites, like polyether ether ketone (PEEK)-reinforced polytetrafluoroethylene (PTFE), exhibit extremely low friction, around 0.1–0.2, when sliding against the same material or other metals (Khare and Anders, 2023). This makes them function effectively as solid lubricants in sliding applications. In general, most polymer materials have friction coefficients ranging from 0.2 to 1 under dry conditions. The work of adhesion in polymers is typically lower compared to ceramics and metals (Ghanem and Lang, 2017). However, their material stiffness and hardness are also lower, and these two effects are usually proportional.

In a general sense, wear denotes the material removal or surface damage arising when two surfaces undergo rolling, sliding, or impacting interactions. These wear phenomena predominantly manifest at a microscopic level, specifically at the surface asperities. As relative motion transpires between two objects or components, material displacement can occur within the interfacial region, consequently inducing alterations in material properties. Nevertheless, it's plausible for this displacement to result in minimal or negligible material loss. This displaced material might either transfer to the opposing surface or fragment into small debris, thereby giving rise to wear. In scenarios where material is transferred from the bulk to the opposing surface, the net mass or volume loss at the interacting surface remains zero, despite wear occurring within the bulk material surface. It's imperative to recognize that wear loss signifies the actual material diminution and can occasionally manifest independently. Wear, as a phenomenon, is not solely determined by material properties but is influenced by the working conditions and environment. It is crucial to consider the specific working atmosphere when evaluating wear at the interface. It is a common misconception that higher frictional forces necessarily lead to an increase in the wear rate. Concerning wear, the common variations include: Abrasive Wear, which arises from the sliding motion of a rugged, unyielding surface against a softer one; Adhesive Wear, which arises due to undesired transference and adhesion of worn fragments from one surface to another; Fretting Wear, which arises from recurrent cyclic abrasion between two surfaces; Erosive Wear, which transpires when solid or liquid particles collide with the surface of an object; Surface Fatigue wear, which occurs when the surface of a substance weakens due to repeated loading cycles; and

Corrosion Wear, which materializes through chemical reactions between worn substances and a corroding medium. Given the array of variations, tribology further elucidates how wear may undergo manifold transformations over time or in response to altering operational circumstances (Swain et al., 2020).

13.3 TRIBOLOGY AND LUBRICATION

Sliding between surfaces involves solid components and is influenced by factors such as the coefficient of friction and wear rate (Khonsari et al., 2021). Surface properties like reactivity, hardness, solubility, and surface energy impact these effects. Newly manufactured surfaces typically exhibit lower wear and friction coefficients compared to clean surfaces. Nonetheless, the presence of foreign substances at the interface has the potential to elevate the coefficient of friction in cases of ongoing sliding. To mitigate wear and friction, the application of lubricants proves effective. Lubrication can be classified into two primary categories: solid lubrication and fluid film lubrication. Solid lubricants like solid films and powders protect surfaces, reduce wear and friction, and find application in low-speed, high-load operations and hydrodynamically lubricated bearings. Hard materials are also used as lubricants in extreme working conditions (Jeyaprakash and Yang, 2020).

Fluid Film Lubrication Regimes: Hydrostatic lubrication maintains a thick layer of lubricant using an external pumping mechanism, resulting in minimal relative motion. The Stribeck curve plots the coefficient of friction against rotational speed and the ratio of absolute viscosity to applied load, identifying different lubrication regimes (Figure 13.1a). Hydrostatic and hydrodynamic lubrication fall under these regimes. Hydrostatic bearings use external pumps to induce fluid pressure and provide a thicker lubricant film. Hydrodynamic lubrication relies on a thin lubricant layer pulled and compressed between bearing surfaces during motion. This mechanism is crucial for hydrodynamic journal and thrust bearings, offering low friction coefficients. Viscosity influences the behavior of physical interactions during sliding, and careful management of lubricant precipitation and film formation helps reduce wear. Adhesive wear may occur during initial and ending processes (Maarof et al., 2019).

Boundary Lubrication: Raising the load results in a reduction of fluid viscosity or velocity, culminating in an elevated friction coefficient. This phenomenon takes place under conditions of limited lubricant supply, where solid surfaces are in close proximity, and the interaction is affected by multimolecular or monomolecular films. Boundary lubricants are crucial in establishing a sheared film on bearing surfaces, effectively diminishing corrosive and adhesive wear. Pertinent attributes of lubricant films, including hardness and shear strength, hold substantial importance. Notably, viscosity exerts minimal influence on wear and friction dynamics (Chen et al., 2023).

Mixed Lubrication: The transition from boundary to hydrodynamic regimes constitutes a mixed lubrication zone where both mechanisms can be active. Solid contacts may occur, but a partial hydrodynamic film remains on a small portion of the bearing surface. Metal surfaces can experience wear debris formation, particle adhesion, metal transfer, and potential seizure. Nevertheless, chemically generated films offer protection against adhesion during sliding, creating a regime of thin film lubrication (Hol et al., 2015).

13.4 TRIBOLOGY AND 3D PRINTING

Understanding the tribological behavior of 3D-printed parts is crucial for optimizing their performance and ensuring reliable operation. Tribological studies provide insights into friction, wear, and lubrication, which are key factors in the functionality and longevity of these parts. By analyzing and controlling these tribological factors, researchers and engineers can enhance the performance of 3D-printed parts, resulting in improved functionality and extended lifespan (Albahkali et al., 2023; Ramezani et al., 2023). One important aspect is material selection. Different materials exhibit varied tribological properties, such as friction coefficients and wear resistance. Through tribological studies, researchers and engineers can identify suitable materials for specific applications, ensuring optimal performance and minimizing wear-related issues. Furthermore, tribological studies contribute to the optimization of the design of 3D-printed parts (Portoacă et al., 2023). By analyzing frictional forces, wear mechanisms, and lubrication requirements, designers can make informed decisions regarding geometry, surface finish, and functional features (Deja et al., 2021). This optimization helps to reduce friction, minimize wear, and enhance overall performance and durability. Different industries may have specific tribological requirements for 3D-printed parts, particularly in safety-critical sectors like aerospace and automotive. Tribological studies enable researchers to tailor materials, coatings, and lubrication strategies to meet industry-specific demands, ensuring compliance with performance and safety standards.

In conclusion, tribological studies are essential for 3D-printed parts as they optimize performance, ensure material compatibility, optimize design, and meet industry-specific requirements. By investigating and controlling tribological factors, researchers and engineers can unlock the full potential of 3D printing technology in various industries, leading to reliable and efficient operation of 3D-printed components.

13.5 3D PRINTING TECHNIQUES

Fused Deposition Modeling (FDM) is a popular 3D printing technique that involves melting and layering of materials. In the late 80s, Scott Crump developed FDM by melting plastics with a hot glue gun and pouring them into thin layers. He later automated the process using polymer filaments and patented it. Chuck Hull is widely recognized as the *"father"* of 3D printing due to his groundbreaking patent for the first AM technique submitted in 1984 (Jordan and Wang, 2019). Although the term "SLA" was mentioned in Hull's patent, the origins of the technique can be traced back to experiments conducted by Hideo Kodama at the Nagoya Municipal Research Institute in 1981. Today, there is a wide range of 3D printing technologies available for professionals and researchers, including FDM, selective laser sintering (SLS), ink jet printing (IJP), stereolithography (SLA), digital light processing (DLP), laser powder bed fusion (LPBF), and binder jetting (BJ). Each method has its own advantages and limitations, and the choice of technique depends on factors such as raw materials, processing speed, resolution requirements, cost, and performance specifications. The general process for 3D printing involves 3D modeling, conversion to .STL format, slicing, and layering of materials. CAD software, 3D scanners, or photogrammetry can be used for 3D modeling. The CAD file is then saved in .STL format, which encodes the 3D model information using triangular facets. The file can be stored in binary or ASCII encoding, representing the coordinates of the vertices and outward unit normal vectors of each facet. The next step is slicing, where the file is converted into G-codes that instruct the printing machine to perform specific actions. Various open-source software programs are available for slicing. The machine reads the G-code line by line and overlaps the 2D layers of materials to create the physical 3D part. Some AM technologies may require additional curing and post-processing to achieve the final usable product (Raj and Dixit, 2022). Actually, ISO/ASTM 52900-2021 classifies whole AM into seven categories (*Additive manufacturing—General principles—Fundamentals and vocabulary*, n.d.).

13.5.1 EXTRUSION-BASED TECHNIQUE

The extrusion-based technique encompasses two technologies, FDM and Direct Ink Writing (DIW), used for 3D model fabrication by extruding materials in the form of strands. FDM is an AM method where a thermoplastic filament like ABS or PLA is heated above its glass transition temperature. The semi-liquid material is then extruded through a nozzle and deposited in layers onto a heated substrate to create a 3D structure (Figure 13.1c). DIW is an AM approach that involves depositing colloids, pastes, gels, or inks onto a substrate to create functional structures. It utilizes a piston to push the viscous material through a micro-nozzle connected to a three-axis positioning stage (Raj et al., 2022c). Curing is achieved through reactive components, heat, or UV light. DIW can be categorized into piston-based, pneumatic, or Archimedes screw systems, each with different extrusion mechanisms (Raj and Dixit, 2022).

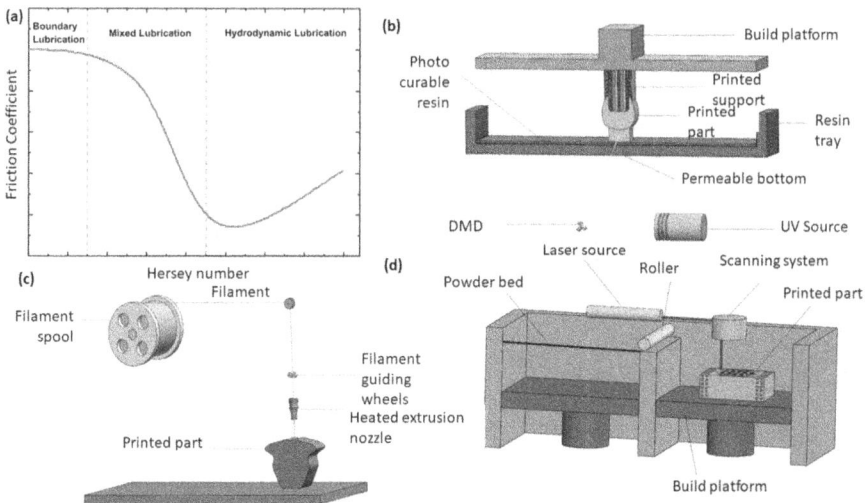

FIGURE 13.1 (a) Representative Stribeck curve; (b) representative mechanism; (c) representative extrusion-based technique; and (d) powder bed fusion schematic. (Adapted with permission from Raj and Dixit, 2022.)

13.5.2 Vat Photopolymerization

Vat photopolymerization is a 3D printing technique that utilizes photopolymerization to solidify liquid polymers into solid objects layer by layer. This method encompasses SLA (Stereolithography) and DLP (Digital Light Processing) technologies (Figure 13.1b), both of which employ photo-curable resins composed of monomers, oligomers, and photoinitiators. The distinction lies in the curing process, where DLP employs UV light from a projector, and SLA utilizes a UV laser beam. In DLP, the UV light cures the entire resin layer at once, while in SLA, the laser beam traces the geometry point by point to solidify the resin layer. The curing time and print resolution can be influenced by parameters such as curing depth, laser power intensity, scanning speed, exposure duration, and the addition of photoinitiators to the resin (Raj et al., 2022b).

13.5.3 Powder Bed Fusion (PBF)

Powder bed fusion (PBF) is a widely used 3D printing technique in industrial AM that fuses powdered material point by point using an energy source like a laser beam or an electron beam. PBF can be applied to both metals and polymers, although material compatibility varies. Various PBF technologies exist, including SLS, LPBF, and Electron Beam Melting (EBM). In SLS, a laser selectively sintered and fuses polymer powder particles to form the object layer by layer. The process involves heating the powder bin and construction area below the polymer melting temperature (Figure 13.1d). LPBF is similar to SLS but used for metals. Unlike SLS, LPBF requires parts to be initially fused to the substrate and often requires support structures for stability and thermal control. LPBF is typically performed under inert gas or vacuum due to the combustible nature of metal powders. Unfused powder can often be reused, although it may degrade over time due to oxidation. EBM is another PBF process for metals. It operates by using an electron beam to melt the metal powder bed under vacuum. Before the 3D printing process begins, each layer of material is typically pre-sintered to ensure proper bonding. Parts are printed within the semi-sintered powder bed, providing support during the build and reducing the need for additional support structures. Overall, PBF technologies offer versatility in printing both metals and polymers, enabling the production of complex and functional parts. The specific technique used depends on the material being printed and the desired properties of the final product (Popov et al., 2021).

13.5.4 Material Jetting

MJ 3D printers utilize viscous materials, spreading small droplets of the material onto the build platform and then curing it using UV light. Once the first layer is completed, the platform lowers, and the printer proceeds to deposit the subsequent layer. This layer-by-layer process continues until the object is fully formed. MJ, includes MultiJet Printing (MJP) and PolyJet or Drop-On-Demand technologies. The working principle of material jetting printers determines the types of materials they can handle. These printers can effectively operate with liquids of specific viscosities, creating objects from tiny droplets. Additionally, the materials used in material jetting are photo-resistant, meaning they solidify under UV light (Gülcan et al., 2021).

13.5.5 BINDER JETTING

Binder jetting (BJ) is a cost-effective and energy-efficient 3D printing technique that uses powdered materials like metals, composites, sand, and ceramics. It creates a fine powder bed by spreading the powder, PBF. Instead of a laser, an industrial printhead selectively deposits a liquid binding agent onto the powder, building up layers based on a CAD file. Plastic parts require curing, while metal parts undergo sintering for completion. Developed at MIT in the 1990s, BJ offers advantages such as lower energy consumption compared to laser-based methods and the use of affordable materials. It enables fast production times, making it ideal for precise and scalable mass production of small, accurate objects. BJ offers benefits like reduced warping at room temperature and cost-effective high-volume production. However, it's important to consider that parts produced through BJ may have moderate mechanical properties and high porosity, which may not meet all requirements (Ziaee and Crane, 2019).

13.5.6 SHEET LAMINATION

Sheet lamination is an AM process that forms 3D objects by stacking and bonding thin sheets of materials like paper, plastic, or metal foil. ISO/ASTM 52900-2021 defines this method, which utilizes welding, adhesive, heat, or pressure for bonding. Laser cutting or CNC machining is employed to achieve the final shape of the object. Sheet lamination can be categorized based on the build material (e.g., paper, plastic, metal, or woven fiber composites) and the forming methods used (e.g., CNC milling, laser cutting, or aqua blasting). Various bonding techniques are utilized in sheet lamination, including adhesive bonding, thermal bonding, and ultrasonic welding. Materials utilized in sheet lamination vary, including paper, polymer, ceramic, and metal, each requiring different binding methods. For instance, paper uses pre-applied adhesive activated by heat and pressure, polymers rely on heat and pressure for melting the sheets together, metal sheets are bound using ultrasonic welding, and fiber-based materials and ceramics combine layers through thermal energy via oven baking (Bhatt et al., 2019).

13.5.7 DIRECT ENERGY DEPOSITION

Directed Energy Deposition (DED) is a 3D printing technique that utilizes a focused energy source, such as a plasma arc, laser, or electron beam, to melt and deposit material through a nozzle. It can be used for repairing existing components or creating new parts. DED is also known as Laser Engineered Net Shaping (LENS), Direct Metal Deposition (DMD), Electron Beam Additive Manufacturing (EBAM), Directed Light Fabrication, or 3D Laser Cladding, depending on the specific method employed. In DED, the material is melted and deposited onto a designated surface, where it solidifies and fuses with the existing materials to form a structure.

The process is performed in a controlled chamber with reduced oxygen levels. Electron beam-based systems operate in a vacuum, while laser-based systems use an inert chamber for reactive metals. To prevent contamination, a shielding gas can be used during metal 3D printing. DED is commonly used with metal parts but can also be applied to polymers and ceramics. It is capable of additively manufacturing various weldable metals, including aluminum, inconel, niobium, stainless steel, tantalum, titanium alloys, and tungsten (Kladovasilakis et al., 2021).

13.6 INDUSTRIAL APPLICATIONS

Traditionally, tribology has primarily been applied to the analysis of common components involved in rolling or sliding motion, such as bearings, gears, cams, brakes, and seals. These components are integral to various machines that involve relative and rotational motion. The initial emphasis of tribology was on improving the performance and longevity of industrial machinery, but its influence has expanded to encompass a wide range of applications.

Tribology research spans across different scales, from macro to nano. While its traditional focus was on the transport and manufacturing sectors, it has become increasingly diverse over the years, branching out into various fields. These include but are not limited to:

13.6.1 CLASSICAL AND OPEN SYSTEMS TRIBOLOGY

This field primarily investigates friction and wear phenomena in machine components like rolling-element bearings, gears, brakes, clutches, and manufacturing processes. On the other hand, Open System Tribology delves into the study of tribological systems that interact with and are influenced by the surrounding natural environment. Various contemporary additive techniques are currently being employed for the production of such components, and numerous research groups worldwide have documented their findings. Research has delved into the tribological aspects of FDM 3D-printed Polycarbonate and Acrylonitrile Butadiene Styrene (PC-ABS) blends. Investigations have been conducted to discern the wear mechanisms in FDM-processed parts (Figure 13.2a). The wear rate of PC-ABS parts produced through FDM escalates with the augmentation of layer thickness and build orientation. Conversely, the wear rate can be significantly ameliorated by enhancing the raster angle and air gap (Ahmed et al., 2017). Additionally, an exploration into the tribological characteristics of 3D-printed stainless steel fabricated via metal-based powder bed fusion has been undertaken. The suitability of this material in both dry and lubricated conditions was assessed. Remarkably, in dry conditions, 3D-printed stainless steel exhibits a lower wear rate compared to traditionally manufactured stainless steel, regardless of the applied load. This discrepancy may be attributed to the heightened hardness of the printed stainless steel or potentially due to variations in surface roughness between the two samples. Notably, in lubricated conditions, 3D-printed stainless steel demonstrates a higher wear rate when compared to traditionally manufactured stainless steel (Kc et al., 2019). Polyphenylene sulfide (PPS) composites reinforced with carbon nanotubes (CNTs) have been successfully produced and subjected to friction and wear testing. The incorporation of CNT into the PPS matrix using 3D printing techniques led to a noteworthy enhancement in tribological performance. The outcomes revealed a substantial reduction in wear rate, dropping from 3.048E−5 mm³/N/m for pure PPS to 0.808E−5 mm³/N/m for the composite containing 0.7 wt% CNTs under lower load conditions. Similarly, at higher loads, the wear rate diminished from 1.14E−3 mm³/N/m for pure PPS to 1.838E−5 mm³/N/m for the composite containing 0.9 wt% CNTs (Pan et al., 2021). The application of vat-photopolymerization-based additive techniques has been utilized in the production of composite materials reinforced with graphene. The investigation delved into the wear characteristics of the samples, with a focus on analyzing the impact of graphene. Notably, the dynamic coefficient of friction in specimens containing graphene exhibited a notable reduction (approximately 50% decrease when printed horizontally, irrespective of the layer thickness) compared to the corresponding samples composed of the unadulterated resin material (Figure 13.2b). This reduction is attributed to the influence of graphene, which could potentially function as a solid lubricant (Hanon et al., 2022). Utilizing micro 3D metal printing, bionic 3D structures were produced. The pin-on-disk trials exhibited the effectiveness of scaly textures in diminishing friction and wear. At lower speeds and higher loads, scaly textures led to a substantial reduction in friction coefficient, achieving a maximum reduction of 51.5%. The wear patterns indicated that textured sections had less conspicuous and uniform wear marks compared to non-textured areas, likely due to the supplementary lubrication effect of the textures. Moreover, the textured surface displayed less severe wear in contrast to the untreated surface

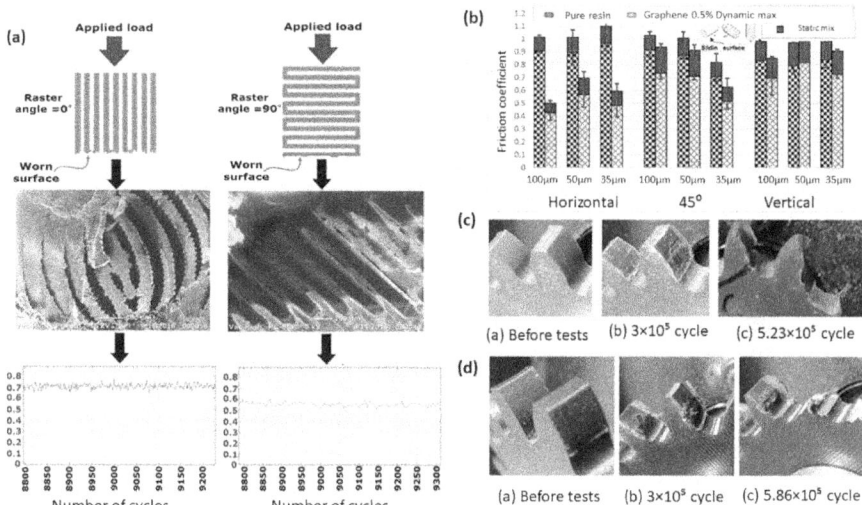

FIGURE 13.2 (a) PC-ABS friction coefficient plots at different raster angles; (b) friction coefficient variation at different orientation 3D-printed samples of graphene composites; and (c) FDM-fabricated PETG and PLA gears tested for wear analysis. (Adapted with permission from Ahmed et al., 2017; Kc et al., 2019; Pan et al., 2021; Hanon et al., 2022; Tunalioglu and Agca, 2022.)

(Liu et al., 2022). Tribological study examined varied surface conditions of additively manufactured H13 steel samples: as-printed, polished, ground and polished, and laser-textured. Tests used base oil-lubricated bearing steel disks in a pin-on-disk tribometer. Grinding and polishing reduced friction; laser texturing and added polishing increased it. Ground-and-polished samples maintained hydrodynamic lubrication at all speeds. Post-processing strongly influenced tribological behavior (Guenther et al., 2021). A theoretical equation determined wear depth in FDM-produced plastic gears using MATLAB for calculation along the line of action. Gear durability was tested through repetitive trials at constant speed until wear damage. Tooth root region displayed highest wear where driving gear engaged. PLA wear depth calculations aligned with experimental, validating equation's applicability. PETG showcased superior wear resistance due to balanced properties (Figure 13.2c and d). PETG outperformed PLA and ABS, emerging as most wear resistant among the materials (Tunalioglu and Agca, 2022). Tribology tests in vacuum on FDM-printed samples demonstrated nylon's self-lubricating ability, exhibiting a low 0.08 friction coefficient at room temperature, but rising at higher temperatures. Nylon remained functional above 130°C, PETG showed good performance up to 80°C with a friction coefficient below 0.1, and ABS performed up to 110°C with increasing friction from 0.12 to 0.29 (Prozhega et al., 2020).

13.6.2 BIOTRIBOLOGY

The additive manufactured parts gaining wide applicability in the modern biomedical industries. From creating custom 3D-printed simple and complex surgical tools, bones, and joints reconstruction to bioprinting artificial tissues and organs are some of the most significant utilities of AM in medical science. This technology produces an extremely accurate and intricate anatomical model that helps to operate complex surgical operations with improved outcomes at lower cost and also cut off the surgical time.

Ensuring enhanced tribological efficiency of AM components employed in surgical instruments and implanted body parts is an essential prerequisite. This is imperative due to the presence of numerous medical apparatuses comprising components that exhibit relative motion, either interacting with native tissue or within biomaterials, often while under load. The friction and wear characteristics of these mobile components not only impact the functionality of the device but also detrimentally influence surrounding natural tissues. This has led to the emergence of the term "biotribology," which encompasses the study of tribological performance within medical and biological systems. Extensive research has underscored the significance of tribology in medical and surgical equipment, spanning artificial teeth, prosthetic joints, dental implants, orthodontic devices, contact lenses, and synthetic limbs.

One of the successful medical devices used in the human body is the artificial joints. Figure 13.3a–c shows some application of artificial joints in human body. These joints not only provide three-dimensional motions to the parts, but also bear a significant amount of load. Some joints such as the hips, knee, and shoulder allow

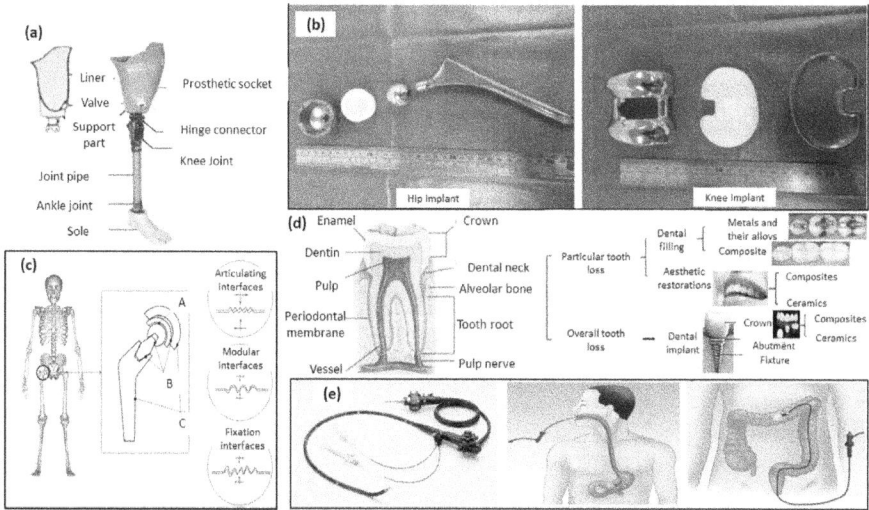

FIGURE 13.3 (a) Artificial limb and prosthetic socket, (b) medical device typically used for a hip and knee implant, (c) interactions involving multiple physical phenomena in the context of a standard medical device like a hip implant, (d) dental repairs and frequently utilized implants in the clinical setting, and (e) gastrointestinal endoscope and diagnosis through upper and lower track. (Adapted with permission from Jin et al., 2016.).

a relatively large motion. Therefore, tribological issues for the proper functioning of such joints are important considerations. Significant effort has been made to improve materials and design for artificial joints resulting in an improved hip and knee implant however still a number of challenges exist.

Bone fractures are the most common issue worldwide (Olson et al., 2015). The primary objective of fracture fixation is to reinstate stability and facilitate the healing process at the fracture locations. The micro-strain present at these fracture sites assumes a critical role (Elliott et al., 2016). Some of the fracture fixation devices include intramedullary nails, modular constructs, fracture plates, and lag screws. Lubrication along with friction is significant factors in governing the relative motion of the lag screw and the transmission of load to the fracture sites (Kummer et al., 2011).

Human teeth are an important part of the human being that is closely associated with facial esthetics and pronunciation. The degradation occurs mainly with aging and other pathologic factors. Dental restorative materials are employed to regain the function, structure, and shape of lost tooth components, and this is achieved through the utilization of dental implants. Among the frequently employed materials for dental implants are metal alloys, composites, and ceramics (Zhou and Jin, 2015). Dental restoration, including filling, crown, onlay, and bridge, is classified as direct and indirect restoration. Direct restoration is made within the patient's mouth, while indirect restoration is made outside and then placed inside. Figure 13.3d shows dental restoration application and commonly used implants in the clinic. The material attributes of the dental composite are closely linked to its wear performance (Lambrechts et al., 2006).

Fixed orthodontic appliance systems can experience wear corrosion, fretting wear, and friction in metals and their alloys due to factors such as the nature of the metals, alloying elements, and oral conditions. Furthermore, ceramic material potentially suffers from brittle fracture, while composite faces the issue of excessive wear posterior composite restoration depending upon the nature of the matrix and interfacial bond strength. Composite resins, which consist of a polymer matrix infused with borosilicate glass or colloidal silica particles, are extensively utilized dental ceramics. These materials are employed for cavity restoration and the replacement of worn tooth tissue (Zhou and Zheng, 2008). A significant drawback lies in the inadequate wear-resistant characteristics of dental composites; nonetheless, novel technologies and approaches have emerged to enhance the wear resistance of these composites (Heintze and Rousson, 2012). Utilizing fine filler particles at a consistent volume fraction was observed to reduce inter-particle spacing and enhance wear resistance (Hahnel et al., 2011). Moreover, silane coupling agents have a crucial function in enhancing the adhesion of the interface between organic resin and inorganic filler, leading to heightened wear resistance.

Dental ceramics find substantial applications in restorative procedures. Nevertheless, they face two primary limitations: the inherent brittleness of ceramics, which can result in brittle fractures, and their higher wear resistance. It's important to note that many ceramic restorations can be abrasive, potentially causing wear on both the natural and artificial dentition of opposing occlusal surfaces (Zhou and Zheng, 2008). In recent years, there has been the development of high-toughness dental ceramics, designed to mitigate the issue of brittle fractures. Zirconia ceramic has been noted for exhibiting superior wear resistance properties compared to traditional ceramics, making it a potential alternative to glass-ceramics for posterior restorations. Past literature also reported that acidulated fluoride in the mouth chemically attacks and degrades the ceramic surface gradually and increases wear rates (Theodoro et al., 2017).

Surgical instruments are generally classified into six classes according to their function while operating such as cutting, holding or grasping, hemostatic forceps, distractors, clamps, and retractors. Gastrointestinal endoscopy and minimally invasive surgery (MIS) are the popular surgical operation associated with some tribological issues occurring at their interface illustrated in Figure 13.3e. The endoscopy is pushed inside the digestive tract of humans with an external force that may lead to complications such as throat abrasion, mucosal tearing, and sometimes bleeding due to repeated insertion, rotation, and pushing operation. Several studies have been made to rectify such serious friction damage. Accoto et al. (2001) were the pioneers in recognizing and addressing the challenge of navigating collapsed and curved digestive tracts with minimally invasive devices. Their work involved assessing the frictional coefficient and its changes by sliding a constant weight block on the surface of the digestive tract at various speeds. Li et al. (2014) conducted an investigation into the friction-induced trauma mechanism within the small intestine during endoscopy procedures involving pulling actions. This study encompassed varying normal forces and friction durations. The investigation observed increase in normal force and friction time increases the total frictional energy dissipation in the small intestine.

Contact lenses, which are extensively employed to improve visual acuity, have gained significant global popularity. The act of placing a contact lens in the eye establishes two interfaces: the lid wiper and ocular surface. These interfaces experience relative motion and loading, underscoring the crucial role of tribological performance in the effective operation of contact lenses during blinking and eye movement. The presence of fluid film lubricant between the contact lens and the eyes ensures minimum friction and smoothen the motion during blinking thus is critically important. The available literature suggests a synergistic lubrication mechanism considering the lens material, tear film, wetting agent, and their interaction essential for the contact lens to perform normally in the human eye. The configuration of the tear film and the constitution encompassing proteins, lipids, and mucin hold paramount importance in both the clinical operation and tribological aspects of a contact lens (Silva et al., 2015). Scientists have also innovated novel materials for soft contact lenses that possess surfaces with elevated water content or incorporate wetting agents like polyvinyl alcohol (PVA), polyvinyl pyrrolidone (PVP), and other substances. This advancement aims to reduce the friction occurring between the surface of the contact lens and the lid-wiper (Pult et al., 2015). The friction coefficients of such material are much lesser than the conventionally used silicone-hydrogel contact lenses. Furthermore, the surface morphology of contact lenses also assumes a crucial role in influencing the tribological behavior of newly designed contact lenses.

Cardiovascular disorders rank as the primary contributor to global mortality within the realm of non-communicable illnesses. This category encompasses afflictions that impact the heart or blood vessel structures, giving rise to conditions like heart failure, arterial disease, and heart valve ailments. In certain cases, when the disease reaches an advanced stage, mechanical intervention utilizing medical equipment such as pacemakers, catheters, heart valves, and vascular prostheses becomes imperative. Given that these devices interact with blood during their movement, their design must prioritize preventing any blood-related harm that could lead to complications like coagulopathy and thrombosis (Jennings and Weeks, 2015). The bearing of the spinning components is the crucial component in the construction of a mechanical circulatory support system. Because of this, it's crucial to maintain a good lubricant film to prevent blood damage and allow enough washout, which is mostly achieved through optimized bearing geometry utilized in medical equipment (Kosaka et al., 2013).

13.6.3 Nanotribology

Nanotribology, the study of friction, wear, and lubrication at the nanoscale, stands as a compelling frontier in the field of materials science and engineering. As technological innovation continues to shrink devices and systems to unprecedented dimensions, the behavior of surfaces and interfaces at the nanoscale becomes increasingly influential (Belak, 1993). Whether in the realm of microelectromechanical systems (MEMS), nanoelectronics, biomedical devices, or emerging nanotechnologies, the intricate interplay of forces at these microscopic dimensions can significantly impact performance, durability, and efficiency (Kim et al., 2007). Traditional tribology, the study of friction and wear on macroscopic scales, has long been central to fields ranging from transportation to manufacturing. However, as time venture into the nanoscale regime, the familiar laws of classical physics no longer hold true. Surfaces that were once considered smooth now reveal a landscape riddled with atomic irregularities and nanoscale features. The forces governing interactions are reshaped by quantum mechanical effects, and phenomena like adhesion, deformation, and surface energy take on novel meanings in this intricate landscape.

In advanced manufacturing sector, the fusion of 3D printing technology with the field of nanotribology has paved the way for transformative innovations in engineering and materials science (Macovei and Paleu, 2022). Nanotribology, which focuses on the intricate interactions of friction, wear, and lubrication at the nanoscale level, presents a realm of challenges and opportunities that align seamlessly with the capabilities of 3D printing. This convergence has led to the creation of bespoke 3D-printed parts, meticulously designed and engineered to exhibit tailored tribological properties at the smallest scales. As a result, these 3D-printed components are poised to revolutionize diverse industries, from biomedical devices to aerospace systems, by ushering in enhanced performance, extended lifespans, and improved functionality through a novel amalgamation of cutting-edge manufacturing and nanoscale science. In this context, exploring the world of 3D-printed parts for nanotribology unveils a frontier of exploration at the intersection of innovation and precision engineering.

The subsequent section will delve into the examination of the incorporation of 3D-printed parts used for nanotribology and its utilization in industrial contexts. This study aims to explain the impact of nanotribology on industries and its role in driving progress in the current era of advanced manufacturing and technology, encompassing both foundational theories and practical case studies. Nanotribology is of paramount importance in shaping the industrial sector as it tackles complex issues pertaining to friction, wear, and lubrication at the nanoscale. The impact of nanotribology in this context is defined by its notable ability to improve efficiency, reduce energy losses, and enhance the reliability of energy systems.

In the realm of nanotribology applications, a range of 3D-printed parts can be classified into various overarching categories. These categories encompass a diverse array of examples that are frequently employed in this field.

The utilization of 3D printing technology enables the fabrication of complex micro- and nanoscale structures, which can be subjected to investigation in terms of their frictional and wear characteristics. The aforementioned structures encompass arrays consisting of micro-pillars, grooves, or alternative surface textures as shown in Figure 13.4. Utilizing the method of two-photon polymerization, micropillars adorned with nanohairs on their upper surfaces were successfully 3D printed. Through observation within a scanning electron microscope, their frictional behavior and deformation tendencies were examined. The study revealed that the structural stiffness of these micropillars significantly influences the initiation of sliding motion in relation to the opposing surface. Notably, elevating the stiffness of the micropillars led to an augmentation in the buckling load of the nanohairs, without sacrificing the inherent hydrophobic characteristics of the surface (Afshar-Mohajer et al., 2023).

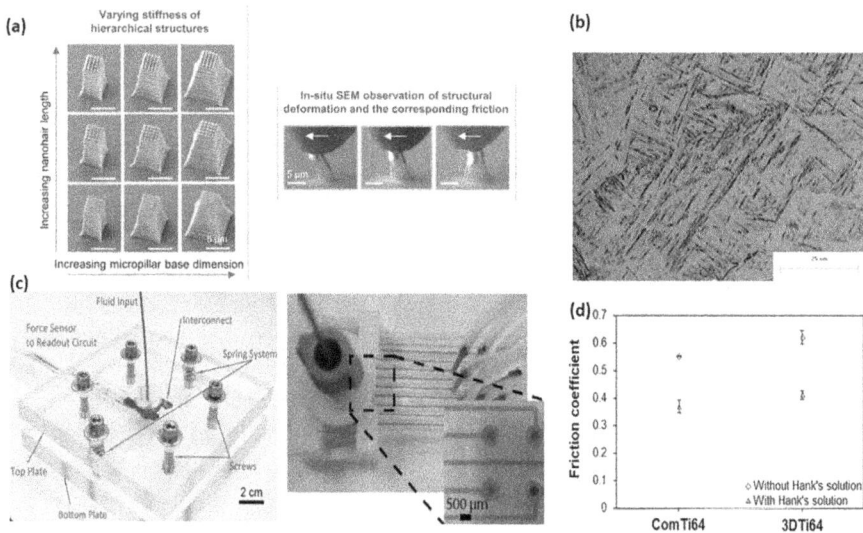

FIGURE 13.4 (a) The geometries and dimensions of the printed structures; SEM images of the structures; (b) 3D-printed Ti64 sample at 25 µm magnification; (c) holder used to apply a precise vertical force on the interconnects, with a spring system counterbalancing the top plate's weight. On one side, there's a 3D-printed interconnect with nine "printed pins," while the opposite side features nine separate individual pins. (d) friction coefficients variations of commercial and 3D-printed Ti64 parts tested against 100Cr6 steel ball. (Adapted with permission from Paydar et al., 2014; Khun et al., 2018; Afshar-Mohajer et al., 2023.)

The study of nanotribological properties of 3D-printed implants and prosthetics in the realm of biomedicine is essential to ensure their seamless interaction with surrounding tissues and materials. An electron beam melting (EBM) procedure was used to 3D print Ti-6Al-4V (Ti64) samples to compare their microstructure, mechanical, and tribological properties to commercial Ti64 samples. The extensively twinned and acicular martensitic structure of 3D-printed Ti64 samples gave them a higher surface hardness than commercial Ti64 samples. The 3D-printed Ti64 samples tested against a 100Cr6 steel counter ball without and with Hank's solution demonstrated superior wear resistance due to their increased surface hardness than commercial Ti64 samples. The lubricating action of Hank's solution during sliding reduced wear on both Ti 64 samples. This study found that 3D-printed Ti64 samples have similar mechanical and tribological properties to commercial Ti64 samples (Khun et al., 2018).

Microfluidic devices, fabricated through the process of 3D printing, offer a means to investigate fluid flow and its influence on nanotribology by providing fine control over channel geometry and surface properties. A study presents a novel approach to addressing the issues associated with the lab-on-chip interface by introducing a quick, prototyped modular microfluidic connection. The connection is a polymer gasket that is co-printed with rigid clamps, hence eliminating the need for adhesives and further assembly. This is achieved using the process of direct multi-material 3D printing, which utilizes a computer-aided design model. This device signifies the initial utilization of multi-material 3D printing in the context of microfluidic interconnects. It possesses the capability to undergo swift redesigning and printing processes, while exhibiting resilience against applied stresses (Paydar et al., 2014).

Thin films and coatings can be fabricated using 3D printing techniques to achieve desired nanotribological characteristics. These coatings have the potential to be applied to various substrates in order to investigate and analyze their tribological characteristics. The investigation focused on the micro/nanotribological and mechanical characterization of TiN sheet and single-crystal silicon prepared using Laser arc hybrid wire deposition (LAHWD)-based hybrid AM process. The hardness, friction, and wear parameters of both materials were assessed using a nanoindenter system. The findings illustrate the resemblances and disparities between the two materials, indicating that TiN film has superior characteristics in terms of hardness and resistance to wear compared to single-crystal silicon. It is probable that the approach derived from this study can be employed to assess the quality of micro/nanofilms. Additionally, TiN film exhibits favorable micro/nanotribological and mechanical properties that make it suitable for applications in MEMS (Cao et al., 2004).

13.6.4 TRIBOTRONICS

Tribotronics is an interdisciplinary field that merges the study of tribology and electronics to investigate the interaction between surfaces and electrical phenomena, focusing on friction, wear, and lubrication. It involves integrating electronic sensors, actuators, and circuits into tribological systems to enable real-time monitoring, control, and optimization of friction and wear processes. The primary objective of tribotronics is to comprehend the electrical signals and phenomena associated with tribological events, such as alterations in friction, contact pressure, and surface conditions. By incorporating sensors and electronics into tribological systems, researchers can capture and analyze these signals to gain insights into the underlying mechanisms and dynamics of friction and wear. The application of tribotronics is extensive and holds promise in various industries. For instance, in the realm of automotive engineering, tribotronics can be utilized to develop intelligent braking systems that continuously monitor and adjust friction levels for enhanced safety and performance. In manufacturing, it enables precise control of friction and wear during machining processes, resulting in improved product quality and tool lifespan. Furthermore, tribotronics finds applications in nanotechnology and MEMS, where it facilitates the measurement and manipulation of surface properties at the micro- and nanoscale. This capability enables the development of advanced materials, coatings, and lubricants with tailored friction and wear characteristics. In summary, tribotronics merges tribology and electronics to explore the electrical aspects of friction, wear, and lubrication. Its potential applications span across industries, empowering the development of intelligent systems, sensors, and control mechanisms to enhance understanding and management of tribological processes.

A 3D-printed flexure hinges-based triboelectric nanogenerator (TENG) demonstrates excellent structural integration and stability. By utilizing the contact electrification between fluorinated ethylene propylene and copper films, the TENG can efficiently convert mechanical energy into electricity. With the flexure hinges serving as mechanical elastomers, the TENG exhibits remarkable stability, as evidenced by minimal fluctuations in the open-circuit voltage even after undergoing 10,000 cycles (Wang et al., 2018). In another study, a soft robotic finger with a triboelectric curvature sensor is directly 3D printed using multi-material printing techniques. The triboelectric curvature sensor (TECS) is situated on the top surface of the finger and consists of an active layer of TECS printed directly onto the reinforced finger body. Another active layer made of elastomers is attached to a stretchable electrode. This triboelectric sensor is capable of measuring curvature up to $8.2\,m^{-1}$ at ultra-low frequencies (Zhu et al., 2020). Additionally, FDM has been utilized for creating optimized digital designs of TENG devices, enabling efficient harvesting of ambient vibration energy. The selection of positive/negative polymers as friction layers in the TENG design significantly improves the output power and energy conversion efficiency. In the vertical contact-separation mode, the triboelectrification efficiency of these TENG devices reaches an impressive 63.9% (Qiao et al., 2018).

13.6.5 Aerospace Tribology

Aerospace tribology is a specialized field that concentrates on the examination of friction, lubrication, and wear as they relate to aerospace systems. It involves investigating the effects of tribological phenomena on different components and systems employed in the aerospace industry. Friction, lubrication, and wear are pivotal factors in aerospace tribology, significantly influencing the performance, reliability, and lifespan of aerospace equipment. This encompasses critical components such as engines, turbines, bearings, seals, and gears found in aircraft, spacecraft, and associated systems. The main objective of aerospace tribology is to optimize the efficiency, safety, and durability of aerospace systems by comprehending and managing tribological factors. By analyzing the interaction between surfaces, lubricants, and environmental conditions, researchers in aerospace tribology strive to develop effective lubrication strategies, minimize wear, reduce energy losses, and enhance overall equipment performance. Research in aerospace tribology also includes the exploration of advanced materials, coatings, and surface treatments that can enhance frictional properties, decrease wear rates, and prolong the service life of aerospace components. This involves conducting experiments, simulations, and field studies to assess and validate tribological models and theories specific to the aerospace domain. Ultimately, aerospace tribology plays a crucial role in ensuring the dependable and efficient functioning of aerospace systems. It contributes to the advancement of aviation and space exploration by addressing the distinctive challenges and requirements of this industry.

Metal AM, such as PBF and DED processes, is widely used to manufacture prototyping and end-use application within the aerospace and aviation industry. The fabricated structure was observed to be stronger and lighter which distinguished it with the part manufactured through traditional methods. The common application of AM in aerospace includes the fabrication of complex structures, frames, and brackets. A joint study between Airbus and EOS group manufactured components with Ti-64-4V alloy using laser-based PBF technique, intended to replace hinge brackets on the aircraft engine that were currently used with a goal of overall mass and cost

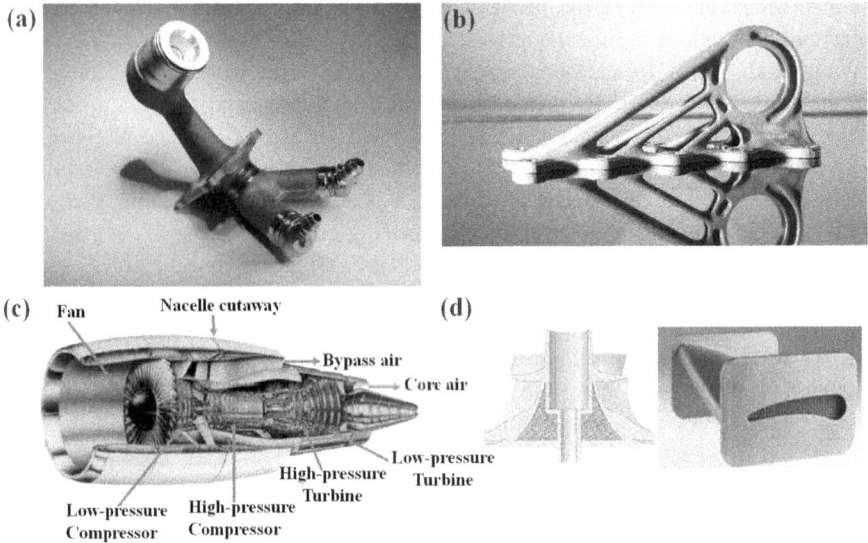

FIGURE 13.5   (a) General electric LEAP engine fuel nozzle, (b) AM of Airbus A350 XWB cabin bracket connector, (c) modern commercial high-bypass engine, and (d) AM-printed impeller with internal lattice. (Adapted with permission from Boyer et al., 2015; Blakey-Milner et al., 2021; Rouf et al., 2022.)

reduction (Tomlin and Meyer, 2011). The wide utility area and potential application of AM part used in aerospace and aviation industry are illustrated in Figure 13.5. The static and dynamic components of aircraft engines are subject to extreme performance even in most harsh environments such as high pressure, temperature, or embrittlement condition. Some example of such components is compressor blades, impellers, turbine blades, and inducers with highly complex geometry (Gebler et al., 2014). Also, the AM techniques extensively applied to repair damaged aircraft components instead of replacing or scrapping them result in significant cost savings for expensive aerospace components (Singamneni et al., 2019). Gobetz et al. investigate the feasibility of using AM technique to manufacture heat exchangers used in aerospace applications and found an effective function under an air-to-air testing setup after successful printing (Rochus et al., 2007). Moreover, large liquid fuel rocket components have also been recently developed using the directed energy method. The selection of material for the aerospace application is based on high strength, low density, durability, heat resistance, damage tolerance, and low weight. Aerospace tribology deals with science and technology related to surface interacting in relative motion. Tribology plays an important role in the operation and service reliability of both aero-engines and air-frames. Koutsomichalis et al. studied the WC-12% tribological coating deposited through plasma spraying. The high hardness and good adhesion with the aerospace component the coating exhibit satisfactory wear resistance (Koutsomichalis et al., 2009). The multilayer material of functionally graded material is also in recent trends an alternative to hard coating technology for aerospace systems operating in extreme environments.

13.7 CONCLUSIONS

The chapter delves into the intricate realm of tribology and lubrication in the context of 3D-printed parts, especially concerning their industrial applications. The influence of various factors on friction and wear is thoroughly explored, underscoring the complexity of these phenomena. The interplay of microstructure, environmental conditions, surface properties, and material interactions shapes friction and wear behavior.

With a focus on open systems, biotribology, nanotribology, tribotronics, and aerospace, the chapter delves into how 3D-printed parts find their place in diverse industrial scenarios. The discussion navigates through the significance of surface properties, deformation behavior, and temperature influence on friction and wear. The various stages of contact, including elastic, viscoelastic, and plastic, are explored, offering insights into their respective behaviors.

Friction, a pivotal element in tribology, is dissected, elucidating its types, classifications, and impacts. The importance of friction in different industrial contexts is highlighted, from its intentional increase for enhanced performance to its detrimental effects on rotating and sliding components. The coefficient of friction is demystified, and its role in monitoring frictional behavior is explained, along with the distinction between static and dynamic friction. Wear, as a corollary to friction, undergoes comprehensive examination. Different types of wear, such as abrasive, adhesive, and erosive wear, are elucidated, emphasizing their dependency on material properties and environmental conditions. The intricate relationship between friction and wear is underscored, paving the way for a more nuanced understanding of tribological phenomena.

The chapter seamlessly bridges the gap between theoretical insights and practical applications. It emphasizes the significance of lubrication in minimizing friction and wear, categorizing lubrication into solid lubrication and fluid film lubrication. An exploration of fluid film lubrication regimes, including hydrostatic and hydrodynamic lubrication, provides insights into their operational mechanisms.

In summary, the study of tribology in the context of 3D printing is paramount for enhancing performance, material selection, design optimization, and industry-specific needs. Tribological insights into friction, wear, and lubrication contribute to the longevity and efficiency of 3D-printed parts. Material selection gains significance as different materials exhibit varying tribological traits. Through such studies, suitable materials can be chosen, minimizing wear-related challenges. Moreover, optimizing designs based on friction, wear mechanisms, and lubrication requirements reduces friction, enhances performance, and ensures durability. Industries with distinct demands, like aerospace and automotive, benefit from tailored materials, coatings, and lubrication strategies to meet safety and performance standards. Ultimately, tribological investigations empower the utilization of 3D printing technology across industries, ensuring reliable and efficient operation of components.

The exploration of tribology in various industrial applications has led to transformative advancements in materials science, manufacturing, biomedicine, and aerospace engineering. The traditional focus of tribology on improving the performance and longevity of machine components has evolved into a multidisciplinary field encompassing diverse sectors. The study of tribology at different scales, from macro to nano, has resulted in significant progress across various domains.

Classical and open systems tribology have been pivotal in investigating friction and wear behaviors in machine components, manufacturing processes, and environmental interactions. 3D printing technology has found a niche in tribological research, enabling the creation of intricate structures for analysis. For instance, the tribological aspects of 3D-printed materials like polycarbonate and stainless steel have been explored, leading to insights into wear mechanisms, friction coefficients, and lubrication requirements. The introduction of nanotribology has allowed for the examination of friction, wear, and lubrication at the nanoscale, influencing fields like MEMS and nanoelectronics. The biomedical sector has embraced the integration of 3D-printed parts with tribology, creating custom implants, prosthetics, and dental restorations. By investigating the tribological behavior of these components, researchers ensure their compatibility with human tissues and enhance overall performance. Additionally, tribotronics, the convergence of tribology and electronics, has emerged as a groundbreaking avenue for real-time monitoring and control of friction and wear processes, offering applications in various industries. Aerospace tribology stands as a critical discipline for ensuring the reliability, safety, and efficiency of aerospace systems. 3D printing has revolutionized manufacturing processes, enabling the production of lightweight and intricate components that meet the stringent demands of aerospace environments. From engine components to heat exchangers, AM has found extensive applications, reducing costs and enhancing performance. The field of aerospace tribology continues to address challenges posed by extreme conditions and complex interactions, driving innovation in material selection, coating technologies, and surface interactions.

In conclusion, tribology has evolved into a dynamic and multidisciplinary field that intersects with various industries, technologies, and scientific domains. Its impact is far-reaching, contributing to improved performance, enhanced functionality, and extended lifespan of components across diverse applications. By understanding and controlling friction, wear, and lubrication, researchers and engineers continue to shape the future of technology and industry.

ACKNOWLEDGMENTS

The authors are grateful for the support and assistance from IIT(ISM) Dhanbad and University Centre for Research and Development, Chandigarh University.

REFERENCES

Accoto, D. et al. (2001) 'Measurements of the frictional properties of the gastrointestinal tract', in *World Tribology Congress*.

Afshar-Mohajer, M. et al. (2023) '3D printing of micro/nano-hierarchical structures with various structural stiffness for controlling friction and deformation', *Additive Manufacturing*, 62, p. 103368.

Ahmed, O. et al. (2017) 'Investigation on the tribological behavior and wear mechanism of parts processed by fused deposition additive manufacturing process', *Journal of Manufacturing Processes*, 29, pp. 149–159. doi:10.1016/j.jmapro.2017.07.019.

Albahkali, T. et al. (2023) 'Adaptive neuro-fuzzy-based models for predicting the tribologi-cal properties of 3D-printed PLA green composites used for biomedical applications', *Polymers*, 15(14), p. 3053.

Belak, J. F. (1993) 'Nanotribology', *Mrs Bulletin*, 18(5), pp. 15–19.

Bhatt, P. M. et al. (2019) 'A robotic cell for performing sheet lamination-based additive manu-facturing', *Additive Manufacturing*, 27, pp. 278–289.

Blakey-Milner, B. et al. (2021) 'Metal additive manufacturing in aerospace: A review', *Materials & Design*, 209, p. 110008.

Boyer, R. R. et al. (2015) 'Materials considerations for aerospace applications', *MRS Bulletin*, 40(12), pp. 1055–1066.

Cao, X. et al. (2004) 'Micro/nanotribological and mechanical studies of TiN thin-film for MEMS applications', *Tribology Transactions*, 47(2), pp. 227–232.

Chabak, Y. et al. (2021) 'Structural and tribological assessment of biomedical 316 stainless steel subjected to pulsed-plasma surface modification: Comparison of LPBF 3D print-ing and conventional fabrication', *Materials*, 14(24), p. 7621. doi:10.3390/ma14247671.

Chen, J. et al. (2023) 'Friction behavior of polycrystalline diamond compact and the evolu-tion of the friction film under different matching materials', *International Journal of Refractory Metals and Hard Materials*, 115, p. 106313.

De Castro, A. et al. (2022) 'An investigation of the parameters for characterization and pre-diction of wear of drum brake friction material', *Journal of Materials Engineering and Performance*, 31, pp. 5712–5725. doi:10.1007/s11665-022-06672-0.

Deja, M. et al. (2021) 'Applications of additively manufactured tools in abrasive machining-A literature review', *Materials*, 14(5), p. 1318.

Deshmukh, K. et al. (2020) Mechanical Analysis of Polymers. In *Polymer Science and Innovative Applications*, pp. 117–152, Elsevier. doi:10.1016/B978-0-12-816808-0.00004-4.

Dwivedi, S. et al. (2022) 'Additive texturing of metallic implant surfaces for improved wetting and biotribological performance', *Journal of Materials Research and Technology*, 20, pp. 2650–2667. doi:10.1016/J.JMRT.2022.08.029.

Elliott, D. S. et al. (2016) 'A unified theory of bone healing and nonunion: BHN theory', *The Bone & Joint Journal*, 98(7), pp. 884–891.

Faghihnejad, A. and Zeng, H. (2013) 'Fundamentals of surface adhesion, friction, and lubri-cation', In *Polymer Adhesion, Friction, and Lubrication*, edited by Hongbo Zeng, pp. 1–57, Wiley. doi:10.1002/9781118505175.ch1.

Fouly, A. et al. (2022) 'Evaluating the mechanical and tribological properties of 3D printed polylactic-acid (PLA) green-composite for artificial implant: Hip joint case study', *Materials*, 14(23), p. 5299.

Gebler, M., Uiterkamp, A. J. M. S. and Visser, C. (2014) 'A global sustainability perspective on 3D printing technologies', *Energy Policy*, 74, pp. 158–167.

Ghanem, A. and Lang, Y. (2017) *Introduction to Polymer Adhesion*, Department of Process Engineering and Applied Science.

Gropper, D., Wang, L. and Harvey, T. J. (2016) 'Tribology international hydrodynamic lubrica-tion of textured surfaces: A review of modeling techniques and key findings', *Tribology International*, 94, pp. 509–529. doi:10.1016/j.triboint.2015.10.009.

Guenther, E. et al. (2021) 'Tribological performance of additively manufactured AISI H13 steel in different surface conditions', *Materials*, 14(4), pp. 1–10.

Gülcan, O., Günaydın, K. and Tamer, A. (2021) 'The state of the art of material jetting-A criti-cal review', *Polymers*, 13(16), p. 2829.

Hahnel, S. et al. (2011) 'Two-body wear of dental restorative materials', *Journal of the Mechanical Behavior of Biomedical Materials*, 4(3), pp. 237–244.

Hanon, M. M. et al. (2022) 'Tribological characteristics of digital light processing (DLP) 3D printed graphene/resin composite: Influence of graphene presence and process settings', *Materials & Design*, 218, p. 110718. doi:10.1016/j.matdes.2022.110718.

Hanon, M. M., Kovács, M. and Zsidai, L. (2019) 'Tribology behaviour investigation of 3D printed polymers', *International Review of Applied Sciences and Engineering*, 10(2), pp. 173–181. doi:10.1556/1848.2019.0021.

Hanon, M. M. and Zsidai, L. (2020) 'Tribological and mechanical properties investigation of 3D printed polymers using DLP technique', *AIP Conference Proceedings*, 2213(March), 020205. doi:10.1063/5.0000267.

Heintze, S. D. and Rousson, V. (2012) 'Clinical effectiveness of direct class II restorations-A meta-analysis', *Journal of Adhesive Dentistry*, 14(5), pp. 407–431.

Hol, J. et al. (2015) 'Multi-scale friction modeling for sheet metal forming: The mixed lubrication regime', *Tribology International*, 85, pp. 10–25.

Hu, D. et al. (2019) 'Effect of spherical-convex surface texture on tribological performance of water-lubricated bearing', *Tribology International*, 134(January), pp. 341–351. doi:10.1016/j.triboint.2019.02.012.

ISO/ASTM 52900 (2021) *Additive manufacturing - General Principles - Fundamentals and Vocabulary*, pp. 1–19. Available at: https://www.astm.org/f3177-21.html (Accessed: 14 May 2022).

Jennings, D. L. and Weeks, P. A. (2015) 'Thrombosis in continuous-flow left ventricular assist devices: Pathophysiology, prevention, and pharmacologic management', *Pharmacotherapy: The Journal of Human Pharmacology and Drug Therapy*, 35(1), pp. 79–98.

Jeyaprakash, N. and Yang, C.-H. (2020) 'Friction, lubrication, and wear', In *Tribology in Materials and Manufacturing-Wear, Friction and Lubrication*, edited by Amar Patnaik, Tej Singh and Vikas Kukshal, pp. 1–17, Intech open.

Jin, Z. M. et al. (2016) 'Tribology of medical devices', *Biosurface and Biotribology*, 2(4), pp. 173–192.

Jordan, R. S. and Wang, Y. (2019) '3D printing of conjugated polymers', *Journal of Polymer Science Part B: Polymer Physics*, 57(23), pp. 1592–1605. doi:10.1002/POLB.24893.

Kc, S. et al. (2019) 'Tribological behavior of 17 - 4 PH stainless steel fabricated by traditional manufacturing and laser-based additive manufacturing methods', *Wear*, 440–441, p. 203100. doi:10.1016/j.wear.2019.203100.

Keshavamurthy, R. et al. (2021) 'Influence of solid lubricant addition on friction and wear response of 3d printed polymer composites', *Polymers*, 13(17), pp. 1–13. doi:10.3390/polym13172905.

Khare, H. S. and Anders, E. A. (2023) 'Effect of counterface material on dry sliding wear of PEEK-PTFE composites', *Tribology Letters*, 71(3), pp. 1–10.

Khonsari, M. M., Ghatrehsamani, S. and Akbarzadeh, S. (2021) 'On the running-in nature of metallic tribo-components: A review', *Wear*, 474, p. 203871.

Khun, N. W. et al. (2018) 'Tribological properties of three-dimensionally printed Ti-6Al-4V material via electron beam melting process tested against 100Cr6 steel without and with Hank's solution', *Journal of Tribology*, 140(6), p. 61606.

Kildashti, K., Dong, K. and Yu, A. (2023) 'Contact force models for non-spherical particles with different surface properties : A review', *Powder Technology*, 418, p. 118323. doi:10.1016/j.powtec.2023.118323.

Kim, S. H., Asay, D. B. and Dugger, M. T. (2007) 'Nanotribology and MEMS', *Nano Today*, 2(5), pp. 22–29.

Kladovasilakis, N. et al. (2021) 'Impact of metal additive manufacturing parameters on the powder bed fusion and direct energy deposition processes: A comprehensive review', *Progress in Additive Manufacturing*, 6, pp. 349–365.

Kosaka, R. et al. (2013) 'Geometric optimization of a step bearing for a hydrodynamically levitated centrifugal blood pump for the reduction of hemolysis', *Artificial Organs*, 37(9), pp. 778–785.

Koutsomichalis, A. et al. (2009) 'Tribological coatings for aerospace applications and the case of WC-Co plasma spray coatings', *Tribology in Industry*, 31(1&2), p. 37.

Kummer, F. J. et al. (2011) 'Sliding of two lag screw designs in a highly comminuted fracture model,' *Bulletin of the NYU Hospital for Joint Diseases*, 69(4), pp. 289–291.

Lambrechts, P. et al. (2006) 'Degradation of tooth structure and restorative materials: A review', *Wear*, 261(9), pp. 980–986.

Li, W. et al. (2014) 'Investigation on friction trauma of small intestine in vivo under reciprocal sliding conditions', *Tribology Letters*, 55, pp. 261–270.

Liu, B., Vollebregt, E. and Bruni, S. (2023) 'Review of conformal wheel/rail contact modelling approaches: Towards the application in rail vehicle dynamics simulation', *Vehicle System Dynamics*, 2023, pp. 1–25. doi:10.1080/00423114.2023.2228438.

Liu, Y. et al. (2022) 'Tribology international designing a bioinspired scaly textured surface for improving the tribological behaviors of starved lubrication', *Tribology International*, 173, p. 107594. doi:10.1016/j.triboint.2022.107594.

Ma, C. et al. (2013) 'The role of a tribofilm and wear debris in the tribological behaviour of nanocrystalline Ni-Co electrodeposits', *Wear*, 306(1–2), pp. 296–303.

Maarof, A. F. et al. (2019) 'Frictional behaviour of trimethylolpropane (TMP) oleate at different regimes of lubrication', *Journal of Mechanical Engineering (*JMechE*)*, 8(1), pp. 1–9.

Macovei, G. and Paleu, V. (2022) 'A review on tribological behaviour of mechanical components obtained by additive manufacturing', *In IOP Conference Series: Materials Science and Engineering*, IOP Publishing, p. 12010.

Neville, A. et al. (2007) 'Compatibility between tribological surfaces and lubricant additives-How friction and wear reduction can be controlled by surface/lube synergies', *Tribology International*, 40(10–12 SPEC. ISS.), pp. 1680–1695. doi:10.1016/j.triboint.2007.01.019.

Nie, N. et al. (2021) 'A review on plastic deformation induced surface/interface roughening of sheet metallic materials', *Journal of Materials Research and Technology*, 15, pp. 6574–6607. doi:10.1016/j.jmrt.2021.11.087.

Olson, S. A. et al. (2015) 'Hot topics in biomechanics: Hip fracture fixation', *Journal of Orthopaedic Trauma*, 29, pp. S1–S5.

Pan, S., Shen, H. and Zhang, L. (2021) 'Effect of carbon nanotube on thermal, tribological and mechanical properties of 3D printing polyphenylene sulfide', *Additive Manufacturing*, 47, p. 102247. doi:10.1016/j.addma.2021.102247.

Paydar, O. H. et al. (2014) 'Characterization of 3D-printed microfluidic chip interconnects with integrated O-rings', *Sensors and Actuators A: Physical*, 205, pp. 199–203.

Phadke, N. et al. (2022) 'Modeling and parametric optimization of laser powder bed fusion 3D printing technique using artificial neural network for enhancing dimensional accuracy', *Materials Today: Proceedings*, 56, pp. 873–878. doi:10.1016/J.MATPR.2022.02.523.

Popov, V. V. et al. (2021) 'Powder bed fusion additive manufacturing using critical raw materials: A review', *Materials*, 14(4), p. 909.

Portoacă, A. I. et al. (2023) 'Optimization of 3D printing parameters for enhanced surface quality and wear resistance', *Polymers*, 15(16), p. 3419.

Prozhega, M. V. et al. (2020) 'Frictional properties of 3D printing polymers in vacuum', *Journal of Friction and Wear*, 41, pp. 565–570.

Pult, H. et al. (2015) 'Spontaneous blinking from a tribological viewpoint', *The Ocular Surface*, 13(3), pp. 236–249.

Purushothaman, P. and Thankachan, P. (2014) 'Hertz contact stress analysis and validation', *International Journal for Research in Applied Science & Engineering Technology*, 2(11), pp. 531–538.

Qiao, H. et al. (2018) '3D printing individualized triboelectric nanogenerator with macropattern', *Nano Energy*, 50, pp. 126–132.

Raj, R. and Dixit, A. R. (2022) 'Direct ink writing of carbon-doped polymeric composite ink: A review on its requirements and applications', *3D Printing and Additive Manufacturing*, 10(4), pp. 828–854 doi:10.1089/3DP.2021.0209.

Raj, R., Dixit, A. R., Łukaszewski, K., et al. (2022a) 'Numerical and experimental mechanical analysis of additively manufactured ankle-foot orthoses', *Materials*, 15(17), p. 6130. doi:10.3390/ma15176130.

Raj, R., Dixit, A. R., Singh, S. S. et al. (2022b) 'Print parameter optimization and mechanical deformation analysis of alumina-nanoparticle doped photocurable nanocomposites fabricated using vat-photopolymerization based additive technology', *Additive Manufacturing*, 60, p. 103201. doi:10.1016/J.ADDMA.2022.103201.

Raj, R., Vamsi, S. et al. (2022c) 'Print fidelity evaluation of PVA hydrogel using computational fluid dynamics for extrusion dependent 3D printing', *IOP Conference Series: Materials Science and Engineering*, 1225(1), p. 012009. doi:10.1088/1757-899X/1225/1/012009.

Ramezani, M. et al. (2023) 'Surface engineering of metals: Techniques, characterizations and applications', *Metals*, 13(7), p. 1299.

Ranjan, N., Tyagi, R., Kumar, R. and Babbar, A. (2023a) 3D printing applications of thermo-responsive functional materials: A review. *Advances in Materials and Processing Technologies*, pp. 1–17. https://www.tandfonline.com/doi/abs/10.1080/2374068X.2023.2205669.

Ranjan, N., Tyagi, R., Kumar, R. and Kumar, V. (2023b) On fabrication of acrylonitrile butadiene styrene-zirconium oxide composite feedstock for 3D printing-based rapid tooling applications. *Journal of Thermoplastic Composite Materials*, 0(0), p. 08927057231186310. doi: 10.1177/089270572311863.

Rochus, P. et al. (2007) 'New applications of rapid prototyping and rapid manufacturing (RP/RM) technologies for space instrumentation', *Acta Astronautica*, 61(1–6), pp. 352–359.

Rouf, S. et al. (2022) '3D printed parts and mechanical properties: Influencing parameters, sustainability aspects, global market scenario, challenges and applications', *Advanced Industrial and Engineering Polymer Research*, 5(3), pp. 143–158.

Silva, D. et al. (2015) 'The effect of albumin and cholesterol on the biotribological behavior of hydrogels for contact lenses', *Acta Biomaterialia*, 26, pp. 184–194.

Singamneni, S. et al. (2019) 'Additive manufacturing for the aircraft industry: A review', *Journal of Aeronautics and Aerospace Engineering*, 8(1), pp. 351–371.

Swain, B. et al. (2020) 'Wear: A serious problem in industry', In *Tribology in Materials and Manufacturing-Wear, Friction and Lubrication*, edited by Amar Patnaik, Tej Singh and Vikas Kukshal, pp. 1–20. Intech open.

Theodoro, G. T. et al. (2017) 'Wear resistance and compression strength of ceramics tested in fluoride environments', *Journal of the Mechanical Behavior of Biomedical Materials*, 65, pp. 609–615.

Tomlin, M. and Meyer, J. (2011) 'Topology optimization of an additive layer manufactured (ALM) aerospace part', In *Proceeding of the 7th Altair CAE technology conference*, pp. 1–9.

Tunalioglu, M. S. and Agca, B. V. (2022) 'Wear and service life of 3-D printed polymeric gears', *Polymers*, 14(10), p. 2064.

Tyagi, R., Singh, G., Kuma, R., Kumar, V. and Singh, S. (2023). 3D-printed sandwiched acrylonitrile butadiene styrene/carbon fiber composites: Investigating mechanical, morphological, and fractural properties. *Journal of Materials Engineering and Performance*, 0(0), pp. 1–14. https://link.springer.com/article/10.1007/s11665-023-08292-8.

Tyagi, R., & Tripathi, A. (2023). Coating/cladding based post-processing in additive manufacturing. In *Handbook of Post-Processing in Additive Manufacturing: Requirements, Theories, and Methods*, edited by Gurminder Singh, Ranvijay Kumar, Kamalpreet Sandhu, Eujin Pei, and Sunpreet Singh, p. 127, CRC Press.

Wang, J. et al. (2018) 'Flexure hinges based triboelectric nanogenerator by 3D printing', *Extreme Mechanics Letters*, 20, pp. 38–45.

Yin, X. and Komvopoulos, K. (2010) 'An adhesive wear model of fractal surfaces in normal contact', *International Journal of Solids and Structures*, 47(7–8), pp. 912–921.

Zhang, S. W. (2013) 'Green tribology: Fundamentals and future development', *Friction*, 1(2), pp. 186–194. doi:10.1007/s40544-013-0012-4.

Zhou, Z. R. and Jin, Z. M. (2015) 'Biotribology: Recent progresses and future perspectives', *Biosurface and Biotribology*, 1(1), pp. 3–24.

Zhou, Z. R. and Zheng, J. (2008) 'Tribology of dental materials: A review', *Journal of Physics D: Applied Physics*, 41(11), p. 113001.

Zhu, M. et al. (2020) 'A soft robotic finger with self-powered triboelectric curvature sensor based on multi-material 3D printing', *Nano Energy*, 73, p. 104772.

Ziaee, M. and Crane, N. B. (2019) 'Binder jetting: A review of process, materials, and methods', *Additive Manufacturing*, 28, pp. 781–801.

Index

For Product Safety Concerns and Information please contact our EU
representative GPSR@taylorandfrancis.com
Taylor & Francis Verlag GmbH, Kaufingerstraße 24, 80331 München, Germany

www.ingramcontent.com/pod-product-compliance
Lightning Source LLC
Chambersburg PA
CBHW060358220326
41598CB00023B/2960

9 781032 509778